MARINE ECOLOGY

To our families:
Angela, Cate, Claire, James, Nick, Richard, Rowena and Toby

MARINE ECOLOGY
CONCEPTS AND APPLICATIONS

Martin Speight

Department of Zoology
University of Oxford

Peter Henderson

PISCES Conservation Ltd
Lymington, Hampshire, UK

WILEY-BLACKWELL

A John Wiley & Sons, Ltd., Publication

This edition first published 2010, © 2010 by Martin Speight and Peter Henderson

Blackwell Publishing was acquired by John Wiley & Sons in February 2007. Blackwell's publishing program has been merged with Wiley's global Scientific, Technical and Medical business to form Wiley-Blackwell.

Registered office:
John Wiley & Sons Ltd, The Atrium, Southern Gate, Chichester, West Sussex, PO19 8SQ, UK

Editorial offices:
9600 Garsington Road, Oxford, OX4 2DQ, UK
The Atrium, Southern Gate, Chichester, West Sussex, PO19 8SQ, UK
111 River Street, Hoboken, NJ 07030-5774, USA

For details of our global editorial offices, for customer services and for information about how to apply for permission to reuse the copyright material in this book please see our website at www.wiley.com/wiley-blackwell

Library of Congress Cataloguing-in-Publication Data

Speight, Martin, 1967–
    Marine ecology : concepts and applications / Martin Speight and Peter Henderson.
       p.   cm.
    Includes bibliographical references and index.
    ISBN 978-1-4051-2699-1 (hardcover : alk. paper) – ISBN 978-1-4443-3545-3 (pbk. : alk. paper) 1. Marine ecology. I. Henderson, Peter. II. Title.
    QH541.5.S3S655 2010
    577.7–dc22
                                                    2009053167

A catalogue record for this book is available from the British Library.

Set in 9.5/12pt Classical Garamond by SPi Publisher Services, Pondicherry, India
Printed and bound in Singapore by Fabulous Printers Pte Ltd

1   2010

CONTENTS

PREFACE

The book *Silent World* was written by Jacques Cousteau, and published by Hamish Hamilton in 1953. In it, Cousteau describes his first encounter with the undersea world using goggles which enabled him to see underwater. He says: "One Sunday morning in 1936 … I waded into the Mediterranean and looked into it through Fernez goggles. I was a regular Navy gunner, a good swimmer interested only in perfecting my crawl style. The sea was merely a salty obstacle that burned my eyes. I was astounded by what I saw in the shallow shingle …, rocks covered with green, brown and silver forests of algae, and fishes unknown to me, swimming in crystal clear water. Standing up to breathe I saw a trolley-bus, people, electric street-lights. I put my eyes under again and civilization vanished with one last bow. I was in a jungle never seen by those who floated on the opaque roof."

Marine ecology has always been fascinating, mysterious, and indeed for many centuries, downright dangerous. Sea monsters such as giant squid and krakens lived in the deep and dragged ships to their doom. Sirens lured unprepared sailors onto rocks, whilst mermaids lured the same sailors into other activities. The bottom of the sea was as far removed from almost everyone as the surface of the moon, and catching fish was a mysterious, hunter-gatherer sort of activity with random and often unpredictable outcomes.

Whilst terrestrial ecologists could walk out into their habitats and ecosystems with a pencil and paper, butterfly net and hand lens, their marine counterparts had to resort to buckets, grabs and cores dangled from boats or piers, somewhat akin to sampling a woodland with a grapnel suspended from a hot-air balloon. The deeper the sea, the bigger the problem, so that anywhere beyond the reach of a depth sounding line or a fishing net was pretty much completely unknown. As Sydney Hickson said in his *Fauna of the Deep Sea* published in 1893, "The bottom of the deep sea was until quite recently (first half of nineteenth century) … terrae incognitae. It was regarded by most persons, when it entered into their minds to consider it at all, as one of those regions about which we do not know anything, never shall know anything, and do not want to know anything" (our parentheses). Ambitious expeditions were nonetheless mounted to explore the sea using the available technology, the most famous of which was the Challenger Expedition that lasted for 4 years beginning in 1872. *HMS Challenger* covered over 120,000 km, surveying, trawling and dredging, and eventually discovering over 4000 new species. Fifty or so years later, another famous and influential marine biological expedition set sail, this time to explore the cold seas of the southern oceans. On board the *Discovery* was the marine scientist Alistair Hardy, later to become Sir Alistair Hardy. Hardy became Linacre Professor of Zoology in Oxford in 1946, and two of his longstanding achievements were firstly the invention of the continuous plankton recorder, and later the publication of the classic two-part book, *The Open Sea*.

Back at the individual level, free diving – holding the breath and reaching far below the surface to collect food or sponges, or attack the enemies' fleets – has been going on since ancient times. Throughout history, we have invented machines to enable us to descend deeper into sea and stay there for longer than a single breath-hold. These new systems enabled us to see a little more of the marine realm first hand, albeit from the bottom of a primitive diving chamber or bell. Aristotle apparently described a diving bell, but it wasn't until the fourteenth and fifteenth centuries that Europeans began to use such apparatus in attempts to raise the valuable bits of shipwrecks, such as cannon and treasure. In 1535, the Italian inventor and explorer, Guglielmo de Lorena, was attributed with the invention of the first proper diving bell, though Leonard da Vinci had produced designs for such a device some years earlier. Leather seals and manual pumps increased the sophistication of diving bells, and by the 1930s, William Beebe was able to descend to nearly 1000 m off Bermuda in his bathysphere. The remaining problem was that such machines had to be lowered by cranes from the surface, and venturing any deeper was very difficult. What followed was the bathyscaphe, a somewhat similar machine, but this time the pressure-proof sphere containing the divers was attached to a large flotation device, allowing the machine to move independently of the surface. The culmination of this development came in 1960, when *Trieste*, a bathyscaphe piloted by Jacques Piccard and Don Walsh, reached the bottom of the Challenger Deep in the Marinas Trench, a depth of 10,916 m. Today, there are many deep submersible vehicles (DSVs), such as *Alvin* owned by the Woods Hole Oceanographic Institute and *Mir* run by the Russian Academy of Sciences, but none can go anywhere near as deep as *Trieste*.

As alternatives to diving bells, at least for shallow waters, the use of individual diving suits became routine in the 1830s when Augustus Siebe formed the company Siebe Gorman to produce the traditional copper-helmeted diving dress. All but a very few of the designs thus far depended on an air supply from the surface, pumped down to the diver under pressure. What was needed by the budding science of marine biology and ecology was a self-contained underwater breathing apparatus (SCUBA) to free the diver from a surface supply.

Various attempts were made to produce a safe and effective SCUBA device, and the first half of the twentieth century saw a series of inventions, such as the oxygen closed circuit systems used by navy frogmen in the Second World War. Though effective enough at depths above 10 m or so, the pure oxygen became toxic at deeper depths and higher pressures, seriously limiting any recreational or scientific applications for the apparatus. The real breakthrough came in 1943, when Jacques-Yves Cousteau and Emile Gagnan invented a demand valve (regulator) that supplied the diver using it with air from steel cylinders on his or her back at the same pressure as that of the surrounding water; the "aqualung" was born.

New institutions for the study of marine biology and ecology were already established. In the USA, Scripps Institute of Oceanography in southern California has its origins in 1903, whilst Woods Hole Oceanographic Institute in Massachusetts was incorporated early in 1930. In the UK, the Marine Biological Association was founded as far back as 1884, and it opened its Citadel Hill Laboratory in Plymouth in 1888. The Monaco Aquarium on the Mediterranean coast was founded in 1910, and in Australia, the Commonwealth Scientific and Industrial Research Organization (CSIRO) set up its Fisheries Investigation Section, later to become the CSIRO Division of Fisheries in 1937. So, marine research around the world was active well before the invention of the aqualung.

Undoubtedly, the aqualung opened the floodgates for the exploration of shallow seas, down to 50 m or so, and we would suggest that detailed marine ecology only really began in the early 1950s as post-war scientists and recreational divers started to explore and study coral reefs and kelp beds alike. Of course the aqualung also enabled much easier exploitation of marine organisms from sponges to scallops, fish to lobsters. SCUBA diving with a speargun was hardly sporting, but very rewarding to some. So, marine ecology in shallow waters at least has burgeoned over the decades since then. For example, a trawl through ISI Web of Knowledge using the search terms "marine and ecology" yielded an average of 5000 or so publications in the 1990s, over 6000 in 2002, over 7000 in 2003, nearly 10,000 in 2004, and more than 11,000 per year in 2005, 2006, and 2007. Human-derived impacts are becoming more far-reaching and serious as the years go by, and climate change such as temperate and sea level rises, is now feared to be having severe and irrevocable effects on shallow marine ecosystems.

In the deep sea, all was thought to be quiet, calm, and possibly boring, until 1977 when scientists from Woods Hole Oceanographic Institute used the DSV *Alvin* to explore areas of underwater volcanic activity near the Galapagos Islands in the eastern Pacific. The enormous diversity of life on newly discovered hydrothermal vents amazed and delighted the scientific world and amateurs alike, and the far-reaching and fundamental research, even down to the origins of life on earth itself, have continued apace. The sheer excitement of vent communities, as well as cold seeps, whale-falls, and so on, is hard to describe.

Critics will no doubt ask the questions "why should Speight and Henderson write such a book?". and "What do they know about marine ecology?". First and foremost, we believe strongly that a textbook for students should be written by teachers, tutors, and lecturers. Research papers are excellent for reporting exciting and challenging new findings at the cutting edge of their fields, but someone has to convert such scholarly works into summaries and syntheses suitable for communication with undergraduates and other students. Secondly, we feel that people who write these textbooks should be good communicators, familiar and practiced with converting sometimes cryptic information into palatable, understandable and indeed enjoyable accounts which will captivate as well as inform.

Martin Speight has been teaching marine biology, ecology, and conservation to university students at undergraduate and postgraduate level for over a quarter of a century. Peter Henderson has done large amounts of university teaching in the field for many years, and also has made his livelihood by examining marine ecological problems and communicating his findings successfully to complete nonexperts. It has never been our intention to steal other people's work, and we have taken great pains throughout the lengthy writing of this book to consult as widely as possible and to seek all approvals and permissions to report the findings of experts in their specific fields. In short, we believe, some will say immodestly, that we are both good teachers and good communicators, and hence well qualified to deliver such a book.

We have strived to base the book on modern primary literature, predominantly post 2000. Some classic work dating back to times before this has of course been required on occasion, but we hope that the book will represent the "state of the art" as perceived at the time of writing this preface. Clearly, research never stays still and we hope to be able to provide new editions as the years go by which will reflect the new findings as they are published. Another problem with this approach, especially when the applied aspects of marine ecology are considered, is that information such as management plans, conservation strategies, and regional or local tactics are not officially published, but merely stay as "grey literature," usually web-based and difficult to verify or attribute. Although we have tried very hard to check such information, and report it as accurately as possible, we apologize to source and reader if we have made mistakes or provided misinformation. Any corrections to this type of error will be gratefully received, and put right in the next edition of the book.

Chapter 1 presents aspects of oceanography and other physical and chemical aspects of the sea which impinge on living things. Chapter 2 discusses the levels of diversity (more realistically, species richness) of marine communities and the various factors which influence them. The remaining chapter structure of the book follows a functional approach as much as possible, rather than describing different types of marine ecosystems separately. Thus, Chapters 3, 4, 5, 6, and 7 discuss various levels of functionality, primary production, herbivory and detritivory, predation and parasitism, competition, succession, and dispersal as major topics in marine ecology. Examples to illustrate concepts have been taken from all parts of the sea as appropriate, from the shallowest intertidal to the abyssal depths. Chapter 8 looks at global fisheries and the problems of sustainable resource use in the sea, and the final two chapters, 9 and 10, consider all aspects of anthropogenic impacts on marine ecosytems, from pollution to tourism, and finally the complex issues of marine conservation and management.

Martin Speight, Oxford
Peter Henderson, Pennington
December 2009

ACKNOWLEDGMENTS

A myriad of people have helped us write the book. The first vote of thanks must go to all the hundreds of marine scientists who have published their findings in learned journals which we have read. We have strived on every single case of a citation and/or graphic reproduction to attribute source and acknowledge authorship. Secondly, we are enormously grateful to all the publishers that have granted us permissions to reproduce published diagrams, tables and photographs, mostly free of charge. Thirdly, we are most thankful for the generosity and kindness of many underwater photographers who have allowed us to use their wonderful pictures. In this context, we must single out Paul Naylor, a hugely accomplished diver and photographer, who has not only provided numerous beautiful photographs in the text, but also the front cover picture of a diverse benthic community in Scotland.

The following list attempts to thank every individual and every institution who have provided information, pictures, permissions, support, and friendship throughout the project. We apologize unreservedly for any omissions:

People – Ian Banks, Joanna Barker, Henry Bennet-Clark, Brian Bett, Monika Bright, Paulyn Cartwright, Josh Cinner, Tim Coles, Ward Cooper, Julian Cremona, Pat Croucher, Robin Crump, Sammy de Grave, Angela Douglas, Dave Fenwick, Magdalena Fischhuber, Peter Funch, Brian Gratwicke, Clive Hambler, Jessica Harm, Rosie Hayden, Dave Harasti, Alex Hayward, Claire Henderson, Scott Henderson, Don Hickey, Jeff Jeffords, Brian Keller, Kathy Kirbo, Raphael Leiteritz, Martin Leyendecker, Pippa Mansell, Kelvin Matthews, Susan Mills, Annalie Morris, Andy Murch, Paul Naylor, Rasmus Neiderham, Andrea Nussbaum, Steve Oliver, Mel Parker, Michael Pidwirny, Jesús Pinada, Camilla Poire, Edi Purwanto, Antares Ramos-Alvares, Joel Rice, Delia Sandford, Scott Santos, Richard Seaby, Charles Sheppard, Jonathan Shrives, Dave Smith, Robin Soames, Angela Speight, Dave Suggett, Carissa Thomas, Hal Thompson, Lizzie Tyler, Ernesto Weil, and Ross Wylie.

Institutions – AGU, American Association for the Advancement of Science, American Society for Microbiology, Annual Reviews Inc., Benjamin Cummings (Addison-Welsley Longman), Cambridge University Press, Cascades Volcano Observatory, Coastal Education & Research Foundation, Commonwealth Scientific and Industrial Research Organization (CSIRO), Coralpedia, Ecological Society of America, Elsevier Science Publishers, Exxon Valdez Oil Spill Trustee Council, Fishbase, Florida Keys National Marine Park (FKNMP), Great Barrier Reef Marine Park (GBRMP), Honduran Coral Reef Foundation (HCRF), Huon Commonwealth Reserve, InterResearch, Joint Nature Conservancy Council (JNCC), Monterey Bay Aquarium Research Institute (MBARI), National Academy of Sciences USA, National Aeronautics and Space Administration (NASA), National Oceanic and Atmospheric Organization (NOAA), Nature Publishing Group, Operation Wallacea, Oxford University Press, Pisces Conservation, Reef Ball Foundation Inc., Reefbase, Rosenstiel School of Marine and Atmospheric Science, Royal Society Publishing, Royal Society of Chemistry, Sinauer, Springer, St Annes College Oxford, Taylor & Francis, United States Geological Survey (USGS), University of Bangor, University of Chicago Press, University of Essex, University of Oxford, University of Southampton, Woods Hole Oceanographic Institute (WHOI), Wiley-Blackwell, and Wiley InterScience.

Chapter 1

The physical template

Introduction

The physical environment determines the most fundamental constraints acting upon life. Life is only possible over a small part of the potential range of physical variables such as temperature that may occur on Earth and all species have evolved adaptations optimized for particular conditions. However, the physical conditions on Earth which all life, including man, are constrained by are not purely the result of physical processes. Life on our planet, and particularly life in the oceans, modifies the physical environment and makes the planet more suitable for life. The physical template we observe is to some extent the product of organisms over millions of years. Since life began in the oceans about 3.5 billion years ago, factors such as salinity, temperature, oxygen and nutrient levels have been shaping the evolution of the myriad of marine organisms alive today and they have in turn been changing these and other variables. Before we examine in detail these organisms and their interactions, it is appropriate to consider the major physical processes acting within the oceans that form the template upon which every ecological community is built.

Marine regions

The sea covers 70% of the surface of the earth and offers greater than 98% of the total space available to life. The Earth from space (Figure 1.1) is clearly a water world; observers approaching from a distance would likely assume all dominant life is marine, simply from the color of the distant planet. Indeed, the preponderance of terrestrial species is a geologically recent phenomenon. Further most of the habitat is in deep water; only about 3% of the world's waters lie over the continental shelf, which have an average depth of around 200 m. The average depth of the oceans is 3200 m and the maximum depth of about 11,200 m is at the bottom of the Challenger Deep in the Marianas Trench near Guam in the western Pacific.

As shown in Figure 1.2, working from the land towards increasing depth, a number of major habitat divisions are recognized. The zone that is influenced by the sea but not always covered in water is the intertidal, or littoral. Next, the sublittoral extends from the extreme low water level down to about 40 m, which is around the safe limit for recreational scuba diving on compressed air. From the edge of the continental shelf the depth increases down the continental slope then slopes more gently down the continental rise to reach the abyssal zone. The continental shelf is the submerged gently sloping border of the land, the width of which varies from 100 m to 1300 km. The continental slope marks the edge of the continents and the region where the seabed slopes at an average angle of 4 degrees to a depth of about 2000 m. The foot of the slope marks the beginning of the abyssal plain.

Marine Ecology: Concepts and Applications, 1st edition. By M. Speight and P. Henderson. Published 2010 by Blackwell Publishing Ltd.

Figure 1.1

Earth from space. Planet Earth taken by Apollo 11, July 16, 1969. (Photograph courtesy of NASA.)

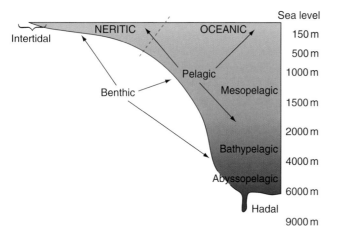

Figure 1.2

Diagram showing location of major marine habitats in relation to depth.

Aquatic habitats are classified by depth and locality within the water body. The term benthic is used to describe living on or within the seabed at any depth. In comparison, the neritic zone extends from the low-tide level to a depth of 200 m, and is thus at or near coastlines in contrast to the oceanic zone which occurs away from land. Pelagic is used to describe the open water habitats, which may lie close to shore and they can also be described as neritic. Pelagic habitats are divided into four depth zones, epipelagic (0–200 m), mesopelagic (200–1000 m), bathypelagic (1000–4000 m),

and abyssopelagic (below 4000 m). The term hadal is used for the deepest parts of the oceans below 6000 m in depth.

The ocean floor is not featureless and the boundaries of the tectonic plates (Figure 1.3) are marked by towering underwater mountain ranges. Figure 1.3 shows the Mid-Atlantic Ridge running down the centre of the Atlantic Ocean, roughly parallel to the shores of Africa and Europe to the east, and the Americas to the west. Similarly, in the Pacific Ocean, approximately 3000 km off the South American coast, there is the East Pacific Rise. This oceanic ridge towers about 2 km from the ocean floor, and stretches from the Gulf of California to the southernmost tip of South America. Submarine ridges owe their formation to the movement of the continental (tectonic) plates. As these plates slowly move away from each other, they leave gaps in the Earth's crust. This allows molten rock from beneath the Earth's crust to move up into the gap, forming a new ocean floor. As the molten rock seeping through these gaps is under pressure, it spews upward, forming a ridge. These ridges cause oceans and seas to be divided into basins. It is in these gaps that hydrothermal vents occur (Figure 1.4), where seawater is superheated by the volcanic activity and discharged in black or white "smokers." This water is rich in dissolved sulfur, iron, and other minerals, and such sites may have supported the first appearance of life on earth.

Mid-ocean ridges are regions of high volcanic activity and are estimated to produce 75% of the total annual output of molten volcanic rock, magma, on earth. It has been estimated that there are more than one million submarine volcanoes and perhaps as many as 75,000 of these volcanoes rise over 1 km above the ocean floor. Some break the surface to form isolated volcanic islands. The Galapagos archipelago in the Pacific Ocean off the coast of South America is a well-known example of a volcanic island group. Ocean trenches are also linked to the boundaries of tectonic plates and are formed as two plates collide and one moves under the other.

Salinity and mineral content

Ocean water has an average salinity of 34.72 parts per thousand (ppt) (or 3.472%) of sodium chloride (NaCl), normally approximated to 35 ppt. This reduces in coastal waters, for example, inshore British or East Coast American Atlantic waters have a salinity around 33–34 ppt. Mixtures of salt and freshwater are termed brackish when the salinity ranges between 8 and 33 ppt and fresh below 8 ppt. In seas where surface evaporation is not balanced by freshwater inputs, salinity can be higher than the average. The Mediterranean Sea has a surface salinity of about 38–39 ppt. Surface water salinity varies across the globe and is, on average, lowest towards the North Pole, probably associated with melting ice-caps, and highest in more tropical latitudes where surface evaporation is greatest (though not necessarily exactly at the Equator due to ocean circulation patterns). Salinity also varies somewhat with depth and lower salinity water is less dense

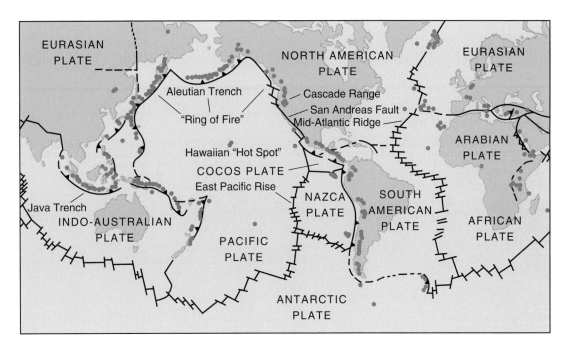

Figure 1.3

The boundaries of the tectonic plates showing the areas of greatest geological activity. (Reproduced with permission of Cascades Volcano Observatory, US Geological Survey.)

Figure 1.4

Hydrothermal venting from sulphur mounds. (Photo courtesy of Submarine Ring of Fire 2006 Exploration, NOAA Vents Program.)

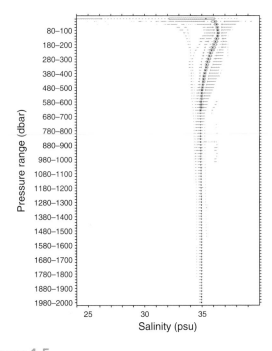

Figure 1.5

Box and whisker plots of salinity within 20 dbar pressure intervals. Data collected from the Gulf of Mexico. (From Thacker 2007; reproduced with permission of Elsevier.)

and will lie over higher salinity water of the same temperature. But as can be seen from Figure 1.5, as depth increases, in this case measured by water pressure measured in dbar or decibars (2000 dbar = 200 bar = 197 atmospheres = 2040 m H_2O), salinity in the open sea (in this example in the Gulf of Mexico) stays fairly constant at just under 35 ppt (Thacker 2007).

As we shall see later in this chapter, temperature and depth are linked, so that deeper water tends to be colder than surface water at least in temperate and tropical seas, though close to deep-sea hydrothermal vents where volcanic activity just beneath the seabed produces superheated seawater from cracks in the Earth's crust, the temperatures can become extremely high. As temperatures reach upwards of 350°C, salinities may drop to less than 10 ppt (1‰ Figure 1.6, Fontaine et al 2007). Maximum salinities of water leaving vents such as black smokers are limited as a result of phase separation, where seawater which enters a vent system becomes separated into a low-salinity vapour phase, which rapidly rises and pours out through the vent chimneys, and a highly saline brine phase, which stays pooled within the vent system and is only released slowly.

The mineral content of seawater is not a simple solution of sodium chloride, but is dominated by 11 chemicals which in order of concentration are chloride, sodium, sulphate, magnesium, calcium, potassium, bicarbonate, bromide, strontium, boron, and fluoride. In addition, there is a large number of trace elements that are listed in Table 1.1. Many of these trace components have important biological functions. For example, calcium is of course a vital building block for exoskeletons, potassium an important fertilizer for marine primary productions, whilst boron is a trace element used for cellular processes by plants such as seagrass. We shall revisit some of these actions in later chapters.

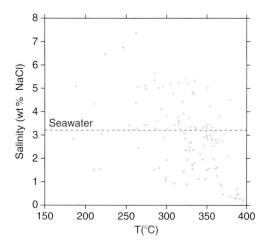

Figure 1.6

The salinity and temperature relationship for high temperature vents. (From Fontaine et al 2007; reproduced with permission of Elsevier.)

Table 1.1 Detailed composition of seawater at 35 ppt salinity in order of abundance (based on values given by Turekian (1968) *Oceans*, published by Prentice-Hall)

	ELEMENT	ATOMIC WEIGHT	CONCENTRATION PPM OR MG L^{-1}
1	Oxygen H$_2$O	15.9994	883000.0000000
2	Hydrogen H$_2$O	1.00797	110000.0000000
3	Chlorine NaCl	35.453	19400.0000000
4	Sodium NaCl	22.9898	10800.0000000
5	Magnesium Mg	24.312	1290.0000000
6	Sulfur S	32.064	904.0000000
7	Calcium Ca	10.08	411.0000000
8	Potassium K	39.102	392.0000000
9	Bromine Br	79.909	67.3000000
10	Carbon C	12.011	28.0000000
11	Nitrogen ion	14.007	15.5000000
12	Fluorine F	18.998	13.0000000
13	Strontium Sr	87.62	8.1000000
14	Boron B	10.811	4.4500000
15	Silicon Si	28.086	2.9000000
16	Argon Ar	39.948	0.4500000
17	Lithium Li	6.939	0.1700000
18	Rubidium Rb	85.47	0.1200000
19	Phosphorus P	30.974	0.0880000
20	Iodine I	166.904	0.0640000
21	Barium Ba	137.34	0.0210000
22	Molybdenum Mo	0.09594	0.0100000
23	Nickel Ni	58.71	0.0066000
24	Zinc Zn	65.37	0.0050000
25	Ferrum (Iron) Fe	55.847	0.0034000
26	Uranium U	238.03	0.0033000
27	Arsenic As	74.922	0.0026000
28	Vanadium V	50.942	0.0019000
29	Aluminium Al	26.982	0.0010000
30	Titanium Ti	47.9	0.0010000
31	Copper Cu	63.54	0.0009000
32	Selenium Se	78.96	0.0009000
33	Stannum (tin) Sn	118.69	0.0008100
34	Manganese Mn	54.938	0.0004000
35	Cobalt Co	58.933	0.0003900
36	Antimony Sb	121.75	0.0003300
37	Cesium Cs	132.905	0.0003000
38	Argentum (silver) Ag	107.87	0.0002800
39	Krypton Kr	83.8	0.0002100
40	Chromium Cr	51.996	0.0002000
41	Mercury Hg	200.59	0.0001500

Table 1.1 (Cont'd)

	ELEMENT	ATOMIC WEIGHT	CONCENTRATION PPM OR MG L^{-1}
42	Neon Ne	20.183	0.0001200
43	Cadmium Cd	112.4	0.0001100
44	Germanium Ge	72.59	0.0000600
45	Xenon Xe	131.3	0.0000470
46	Gallium Ga	69.72	0.0000300
47	Lead Pb	207.19	0.0000300
48	Zirconium Zr	91.22	0.0000260
49	Bismuth Bi	208.98	0.0000200
50	Niobium Nb	92.906	0.0000150
51	Yttrium Y	88.905	0.0000130
52	Aurum (gold) Au	196.967	0.0000110
53	Rhenium Re	186.2	0.0000084
54	Helium He	4.0026	0.0000072
55	Lanthanum La	138.91	0.0000029
56	Neodymium Nd	144.24	0.0000028
57	Europium Eu	151.96	0.0000013
58	Cerium Ce	140.12	0.0000012
59	Dysprosium Dy	162.5	0.0000009
60	Erbium Er	167.26	0.0000009
61	Ytterbium Yb	173.04	0.0000008
62	Gadolinium Gd	157.25	0.0000007
63	Ruthenium Ru	101.07	0.0000007
64	Praesodymium Pr	140.907	0.0000006
65	Beryllium Be	9.0133	0.0000006
66	Samarium Sm	150.35	0.0000005
67	Thorium Th	232.04	0.0000004
68	Holmium Ho	164.93	0.0000002
69	Thulium Tm	168.934	0.0000002
70	Lutetium Lu	174.97	0.0000002
71	Terbium Tb	158.924	0.0000001

Defining and measuring salinity

Salinity is expressed as either parts per thousand (ppt) or on a practical salinity scale (PSS) often termed practical salinity units (psu). For most purposes and waters there is little numerical difference between ppt and psu measurements. Originally salinity was defined to be the total amount of dissolved material in grams in one kilogram of seawater. This is not useful in practice because the dissolved material is impossible to measure. Because salinity is directly proportional to the amount of chlorine in seawater, and chlorine can be measured accurately by a simple chemical analysis, salinity, S, was redefined using chlorinity, Cl, as

$$S = 1.80655 \text{ chlorinity}$$

where chlorinity is defined as the mass of silver required to precipitate completely the halogens in 0.3285234 kg of the seawater sample.

Oceanographers now use conductivity meters to measure salinity, where the passage of an electrical current through water is related to the amount of salts dissolved within it. The equation relating conductivity to salinity is termed the practical salinity scale (PSS). With careful calibration, an accuracy of 0.002 PSS and a precision of 0.001 PSS can be achieved. Biologists working in coastal and estuarine waters are more likely to use refractometers which measure the salt content by the change in direction of light as it passes across a film of water placed on the instrument. The accuracy at best is 0.1 ppt.

Estuaries and sediments

Within estuaries there are salinity gradients ranging from 0 in the river to 35 ppt at the seaward limit. In the water column, salinity varies with the tide, wind and river flow creating a rapidly and constantly varying habitat for organisms that maintain a fixed position on the seabed. Because saline water has a higher density than freshwater there is a tendency for marine waters to flow in along the bottom and freshwater to flow out on the surface. The intrusion of higher salinity waters along the bed of an estuary is often termed the salt wedge. These flows and changes in salinity also cause the flocculation of clay and the deposition of sediments. Flocculation occurs when very small clay particles combine into groups to form larger crumbs or flocs which sink to the bottom, removing significant amounts of metal ions from the water column. As shown in Figure 1.7, the dramatic changes in water salinity observed in estuarine water do not occur within the bottom sediments. Within a few centimeters below the sediment surface, salinity concentrations remain fairly constant no matter what is happening in the water above. This relative stability within the sediments is important for bottom-living organisms that are unable to tolerate changes in salinity.

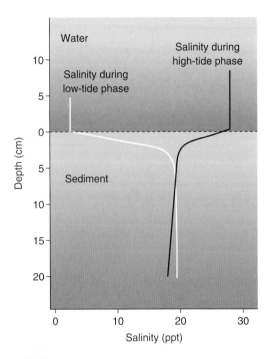

Figure 1.7

Variation in the salinity within the water column and within the bottom sediments of an estuary.

Salinity tolerance

Organisms are classified by their ability to withstand variation in salinity. Obligate freshwater organisms do not live in waters that exceed 8 ppt and obligate fully marine organisms, which will not tolerate salinities below about 30 ppt, are termed stenohaline. Echinoderms such as starfish and sea urchins are stenohaline, predominantly due to their unique water vascular systems which will only function if their internal body fluids are isosmotic (having an equal osmotic pressure) with the surrounding seawater. Both freshwater and stenohaline marine species cannot survive in the variable salinities of estuaries. Animals able to withstand wide salinity variation are termed euryhaline, and include many of the familiar Crustacea such as shore crabs (Figure 1.8) which can be found in all estuaries, salt marshes, and rock pools, and fish such as salmon, flounder, shad, and eel. Many fish and lamprey, including river lamprey, salmon and shad, undertake most of their growth in the sea and only return to freshwater as adults to breed. These species are termed anadromous. Species such as eels (*Anguilla* spp.) (Figure 1.9) start their life at sea and may enter freshwater to grow, only returning to the sea to spawn. These species are termed catadromous. These fish are discussed in more detail in Chapter 8. In addition to reproductive movements between marine and freshwaters, there are also many species of marine fish that use estuaries as nursery grounds (Elliot et al 2007) as they offer rich feeding and sheltered habitat such as salt marsh.

Salinity variation lies at the core of estuarine biology, acting as a physiological barrier for species lacking the

Figure 1.8

The shore or green crab, *Carcinus maenas*. (Photograph courtesy of Paul Naylor.)

Figure 1.9

European eel, *Anguilla anguilla*, a catadromous species of fish that during its lifecycle moves from freshwater to the sea and back to freshwater. (Photograph courtesy of Richard Seaby, Pisces Conservation Ltd.)

physiological ability to adapt. Euryhaline animals use several different strategies to adapt to salinity change. Among the vertebrates, blood osmotic concentrations are regulated within a narrow range by hormonal controls of ion fluxes and the accumulation of organic chemicals (amino acids and their derivatives) called osmolytes, which adjust the water content of cells and maintain their volume under varying environmental salinity levels (Pequex et al 1988). Invertebrates show several adaptive strategies, but they can be roughly classified as conformers, regulators, or a mixture of the two. The common shore crab *Carcinus maenas* (Figure 1.8) demonstrates both invertebrate approaches. At salinities above 25 ppt, the blood osmotic concentration tracks that of the ambient water, it is a conformer. At salinities below 25 ppt, it uses physiological mechanisms to regulate blood salt levels. This regulation can be maintained down to salinities of 8 ppt; it cannot survive in freshwater.

Man also discharges hypersaline water into estuaries and the ocean. The effects of these artificial elevations are discussed on page 190.

Depth, pressure, and topography

It is notable that more than 60% of the earth surface is more than 2 km below sea level, and physical conditions at this depth differ greatly from those on the surface. The pressure at the surface of the sea is approximately 1 atmosphere (depending on weather conditions), and increases by roughly 1 atm for every 10 m increase in depth. 1 atmos-

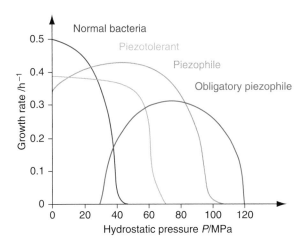

Figure 1.10

Definitions of the relationships between growth rate of microorganisms and pressure. Atmospheric pressure (surface) = 0.1 MPa; 120 MPa = 1200 atmospheres or 12,000 m. (After Margesin & Nogi 2004; reproduced with permission of Chemical Society Reviews.)

phere (atm) = 14.69 pounds per square inch (psi), or 1.03 kilograms per square centimeter (kgf/cm^2), or 0.1 megapascals (MPa). At 1000 m depth, the pressure is just over 100 atm, 1472.6 psi, 103.5 kgf/cm^2, or 10.2 MPa. These pressure conditions have resulted in the evolution of organisms specially adapted to living in deep water. Interestingly, it has been suggested that all life on earth may have originated in the stable, calm, protective depths of the deep ocean, maybe 3.8 billion years ago (Daniel et al 2006). The external pressure (and temperature) affects membrane and enzyme systems (Carney 2005) resulting in a vertical zonation of species, adapted to different (but relatively constant) pressures (Blankenship et al 2006). An example of this type of zonation is shown by bacteria (Figure 1.10, Daniel et al 2006 after Margesin & Nogi 2004). Organisms specially adapted to life in very deep water at very high pressures are termed piezophilic, and organisms such as these bacteria are specialized to grow optimally at particular depths and pressures. Similar species are partitioning the depth resource resulting in the avoidance of interspecific competition (see Chapter 6).

Light and irradiance

Only a small fraction of the sunlight incident on the sea surface is reflected, the greater proportion entering the water. The rate at which sunlight is attenuated determines the depth that is lit and heated by the sun. Attenuation is due to absorption by pigments and scattering by dissolved molecules and suspended particles. The rate of attenuation depends on the wavelength of the light. Blue light is absorbed the least and red light is absorbed most strongly. Thus, as divers move down through clear ocean water they perceive an environment that becomes increasingly blue; bright colors, especially the reds and yellows, quickly fade to grey. The change in the light spectrum with depth is shown in Figure 1.11 and the

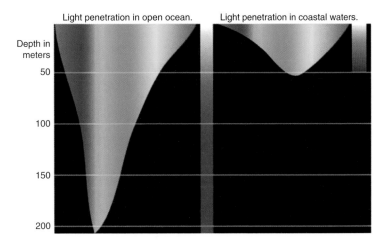

Figure 1.11

Light penetration with depth in open ocean and coastal waters. (Courtesy of Kyle Carothers, Ocean Explorer, NOAA.)

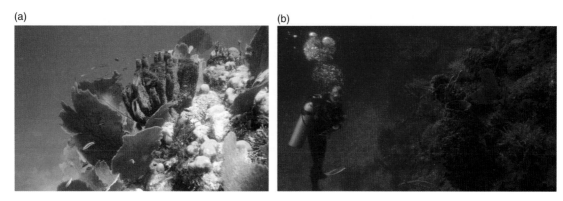

Figure 1.12

(a) Bay Islands (Honduras) coral reef at 2 m depth. (b) Bay Islands (Honduras) coral reef at 20 m depth. (Photographs Martin Speight.)

photographs in Figure 1.12 show sections of the same coral reef on the island of Utila in the Caribbean off the mainland of Honduras taken using natural light only with the same camera on the same dive. Note that the vast majority of wonderful photos of marine life showing striking colors are taken with powerful flash guns (strobes). These colors are normally invisible to the local fauna and visitors alike. The physical mechanisms for differential light absorption are complex. Put simply, pure water is itself very slightly blue because water molecules absorb light at the red end of the visible spectrum (Braun & Smirnov 1993), In fact, if the absorption coefficient is constant, the light intensity decreases exponentially with depth:

$$l_x = l1 \, \exp^{(-cx)}$$

where l is the original radiance or irradiance of light, and l_x is the radiance at depth x and c is the absorption coefficient.

In addition, coastal waters are typically more turbid (less clear) than offshore ocean waters. They contain pigments from land (sometimes called gelbstoffe which just means yellow stuff) and suspended sediments from rivers and the action of waves on the seabed in shallow water. Very little light penetrates more than a few meters into these waters. In some particularly turbid estuaries where high tidal currents result in levels of suspended solids as high as 1 g l^{-1} or more, light may penetrate less than 1 m. Crucially of course, light fuels the primary production of shallow seas via photosynthesis, so that if it is unable to penetrate far into the water, primary productivity will be highly constrained. Further, photosynthetic organisms such as macroalgae (seaweeds), microalgae (phytoplankton), and symbiotic algae such as zooxanthellae in coral polyps and other cnidarian tissues also respire, so that if their energy capture by photosynthesis is less than that used by respiration, there is a net loss of production. The depth at which respiration losses equal photosynthetic gains is called the compensation depth, where light penetration is just sufficient for production to match that lost by respiration (see Chapter 3). Above this depth, light can influence the distribution of organisms,

and/or their abilities to survive and grow, as shown in the example of corallimorph, *Rhodactis rhodostoma* (Cnidaria: Anthozoa) from Red Sea coral reefs (Kuguru et al 2007). As mentioned above, almost all corals, and many other marine organisms, contain intracellular symbionts, dinoflagellates called zooxanthellae in the genus *Symbiodinium*, which photosynthesize using nutrient chemicals from their host. The abundance of zooxanthellae within polyps, and the quantity of chlorophyll *a* pigment they hold, increases significantly with depth. Both these changes are responses to reduced light levels with depth. Because of this response animals like *Rhodactis* are able to exist successfully over a range of depths and varying light levels. It seems that different strains of zooxanthellae with different responses to irradiance levels occur in polyp tissues at different depths. In contrast to the limitations caused by low light levels, too much light (high irradiance) can have severe affects on marine organisms. Coral bleaching is one of the most serious global threats to shallow marine tropical ecosystems and, in part at least, seems to be a function of intense light levels, especially from the ultraviolet end of the spectrum. In very high light levels, the zooxanthellae either lose their chlorophyll, and/or die. Either way, the photosynthetic ability of the symbionts declines catastrophically, to the detriment of the host animal. This condition may not be irreversible. In the case of *Rhodactis* at least, removing the stressing effects of light (and temperature) enables the zooxanthellae to regain their photosynthetic ability.

Temperature

Figure 1.13 shows the variation in seawater surface temperature (SST) with latitude from less than 0°C (<32°F) close to the poles to over 30°C (86°F) in the tropics. This latitudinal gradient is linked to variation in the light energy received per unit area. However, surface temperatures are not perfectly correlated to the received energy because of

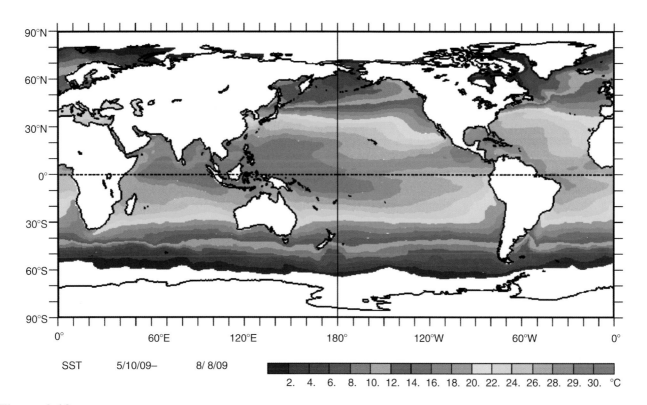

Figure 1.13

Global variations in sea surface temperatures (SST). (Courtesy of NOAA – www.cdc.noaa.gov/map/images/sst/sst.gif)

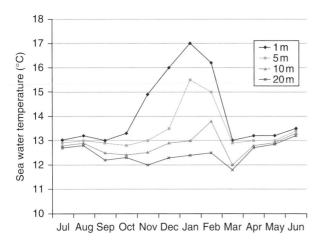

Figure 1.14

Annual variation in sea temperature according to depth at Las Cruces in central Chile. (From Narváez et al 2004; reproduced with permission of Elsevier.)

ocean currents. Notice, for example, the incursions into the otherwise warm areas along the west coasts of South America and Africa, caused by cold currents from the Antarctic (Humboldt and Benguela currents respectively).

Deep ocean waters are fairly constant in temperature, ranging from about 0°C to 4°C (32°F to 39°F), though high pressures at depth cause slight adiabatic warming because of compression. However, deep-sea hydrothermal vents are

a notable local exception as we mentioned earlier in this chapter: water from these can raise local temperatures to well over 100°C (212°F), and may exceed 400°C at the point of emergence. In the shallow temperate zones there are considerable seasonal temperature variations. Some of the most extreme occur on the North American East Coast. In the River Hudson Estuary near New York, for example, surface temperature can vary from below 0 to 30°C. As shown in Figure 1.14 coastal waters do not show the same degree of variation and this variation declines with depth (Narváez et al 2004). In this example from the southern hemisphere, notice that the coldest seawater can be experienced not in mid-winter, but in spring, due to the time lag in the cooling and heating of the huge mass of water. In general, most marine organisms living in deep water experience relatively small variations in temperature, when compared to terrestrial life, they are not well adapted to extremes of temperature. The only exceptions are organisms specially adapted to changing conditions such as those of littoral habitats. An example of an animal with a particularly narrow and limited temperature adaptation is the mussel *Bathymodiolus childressi* (Mollusca-Bivalvia: Mytilidae) (Figure 1.15), which occurs around cold seeps in 750 m of water in the Gulf of Mexico (Berger & Young 2006). Cold seeps were only discovered in the 1980s, and are places where water from the underlying bedrock flows out, rather like an underwater spring. This water is rich in methane and sulfides which provide the chemosymbiotic bacteria in the mussel tissues with fuel for primary production (see Chapter 3). Unlike the very hot hydrothermal vents, cold seeps are at the same temperature as the surrounding water, perhaps 2 or 3°C. Under these stable conditions, *Bathymodiolus* is unable to survive in water warmer than 20°C for very long.

Water temperature at the sea surface can vary considerably, and studies of sea surface temperature (SST) are an important research topic. SST can now be measured using high-resolution satellites such as those deployed by NOAA (National Oceanic & Atmospheric Administration), and NASA (National Aeronautics & Space Administration), both in the USA (Mesias et al 2007). Clearly, any increase in temperature will have some influence on the metabolic rate of most marine organisms, since the vast proportion of spe-

Figure 1.15

Deep sea mussel community with squat lobsters and shrimps. (Photo courtesy of Submarine Ring of Fire 2006 Exploration, NOAA Vents Program.)

cies living in the sea are poikilothermic ("cold blooded"), and as we shall see in detail in Chapter 3, oceanic primary productivity is closely linked to water temperature, though not necessarily in a simple linear manner. The SST can influence the whole structure of marine communities, as shown for example in intertidal habitats in California (Blanchette et al 2006). The percentage cover of filter feeders such as barnacles, *Chthalamus* and *Balanus* spp. (Crustacea: Cirripedia), and mussels, *Mytilus* spp., (Mollusca-Bivalvia: Mytilidae) increases linearly as mean SST increases, linked to the increasing numbers of juveniles settling (so-called recruitment rate) with increasing SST. However, the cover of primary producers such as seaweeds decreases with increasing SST, probably linked to increased numbers and activity of herbivores (see Chapter 4).

SSTs that exceed normal variations, or are atypical at various temporal scales, may indicate changes which can have serious, even catastrophic, consequences for marine ecosystems and indeed global climate patterns. One illustration of these SST anomalies is shown in Figure 1.16 (Behrenfeld et al 2006). The diagram shows positive (pink) and negative (blue) anomalies by comparing SSTs from 1999 to 2004. Changes in SSTs to warmer conditions can be seen parts of the Arctic, Atlantic, Indian and especially Pacific oceans, as well as the Caribbean. One of the most threatening physical factors in marine ecology today is that of elevated SSTs on coral survival. Figure 1.17 illustrates a clear relationship between SST anomalies and coral bleaching in the Caribbean (McWilliams et al 2005). It seems that as little as a 1°C increase in SST during the hottest months of the year can cause bleaching, when the symbiotic zooxanthellae either lose their chlorophyll or die. The optimum temperature for most hard (scleractinian) corals is between 25 and 29°C, and even increases in water temperature to 30 or 31°C can cause serious losses of zooxanthellae

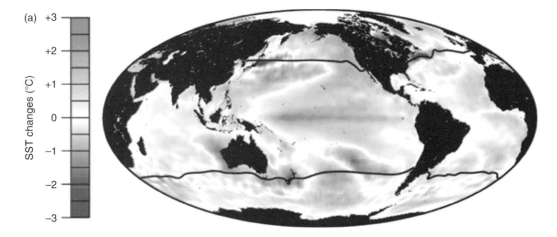

Figure 1.16

Global changes in annual average sea surface temperatures (SSTs) for the period 1999 to 2004. (From Behrenfeld et al 2006b; reproduced with permission of *Nature*.)

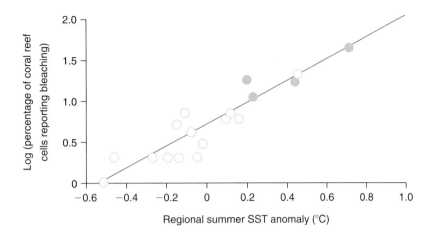

Figure 1.17

The relationship between the regional SST anomalies and the percentage of coral bleaching. (From McWilliams et al 2005; reproduced with permission of *Ecology* – ESA.) Each data point represents 1 year. Solid circles represent years described in the literature as mass-bleaching events; open circles represent other years.

(Sammarco et al 2006). If the symbionts are unable to recolonize, or the warm conditions persist for too long, corals may die on a massive scale. It may be that variations in temperature are more destructive than steady but stable increases. The above authors looked at SSTs using discriminant function analysis (DFA) to group their data on coral bleaching events on reefs around Puerto Rico. Three groups were identified: cool water with no bleaching; warm water also with no bleaching; and warm water with bleaching. The coefficient of variation (CV) of the data measures variability (degree of fluctuation), and the likelihood of bleaching in warm water was found to increase at lower temperatures as the temperature CV increased. Without doubt, climate change and global warming (see later in this chapter) are exerting pressures on some of our most precious marine ecosystems. As we suggested above, all may not be lost; some coral species seem able to thermally acclimatize to increasing water temperatures, and their symbionts, the zooxanthellae, are able to exchange temperature-tolerant genotypes. Berkelmans & van Oppen (2006) suggest that "though such mechanisms might not enable corals to survive all of the SST increases predicted for the next 100 years, it may buy them time."

Man also discharges heated water into estuaries and the ocean. The effects of these artificial temperatures are discussed in Chapter 9 (see p. 184).

Oxygen

In the open ocean the oxygen content at the surface is relatively high (about 6 ml l^{-1}) and is replenished from the air. Deeper in the water the oxygen content begins to decrease with depth until at about 1000 m (3082 feet) the value reaches a minimum. The reason for the decrease is the con-

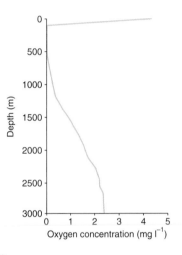

Figure 1.18

The variation in oxygen concentration with depth in the eastern tropical Pacific Ocean at 13°23'N, 102°27'W. (Modified from Wishner et al 1990.)

sumption by bacteria of the rain of organic debris (marine snow) falling through the water. The exact amount of oxygen at the minimum varies with location in the ocean. The oxygen concentration profile for the Eastern tropical Pacific Ocean, which is noted for the severity of the oxygen minimum, is shown in Figure 1.18. This minimum is known to reduce the abundance and diversity of life in the midwater region (Wishner et al 1990). The deep water in the ocean starts out at the surface in polar regions and when it sinks it carries dissolved oxygen from the surface (see p. 16 for information on currents).

In inshore, shallow waters oxygen concentration can vary greatly both spatially and temporally. It is not uncommon for bottom waters in some parts of estuaries to be almost anoxic because of oxygen consumption by bacteria and other micro-organisms. In estuarine and shallow coastal waters the oxygen concentration is one of the key physical variables determining the abundance and diversity of life. While hypoxic and anoxic waters occur naturally, there are clear indications that oxygen deprivation is increasing and that this is linked to the activities of man. Diaz (2001) in a review of hypoxia concluded "that many ecosystems that are now severely stressed by hypoxia may be near or at a threshold of change or collapse (loss of fisheries, loss of biodiversity, alteration of food webs)." He also noted that several large systems for which we have reliable nineteenth century oxygen concentration data (including the Kattegat, between Denmark and Sweden) and did not then suffer from hypoxia, now experience severe seasonal hypoxia. Reports of a decline in ocean oxygen levels are generally becoming more frequent, and oxygen concentration decline is likely to be an important area of concern for the foreseeable future.

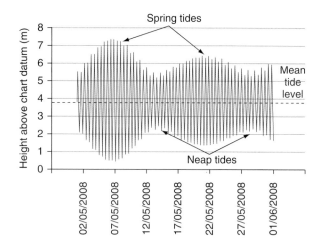

Figure 1.19

Typical series of tidal cycles over a month from Milford Haven in Southwest Wales. Chart datum (*y*-axis) is mainly used on nautical charts and is the lowest possible astronomical tide which may never actually be achieved over many years. (Data from 'Tide Plotter', Belfield Software.)

Tides

Tides are the periodic rise and fall of the sea. They are one of the most important physical features for life in coastal waters, creating the productive but challenging conditions within the littoral zone and the currents used by animals and plants for dispersal (see Chapter 7). The most important tidal waves are caused by the gravitational interaction between the Earth and the Moon (lunar waves), with other components such as the interaction between the Earth and the Sun (solar waves) being significant but of lower magnitude. The gravitational attraction of the Moon causes the oceans (simply a very large volume of incompressible fluid) to bulge out in the direction of the Moon. Another bulge occurs on the opposite side, since the Earth is also being pulled toward the Moon (and away from the water on the far side). As the Earth is rotating, there are about two high tides per day, but the Moon actually takes about 24 hours and 50 minutes to return to the same position in the sky from one day to the next. Thus in general, each high tide is 12 hours and 25 minutes later than the one before it.

Spring tides are especially strong tides that occur when the Earth, Sun, and Moon are aligned and the gravitational pull of the Moon and the Sun are working together. This alignment occurs at the full and new moons, so that there are two spring tides every month. Note that the term "spring" has nothing to do with the seasons. In contrast, the smallest tidal ranges over the lunar cycle, called neap tides, occur when the sun and moon are pulling in opposite directions. Not all springs and neaps are of equal extent however, since the Moon comes closer to the Earth at certain times of the year. When the Moon is closest to the Earth, it is said to be at apogee, and when it is furthest away, it is at perigee. If the Moon at apogee is directly in line with the Sun, then an extra pull on the oceans occurs and so produces extreme spring tides. The highest and lowest tides of

a year take place a day or two after the nearest new or full moon to the spring (now a season) and autumn equinoxes in March and September. To summarize, Figure 1.19 shows typical tidal cycles in South Wales over a month, indicating that the tides go in and out twice a day, that springs and neaps occur twice a month, and that the extent of a spring tide also varies over a few weeks. Most importantly, notice that whatever the peaks and troughs, or high tide and low tide extents, the average of a tidal cycle (between high and low tide on a particular cycle at a particular place, is always the same. We shall return to the ecological significance of mean tide level (MTL) below.

Tides differ in periodicity and the rate of rise and fall between localities because the tidal wave is reflected from the continental edges creating interference patterns and can be funneled within inlets creating exceptionally large tidal ranges. The tidal range varies dramatically between localities (Figure 1.20, Kowalik 2004). The M_2 tides are depicted in Figure 1.20; these are the principal lunar component of total tidal cycles with an absolute periodicity of 12.42 hours. The Figure shows that the highest and lowest M_2 tides can be experienced on the Atlantic coast of Western Europe and North Africa, on the Indian Ocean coasts of East Africa, and on the Pacific coasts of Alaska, British Columbia and Washington State, Columbia, and Ecuador. The largest tidal ranges of more than 15 m occur in the Bay of Fundy, Canada, in estuaries in Northern France, islands in the western English Channel (Figure 1.21) and in the Bristol Channel, UK. These huge tides are created by the flow of tidal waves into funnel-like water bodies. The exceptional long narrow funnel of the Bay of Fundy results at Burntcoat Head in a tidal range of 16.1 m, the greatest on the planet. In contrast, Eureka, on Ellesmere Island, Canada

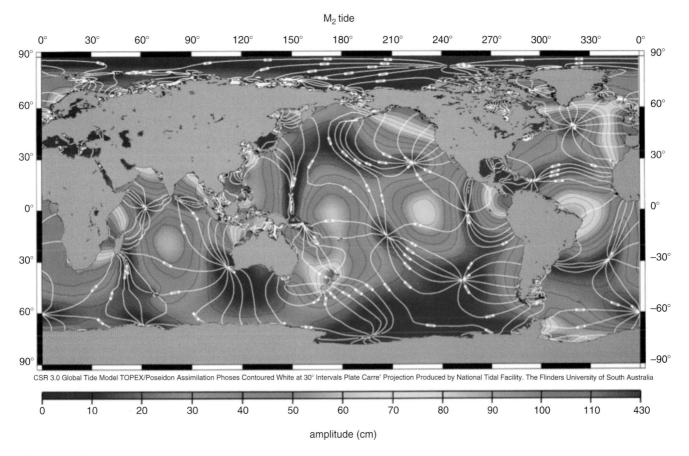

M$_2$ tide

CSR 3.0 Global Tide Model TOPEX/Poseidon Assimilation Phoses Contoured White at 30° Intervals Plate Carre' Projection Produced by National Tidal Facility. The Flinders University of South Australia

amplitude (cm)

Figure 1.20

The geographical variation in tidal height. This figure shows the M$_2$ tidal component which is the dominant tidal component caused by the movement of the Moon. (From Kowalik 2004; reproduced with permission of Institute of Oceanology PAS.)

Figure 1.21

Low and high tides in Jersey. (Photographs courtesy of Jonathan Shrives.)

probably has the smallest tidal range of only 0.1 m. In the mouths of some rivers, the incoming tide meets the out flowing current and builds up forming tidal bores. These are fast-moving currents that travel as a wave front or wall of water. They can produce spectacular waves that in the River Severn, England and the Amazon estuary, Brazil, can be used by surfers.

Mean tide level (MTL) has particular significance for organisms living between the tides. Any organism living on a rock, in a pool or in sediment at this point will spend

50% of its time away from the direct influence of the sea. Above MTL, life for marine organisms becomes more and more difficult, requiring complex physiological, morphological and/or behavioral adaptations to cope with living in a terrestrial environment for increasing periods of time. Of course it is perfectly possible and indeed normal on all but the most sheltered shores for the sea's influence to extend much further than the height of an extreme high tide by virtue of splash driven by winds and waves. We discuss the phenomenon of exposure in this context in the next section. Note finally that although tides are usually thought of as operating at the sea's surface, they can also occur in the deep ocean as internal tides (Garrett & Kunze 2007). Internal tides are produced by the interaction of deep currents with the varying seabed topography, and can cause the vertical displacement of water by tens or even hundreds of meters, enabling mixing of water masses. Mixing in a fluid such as seawater increases dramatically over a region of structurally complex seabed as compared with a homogeneous, smooth topography, and this turbulence can extend for many meters above the seabed. These tides are not much influenced by astronomical bodies, but mainly by pressure and topography, hence their name of barotropic tides. It is not hard to imagine the great potential for sediment suspension, and the nutrient and propagule mixing in the deep-sea derived from internal tide generation.

Waves

Wind causes surface waves. The wind transfers energy to the water, through friction between the air molecules and the water molecules. Waves of water do not move horizontally, they only move up and down. The wave height is the distance between the wave crest and trough. This can vary dramatically from negligible to extreme, with a maximum of probably in excess of 30 m, although such monster waves have rarely been measured. In 1998, a buoy moored 500 km southeast of Cape Breton recorded a maximum wave height of 27 m when the eye of Hurricane Danielle passed nearby. In September 2004 Scientists at the Stennis Space Centre measured a record ocean wave of 27.7 m in height when the eye of Hurricane Ivan passed over moorings deployed in the Gulf of Mexico. The highest average wind speeds occur in the Southern Ocean where wave heights frequently exceed 6 m. The distance between wave crests is termed the wavelength and the maximum depth at which the wave motion is experienced is half the wavelength. It follows therefore that the deeper the water for a wave of a given length, the less affected organisms and habitats will be. For example, with a wavelength (distance between one wave and the next following it) of say 30 m, a diver or any other object would hardly feel the movement or swell at all. As all SCUBA divers

(a)

(b)

Figure 1.22

Rocky shore communities under two exposure extremes. Both habitats are in close proximity on the North Somerset coast, England. High exposure shows domination of species that attach tightly such as barnacles, low exposure shows luxuriant macroalgal growth and few barnacles. (a) Very exposed to wave action. (b) Very sheltered from wave action. (Photographs courtesy of Richard Seaby, Pisces Conservation Ltd.)

know, it is the ascent to the choppy or even violent surface that can be the worst part of a dive.

The degree of exposure of the coast to wave action determines the nature of the substrate and the community of plants and animals. There is a clear change in the temperate rocky shore community that can be related to wave action. W.J. Ballantine invented a system of exposure rating in the UK in 1961 which depends on the distribution and occurrences of common sessile or sedentary intertidal organisms. In a very sheltered region (exposure scale 7 or 8), large, luxuriant species such as brown seaweeds (fucoids or wracks) dominate, whereas on exposed shores (exposure scale 1 or 2), where large weeds would be washed away, encrusting species are most common such as barnacles and a few small, tightly attached macroalgae (Figure 1.22a,b). This concept of exposure in relation to shallow-water community structures also has parallels in the tropics.

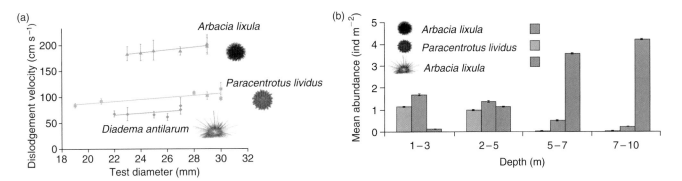

Figure 1.23

(a) Mean "velocities of dislodgement" for each individual of three sea urchin species calculated from hydrodynamic experiments. Error bars represent ±s.e. of means. (b) Mean abundances of each sea urchin species at each depth stratum across the study area. Error bars represent ±s.e. of means; $n = 140$ for each species. ((a,b) From Tuya et al 2007; reproduced with permission of Elsevier.)

For example coral reef structure is influenced by wave action caused by storms (Hubbard and Dennis 1989). For example, Caribbean reef type is determined by the wave energy and three types of habitat can be defined:

- TYPE I: algal ridges with reef crests dominated by coralline algae rather than coral. The exposure to frequent storm damage breaks corals and provides coral substrate for algae. High wave energy reduces fish grazing which would otherwise inhibit algal growth.
- TYPE II: branching elkhorn coral, *Acropora palmata*, dominates. There is high wave energy, but less frequent storms.
- TYPE III: only scattered coral cover with open "pavements" and a relatively low diversity community. Frequent storms disrupt the reef-crest, but low wave energy conditions between storms permits grazing, reducing the deposition of thick algal crusts, but also discouraging coral recruitment.

On an even more general scale, whole regions of coral reefs can be recognized (see also Chapter 2). Caribbean reefs, for example, have many more soft corals than those in Indonesia or Australia, since the ones in the Caribbean have evolved over thousands of years with annual storms and even fairly regular hurricanes (especially in recent years), which soft corals are better able to withstand than hard species.

The distribution of individual species can be determined by wave action and exposure. Experiments by Tuya et al (2007) with three species of sea urchin showed that whilst the abundance of *Diadema antillarum* increased with depth (Figure 1.23a,b), two other species, *Arbacia lixula* and *Paracentrotus lividus*, showed a reverse zonation. These distributions were directly related to the dislodgement water velocity for each; *Diadema* with the biggest body and largest spines being most easily dislodged by waves and tides and hence occurred only at deeper, quieter, depths.

Ocean currents

Ocean circulation includes both horizontal and vertical flows that are important for the movement of heat over the planet. Vertical flows are the movement of water up or down the water column, which may take place over many hundreds if not thousands of meters of depth. Ocean circulation, both horizontal and vertical, is induced by the wind acting on the sea surface, and by buoyancy changes caused by the alterations in salinity and especially density. Warm air at the Equator rises leaving a less dense, low pressure region into which air flows from both the north and the southern hemispheres producing the trade winds well known to ocean explorers. At about 30 degrees north and south, this warm air cools and sinks again, some returning to the Equator, but the rest heading towards the poles as westerly winds in both north and south hemispheres. Finally, cold air at the poles sinks and flows away from the poles to meet the westerlies. All these winds cause a frictional drag on seawater at the surface, and move it with them; typically, a surface current is around 2 or 3% of the speed of the wind which blows over it. Buoyancy differentials between surface and deeper waters are capable of inducing overturning currents (they are termed overturning because they bring bottom water to the surface and visa versa) that reach from the surface to the seabed. Cooling and evaporation both make surface seawater denser and therefore reduce buoyancy, so that surface water tends to sink at the poles. In contrast, solar heating and rain reduce surface density and therefore increase buoyancy. The rotation of the Earth rotates current flow to the right of the wind direction in the northern hemisphere, and to the left of the wind direction in the south, via Coriolis forces. Coriolis was a French scientist who described the ways in which winds flow from high to low pressure areas, and he discovered that because of the planet's rotational spin, this air flow does not occur in a straight line but is bent relative to an observer on Earth. Note that Coriolis forces are zero at the Equator.

As currents on the surface are shifted by these forces, so frictional coupling with slower, deeper water layers drags subsurface currents along with the surface ones, but since Coriolis forces also act on these deeper currents, the latter are deflected further around in a spiral fashion which moves further left or right depending on hemispheres. This produces a phenomenon known as Ekman transport, named after the Swedish physicist, which adds to Coriolis forces moving currents further right or left relative to wind direction. The lowest layers of seawater may be rotated up to 90 degrees compared with those at the surface.

The major horizontal currents

The sum of wind, buoyancy, Coriolis effects and other forces produce the major ocean surface currents, which are listed in Table 1.2 and shown in Figure 1.24. These currents circulate in paths called gyres, which rotate in a clockwise direction in the northern hemisphere and a counter-clockwise direction in the southern hemisphere. One of the most well known gyres is the Gulf Stream that flows across the Atlantic Ocean

Table 1.2 The major ocean surface currents

NAME	LOCALITY	TEMPERATURE
Agulhas Current	Indian	Warm
Alaska Current	North Pacific	Warm
Benguela Current	South Atlantic	Warm/Cool
Brazil Current	South Atlantic	Warm
California Current	North Pacific	Cool
Canaries Current	North Atlantic	Cool
East Australian Current	South Pacific	Warm
Equatorial Current	Pacific	Warm
Gulf Stream	North Altantic	Warm
Humboldt (Peru) Current	South Pacific	Cool
Kuroshio (Japan) Current	North Pacific	Warm
Labrador Current	North Atlantic	Cool
North Atlantic Drift	North Atlantic	Warm
North Pacific Drift	North Pacific	Warm
Oyashio (Kamchatka) Current	North Pacific	Cool
West Australian Current	Indian	Cool
West Wind Drift	South Pacific	Cool

from the southern states of the USA to Western Europe including the UK. The most important consequence of this is that this part of Europe has the warmest climates of anywhere on the globe at this latitude. Without the Gulf Stream the UK would be a very much colder place in winter. This surface current moves at an average of 3 or even $4\,km\,h^{-1}$ in a narrow band perhaps only 50–75 km wide, and it is quoted as transporting more than 30 million cubic meters of water per second (Lund et al 2006), with the transportation potential of heat and solids of almost unimaginable quantities.

Horizontal currents also occur in the deep-sea. Indeed, the Earth's climate is regulated to a great extent by the movement of large, deep-water masses such as the Antarctic Bottom Water (AABW) and the North Atlantic Deep Water (NADW). These currents are termed global "motors" for the exchange of large masses of water (Schlüter & Uenzelmann-Neben 2007).

Vertical currents and the global conveyer belt

Vertical motions in the ocean are driven by small differences in water density due to differences in salinity and/or differences in temperature. These water movements are termed thermohaline circulation. Increased salt content increases the density of water, and above 4°C the density of water decreases with increasing temperature. However, below 4°C the density of pure water starts to decline with decreasing temperature and this property is very important in Arctic and Antarctic waters. It is easy to remember this important feature of cold water, as icebergs float. The fact that ice is less dense than water is very important for the biology of temperate regions, if the reverse were true, lakes and seas would freeze from the bottom upwards and it would be impossible for fish and other organisms to survive in waters in the far north or south as some water bodies would freeze completely.

As was discussed earlier in this chapter, the oceans do not have a uniform salinity. As water flows towards the poles from the Equator, it passes the subtropical high-pressure zones where there is little rain but high levels of sunshine producing high rates of evaporation. Evaporation increases the salt content of the surface water raising the density. The influence of evaporation is particularly apparent in the Mediterranean Sea that receives relatively low inputs from rivers but has a large evaporative loss due to high levels of solar radiation. As a result, dense water is created in the basin that flows out of the Straits of Gibraltar close to the seabed. This is replaced by an inflow of less salty Atlantic Ocean water at the surface.

In contrast to the Mediterranean there are also regions where reduced surface salinities are generated. In regions where rainfall is high, such as the Intertropical Convergence

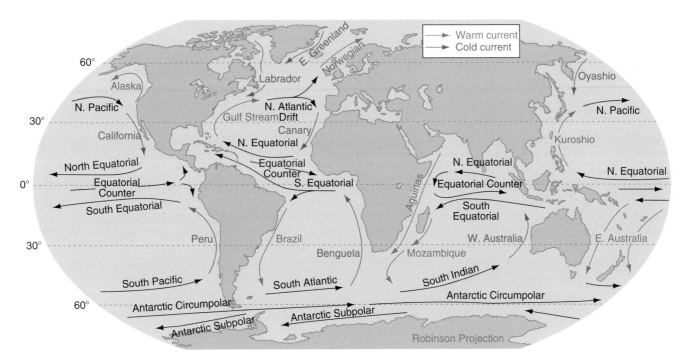

Figure 1.24

The major ocean currents. Surface currents are shown in red and bottom currents in blue. (From http://www.geni.org/globalenergy/library/renewable-energy-resources/oceanbig.shtml; reproduced with permission of Michael Pidwirny.)

Figure 1.25

A diagram of the global conveyer belt – the circulation pattern which moves water heat and organisms around the globe. (From Haupt & Seidov 2007 after Brasseur et al 1999; reproduced with permission of Elsevier.)

Zone in the central Pacific Ocean, low salinity water floats on top of the more saline ocean water. Similarly melting ice in polar regions reduces the density of surface waters both because of the lack of salt and the reduction of the surface temperature below 4°C. It is currents caused by density differences that link the surface and abyssal ecosystems. These various interactions result in three-dimensional ocean circulations. In the North Atlantic, for example, water flowing north at the surface passes through the subtropical high-

pressure zone where density increases. As it continues north the surface water cools, causing a further increase in density. Finally, to the north of Iceland, the density increases sufficiently as freshwater freezes out to cause the water to sink to the ocean bed and then flows south close to the seabed. This North Atlantic current is one part of a global pattern of ocean circulation, called the conveyer belt (Figure 1.25), which circulates throughout the entire expanse and depth of the world's ocean system. Arbitrarily starting in the Arctic, cold, dense, surface water sinks and flows south along the Atlantic Ocean bottom. The area of greatest downwelling is off Greenland. This dense water flowing south combines with sinking Antarctic water and flows around Africa into the Indian Ocean and onwards to Australia into the Pacific. In the Pacific basin it warms and wells up toward the surface. From there, surface currents move in the opposite direction towards the Atlantic and the cycle starts again. One of the crucial factors which keeps the conveyer belt moving is the slight salinity differential between the Pacific and Atlantic Oceans (Haupt & Seidov 2007). It may take 1000 years or more to complete one global cycle, but there is no doubt that the global conveyer is a vital basis for the world's food chains. It transports nutrients and respiratory gases as well as warmth from areas rich in these essentials for life to those where one or more are in short supply. Indeed, there have been fears that climate change might weaken the conveyer belt by warming Arctic waters or altering salinity differentials, and if this were to occur the impacts on life on our planet could be enormous.

We have already mentioned upwelling in a rather different context, but this phenomenon also has crucial consequences for marine primary productivity, especially near to coasts on Continental Shelf systems. Winds blowing parallel to a shoreline influence surface currents via Ekman transport (see above), whilst other winds blowing from the land out to sea drag surface waters with them. Either way, seawater under these influences tends to flow away from land, causing deep water to upwell to replace it. This deep water brings vital nutrients accumulated at the bottom of the sea to the surface. Even without wind effects, deep currents will bring nutrient-rich bottom water up into the shelf regions because of seafloor topographies. Therefore for a number of reasons near-shore productivity is enhanced (Phillips 2005). The same strong winds which in these regions produce upwelling also

tend to disperse or "export" nutrients once they have reached the surface layers, but sufficient nutrient retention seems to occur to fuel subsequent algal blooms (Roughan et al 2006). Once phytoplanktonic primary productivity is enhanced in this way, a cascade of effects occurs further up marine food chains. Planktonic marine larvae of many species near to shore may be transported in upwelling currents, and indeed may be exchanged with offshore water in high speed currents (Shanks & Brink 2005). Upwelling-driven production influences even the largest animals in the sea, so that the migrations of blue whales, *Balaenoptera musculus*, for example, may be affected by seasonal patterns in this productivity (Croll et al 2005). Not only does upwelling significantly affect the abundance of primary producers and their consumers, it can also dictate the structure of communities. Herrera & Escribano (2006) studied the species composition of phytoplankton (mainly diatoms and dinoflagellates) off the coast of Chile, and found clearly distinctive communities using principle components analysis (PCA) between upwelling and nonupwelling conditions.

Local currents

While the great ocean currents influence all life on Earth, there are local currents which extend a mere few meters or, indeed, a few centimeters, that are also of great ecological importance, just on a much smaller scale. Tidal currents, which range in speed from zero to above 5 knots (260 cm s⁻¹), are the most important local currents and together with wind and wave action mold the physical features of littoral and sublittoral habitats. Of course, many marine organisms are adapted to tidal currents and can use them to advantage. For example, as is shown in Figures 1.26a,b and 1.27, the lugworm, *Arenicola marina*,

(a)

(b)

Figure 1.26

(a) Lugworm (*Arenicola marina*) casts at low tide on a sand/mud beach (Photograph courtesy of Paul Naylor.) (b) Lug worm. (Photograph courtesy David Fenwick, www.aphotomarine.com)

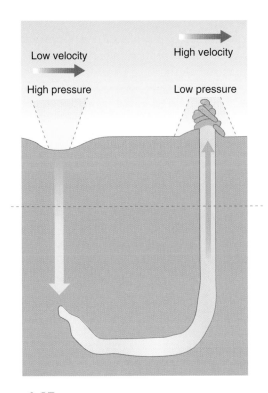

Figure 1.27

The U-shaped tube of the lugworm, *Arenicola*, showing the effects of water movement on the pressure at the two entrances to the burrow. The pressure differential results in a flow through the burrow.

lives in a U-shaped tube in sand or mud. The waste opening of the tunnel is raised above the seabed by only a centimeter or so, but the other opening is flush with the sand surface. When the tide is flowing this tiny, but physically (and ecologically) significant, height differential creates a pressure difference by the Bernoulli effect, which in turn causes water to flow into the lower opening and so helps to supply the worm with fresh seawater and hence food and respiratory gases (Figure 1.27).

Current speeds vary with distance from the seabed and there is a region termed the benthic boundary layer close to the seabed in which currents are appreciably reduced in speed. The benthic boundary layer can itself be subdivided. In the zone up to about 2 mm from the seabed frictional forces greatly reduce current speeds. From about 2 mm to 1 m from the bed there is the logarithmic layer in which the speed increases linearly with the logarithm of the distance from the bed. Between about 1 and 9 m above the bed marks the top of the benthic boundary layer. Above this zone the current flows at a constant speed. This reduction in current velocity close to the seabed influences the distribution of filter feeders. Active feeders (animals that create their own feeding currents such as bivalves, sponges, and sea squirts) are most abundant in the lower current speed benthic zone and passive filter feeders (ani-mals that cannot generate their own currents such as cnidarians) higher in the water column where currents are faster. For example, Figure 1.28 shows the vertical distribution of four species of sessile filter feeders on the hydroid *Nemertesia* (Figure 1.29). The active filter feeders, *Electra* and *Scruparia*, are both more abundant closer to the seabed than the passive filter feeders, *Plumularia* and *Clytia*, where though the current velocity is slow relative to higher above the seabed, these two species can still feed because they can create their own water currents (Hughes 1978). This of course demands some expenditure of energy, so the passive species may be able to outcompete the active ones as long as there is sufficient current to supply them with food, hence their location higher up the living substrate.

Suspended sediments

Sediment, which is denser than water, is lifted and held in the water column by water movement. When the seabed is composed of soft sediments such as mud or fine sand, the quantity of suspended particulate material (SPM – frequently expressed as turbidity) increases with the current speed. This is because faster flows lead to greater turbulence and higher rates of vertical mixing and re-suspension. As currents slow down, particles suspended within them will drop out, the largest first. Simple fluid dynamics has fundamental consequences for marine habitat creation. Imagine an island around which tides and currents are flowing. The velocity of the currents at particular locations will be determined by the topography, so that water speeds up as it flows around a headland and slows down in a sheltered bay. Faster currents mean that small particles stay in suspension and only the large ones such as rocks or boulders are left in position. In slow current regions, even very small particles such as sand and mud drop out of suspension. This process therefore creates and maintains physical habitats in the sea, which in turn dictates the types and ecologies of organisms able to live in these places. Slow currents mean soft sediments containing burrowers and detritus feeders; fast currents support encrusting species that filter feed, and so on (Figure 1.30a,b). Note also that this basic system can work over a range of spatial scales, so our island in the diagram may be miles across or alternatively, a few meters or even centimeters. The same old rules apply.

Estuaries with high tidal currents can be exceptionally turbid. For example, the Bristol Channel, UK, has a maximum tidal range of about 15 m and tidal currents and tidal current speeds generally exceeding 1.5 m s^{-1} at springs and 0.75 m s^{-1} at neaps, so that a suspended particle can move up to 25 km over a flood or ebb tide. These high currents over a muddy bed result in suspended sediment loads as high as 4 g l^{-1}. It is not only tidal currents that generate

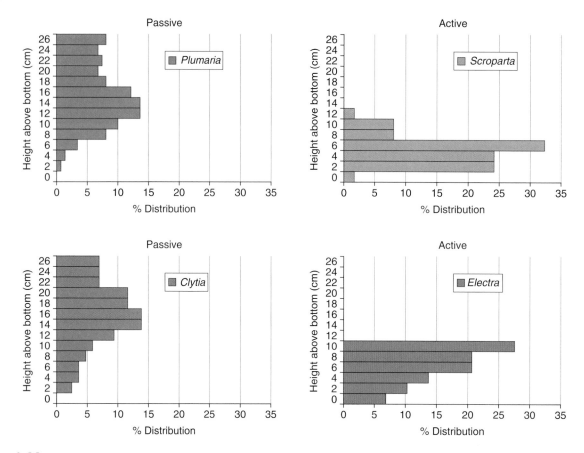

Figure 1.28

The differential vertical distribution of active and passive filter feeders growing on the hydroid *Nemertesia*. (From Hughes 1978.)

suspended sediment loadings, these are also influenced by wind and wave action, and suspended sediment loads can vary greatly both seasonally and also from day to day. Figure 1.31 shows the variations in both suspended particulate matter, SPM (in mg l⁻¹), and the size of suspended particles off the coast of Belgium over 24 hours, according to current velocity and variations in depth due to tidal fluctuations (Fettweis et al 2006). Several relationships can be identified. High SPM is clearly linked to high current velocities, though there seems to be a lag in the system such that peak SPM occurs an hour or more after peak velocity is reached. Water depth appears less significant.

As we have already suggested, the amount of suspended solids in the water column has a considerable influence on the community of plants and animals living in an area. The higher the turbidity the lower the light penetration and light levels can be reduced to such a low level that sublittoral plants do not occur. This is the case in parts of the Bristol Channel in the UK. High tidal currents can also produce regular sand and mud storms as the bottom sediment is taken into suspension by the flow and then resettles at slack water. This can result in a greatly impoverished ben-

thic infauna (organisms that live within the sediments) because of the smothering and a lack of oxygen. In many localities the effects are not so extreme, but suspended sediment loads can still have a great effect on the species present. Even the deep-sea experiences remarkable fluctuations in suspended loads.

The ecological effects of elevated suspended sediment depend primarily on two factors: the size range of the sediment particles, and the food content of the suspended sediment. The percentage organic content of SPM tends to decrease as the suspended solids loading increases (Yukihira et al 1999). This change has a considerable effect on filter feeders. If the food content decreases, animals have to expend more energy to collect their food, or alternatively, live their lives more slowly. Even closely related species may vary in their suspended sediment tolerance. For example, black lip, *Pinctada margaritifera*, and silver lip, *Pinctada maxima*, pearl oysters occupy quite different habitats. *P. margaritifera* is typically found in coral reef waters which are oligotrophic and of low turbidity. In contrast, *P. maxima* inhabits mud, sands, gravels and seagrass beds and is most abundant in water with relatively

Figure 1.29

The hydroid *Nemertesia antennina*. (Photograph courtesy of Paul Naylor.)

(a) (b)

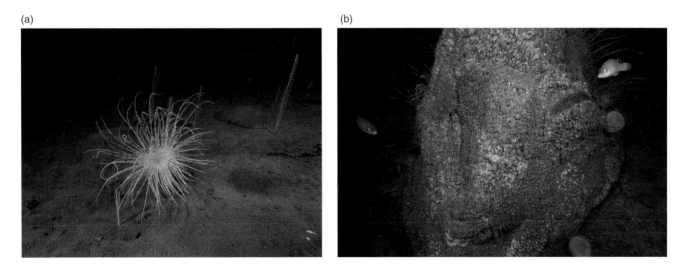

Figure 1.30

(a) Low current velocity community, with fireworks anemone and sea pens, Duich, W Scotland. (b) High current velocity community, with jewel anemones and cuckoo wrasse, Manacles, SW England. (Photographs courtesy of Paul Naylor.)

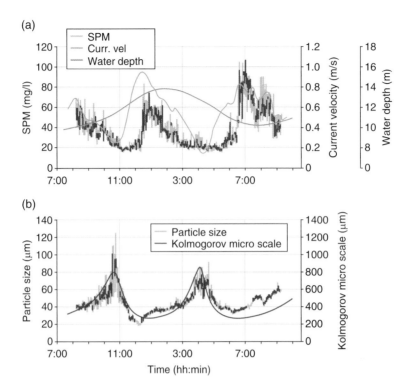

Figure 1.31

Zeebrugge site survey 2003/22. Through tide measurements from September 8, 2003, 8.00p.m. until September 9, 2003, 9.00a.m. (a) SPM concentration, water depth and vertical averaged current velocity. (b) Averaged particle size and Kolmogorov microscale of turbulence. Measurements have been taken at about 3 m from the bottom. (From Fettweis et al 2006; reproduced with permission of Elsevier.)

high sediment and nutrient loadings. Studies on the filtering capacity of these two oysters has shown that the species from clear water (*P. margaritifera*) is able to retain much smaller particles and to absorb energy from the smallest of them, as compared with the turbid water species (*P. maxima*) which is unable to retain small particles and gains maximum energy from medium size particles. This is an example of resource partitioning to avoid interspecific competition between similar species, which we discuss in more detail in Chapter 6.

Some organisms do well in conditions of high SPM and elevated turbidity, such as copepods (Islam et al 2005). However, environmental concerns have been expressed about increased suspended sediment concentrations resulting from all sorts of human impacts, such as paddlers on reefs, boat traffic, and dredging and disposal plumes. One of the worst-case scenarios of this type involves the logging of tropical forests resulting in increased erosion that dumps huge quantities of soil on fringing coral reefs (see Chapter 9). These unnatural events harm fish eggs, reduce growth and survival of larval and early juvenile fish, stunt the growth and reduce survival of bivalves and corals. Deposited sediments may simply smother sessile and sedentary animals and plants, or the suspended material may reduce feeding efficiency and impair the function of gills and other respiratory surfaces.

The activities of man also cause elevated suspended solids levels. The effects of these are discussed in Chapter 9.

Climate change

Physical conditions in the sea may be thought of as fairly stable and predictable, especially when compared with many terrestrial locations, but the rather obvious effects of climate fluctuations for example on land are also reflected in the sea. Climate change is the best known cause of physical change. There are many actual or potential consequences of climate change, global warming and so on for marine ecosystems, and a thoughtful review of this chapter may indicate what might happen if temperatures rise or storms increase in frequency and intensity. Two of the most important aspects of climate change are increases in sea level, and changes to the infamous El Niño.

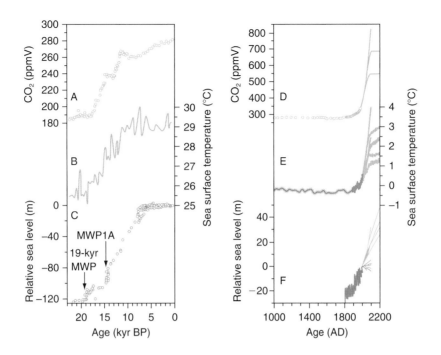

Figure 1.32

Time series of key variables encompassing the last interval of significant global warming (last deglaciation) (left) compared with the same variables projected for various scenarios of future global warming (right). (A) Atmospheric CO_2 from Antarctic ice cores. (B) Sea surface temperatures in the western equatorial Pacific based on Mg/Ca measured in planktonic forminifera. (C) Relative sea level as derived from several sites far removed from the influence of former ice sheet loading. MWP = meltwater pulse. (D) Atmospheric CO_2 over the past millenium (circles) and projections for future increases (solid lines). Records of atmospheric CO_2 are from Law Dorne, Antarctica and direct measurements since 1958 are from Mauna Loa, Hawaii. Also shown are three emission scenarios of atmospheric CO_2 over the course of the 21st century and subsequent stabilization over the course of the 22nd century. (E) Temperature reconstruction for the Northern Hemisphere from 1000 to 2000 AD (grey time series), global temperature based on historic measurements, 1880 to 2004 (blue time series), and projected warming based on simulations with two global coupled three-dimensional climate models with the use of three emission scenarios (orange time series). (F) Relative sea-level rise during the 19th and 20th centuries from the tide gauge record at Brest, France (green time series), projections for contributions from combined Greenland and Antarctic ice sheets (dark blue time series), and projections from sea-level rise from thermal expansion based on climate simulations shown in (E) (light blue time series). (From Alley et al 2005; reproduced with permission of *Science* – AAAS.)

In Figure 1.32 we can look back in history as much as 20,000 years, and notice that on a large time scale, increases in atmospheric CO_2 concentration, sea surface temperature and sea level all follow the same pattern. 20,000 years ago, the atmosphere contained much less CO_2, and the sea was appreciably colder and 120 m or more lower than now (Alley et al 2007). The graphs also suggest that little has changed in the last 5000 years. However, on a time scale of a mere 1000 years, we notice that the last 200 years have seen an almost exponential increase in CO_2 levels, SSTs, and a steep but linear increase in sea levels. By the year 2200, sea levels are predicted to be around 0.5 m above present levels, which roughly equates to an annual sea level increase of 2 or 4 mm, but these types of predictions are notoriously unreliable. What we can say is that the rate of change in global sea level is related to temperature anomalies (Figure 1.33) (Holgate et al 2007), and it seems that apart from a slight dip in the 1950s and 1960s, annual mean global temperature has risen steadily, and is now about 0.5°C warmer than it was 100 years ago. This anomaly is reflected in a roughly three-fold increase in the rate of sea level change.

Though rising seas levels will have serious consequences for low-lying coastal areas on land, it would seem likely that the majority of marine species will be little affected. However, temperature changes can have a much greater impact on marine systems, and the phenomenon of El Niño is a prime example. Essentially, sea surface temperatures are significantly enhanced in El Niño years, so that a band of unusually warm water extends virtually all the way across the Pacific Ocean from west to east. Figure 1.34 provides more details of the mechanism, where weakened trade winds allow warm water to move eastwards. These effects can have general impacts on the planet's weather. One of the most serious effects can occur over large parts of Southeast Asia which become dry during an El Niño; rain forests do not do well in dry conditions. The most rapid climate change of this type is predicted to take place

(a)

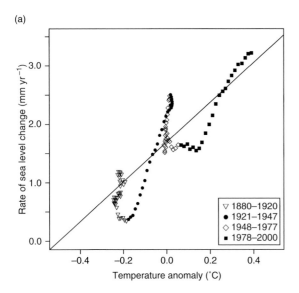

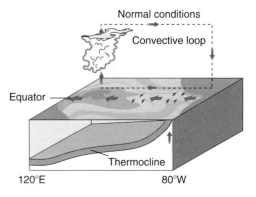

Normal conditions

Convective loop

Equator

Thermocline

120°E   80°W

(b)

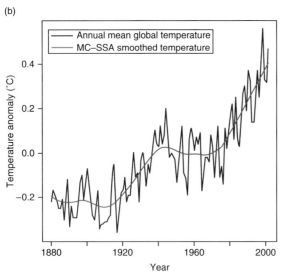

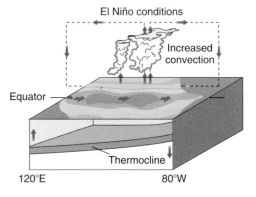

El Niño conditions

Increased convection

Equator

Thermocline

120°E   80°W

El Niño conditions

Weakened trade winds allow warm water to move eastwards

Thick upper-ocean layer keeps nutrient-rich water from upwelling along the coast of the Americas

Ocean heat released into the atmosphere increases cloud formation and alters path of jetstream

Figure 1.33

(a) The relationships of the rate of global mean sea-level rise to global mean sea-level surface temperature with the data divided into four epochs, each showing a different relationship between the variables. (b) The global mean surface temperature record, annual data and data smoothed using the MC-SSA method. The four epochs described in (a) relate to the four sections of the temperature record that can be clearly seen. (From Holgate et al 2007; reproduced with permission of *Science* – AAAS.)

Figure 1.34

Marine and atmospheric conditions in the mid-Pacific under normal and El Niño conditions. (Courtesy of NOAA/PMEL/TAO Project Office, Dr. Michael J. McPhaden, Director.)

in the Southern Ocean (Trathan et al 2007), which comprises much of the Antarctic and Polar seas. El Niño effects here have been called "El Niño–Southern Oscillations" by virtue of their variations, known as ENSO for short, and they have been blamed for a series of catastrophes such as the reduction in krill populations in the south Atlantic with all the associated food chain effects on marine megafauna such as whales.

In the sea, one of El Niño's biggest influences is on primary productivity (see also Chapter 3). Rising convection

currents occur in mid-ocean, resulting in significantly altered wind patterns, jet streams and cloud formation, but most importantly, nutrient-rich water is prevented from upwelling in various parts of the world with huge impacts on marine productivity. One of these areas occurs along the west coast of the Americas, and a lot of research work has centred on Peru and Chile where fisheries for anchovy and sardine are some of the biggest in the world (see also Chapter 8). In Peru, El Niño events were recorded in 1972/73, 1982/83, and 1997/98 (Niquen & Bouchon 2004). Figure 1.35 summarizes the major changes in environmental conditions on pelagic communities over the various ENSO periods, and as can be seen the results are complex. Commercially, the most significant effects of El Niño are on fishing stocks, and two species show opposite

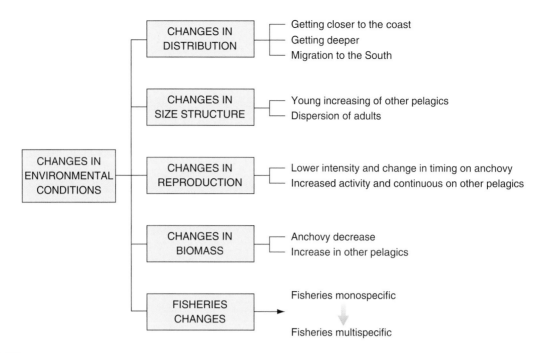

Figure 1.35

Effects of El Niño events from the 1970s to 1990s on pelagic resources in Peruvian waters. (From Niquen & Bouchon 2004; reproduced with permission of Elsevier.)

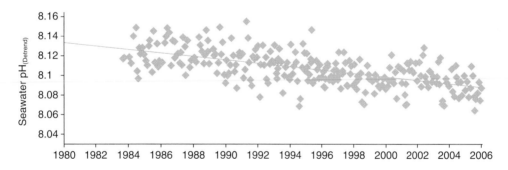

Figure 1.36

Trends in oceanic surface water pH in the northeast Atlantic near Bermuda. (From Bates & Peters 2007.)

reactions. Anchovy suffered huge declines in stocks in all three decades, whereas sardine stayed fairly constant in the 1970s and 1990s but showed large increases in the 1980s. In neighboring Chile, Escribano et al (2004) reported the total anchovy catch remained stable in 1997 at a maximum of around 1 million tons. In 1998, however, the anchovy catch declined to 400,000 tons, before recovering to 1.2 million tons in 1999. These sorts of fluctuations can have serious impacts on the socioeconomics of a region.

Finally, one of the most potentially serious physical changes for the future of our oceans involves acidity. As CO_2 levels in the atmosphere rise, the pH of seawater is likely to fall (become more acidic as the hydrogen ion concentration increases). Acid rains and direct absorption of CO_2 by surface seawater may add to the problem, though its extent and consequent ecological effects are still uncertain. For example, Bates & Peters (2007) studied the changes in pH of surface seawater near Bermuda, and were able to detect a slight decrease over 25 years (Figure 1.36). The authors concluded that surface seawater pH in this part of the Atlantic Ocean decreased by 0.0017 (±0.0001) pH units per year. In this study, there was no obvious impact on coral growth, but since coral skeletons are made of a form of calcium carbonate called aragonite, then the ability of hard corals to lay down skeletons will be reduced if seawater becomes too acidic. On a small scale, however, Jokiel et al (2008) were able to detect a reduction in calcification of skeletal material in the hard coral *Montipora capitata* held in $1 \times 1 \times 0.5\,m$ fibreglass tanks treated with diluted hydrochloric acid at a concentration designed to simulate the increase in seawater acidity expected during the twenty first century. Overall, this experiment found a 20% reduction in calcification rate, but there appeared to be no effect of increased acidity on gamete production in the coral. It remains to be seen whether or not declining pH in our global seas will indeed add significantly to the already large list of troubles ahead for our coral reefs (see also Chapter 9).

Conclusions

Clearly there are numerous physical factors and conditions that influence life in the sea. Depth, temperature, salinity, turbidity, velocity, pressure and so on, all combine to produce a myriad of physical and chemical habitats which provide a huge number of combinations and permutations. The oceans are not at all the bland, homogeneous expanse that we might think, and with this in mind, it should be no surprise that the diversity of life in the sea is as rich and varied as the physical conditions in which it lives. Biodiversity is the subject of Chapter 2.

Chapter 2

Marine biodiversity

Introduction

The term biological diversity or biodiversity, a measure of the total genetic and ecological diversity, has become a common expression in ecological literature over the last 20 years (Wilson 1988; Reaka-Kudla 1997). The quantification of biodiversity has been driven by increased interest in conservation and concern that diversity is declining throughout the planet, and for this reason there has been more effort in recent years to identify biodiversity hotspots on land and in the sea where our conservation efforts should be focused (see Chapter 10).

In the ecological literature the term diversity is frequently used to summarize two community attributes, species richness (the number of species in a particular habitat) and relative species abundance. A myriad of mathematical indices are available to measure and compare species diversity (Magurran 1988). The number of species (the richness) can be exactly the same in two habitats for which the species diversity indices are greatly different, because of differences in relative species abundance. Only by creating and applying reliable measures of diversity can we measure how it varies both spatially and temporally and thus recognize the influences that create, maintain and destroy diversity. In this chapter we will generally measure biodiversity in terms of the number of species, genera, families and phyla present. While it is common to use number, it can also be appropriate to measure diversity using biomass as a measure of taxonomic abundance.

The comparative richness of marine habitats in terms of animal phyla

As Figure 2.1 illustrates, many animal phyla can be found coexisting in just one small region of the sea, and the use of higher taxa (genera, families, phyla, etc.) is often more efficient than using species when trying to compare diversity in many marine environments (de Voogd & Cleary 2008). The phylum is the highest taxonomic level within the animal kingdom and each phylum has a distinct body plan. Angel (1993) reported 31 animal phyla from marine habitats (Figure 2.2), of which 15 were endemic to the sea. However phyla are constantly under revision, so by 2009, we have additions, including the Cycliophora which was first described in 1995 for the single species *Symbion pandora* (Figure 2.3) (Funch & Kristensen 1995). These animals have a complex lifecycle, with the adults apparently living nowhere else but on the mouthparts of the Norway lobster (scampi), *Nephrops norvegicus*. Two more new phyla are the Orthonectida and Dicyemida (Slyusarev 2008), and these recent discoveries suggest tantalizingly that the oceans may hold further major groups of organisms yet to be described. Note, however, that we have lost several others through molecular taxonomy (e.g. the Pogonophora have

Marine Ecology: Concepts and Applications, 1st edition. By M. Speight and P. Henderson. Published 2010 by Blackwell Publishing Ltd.

Figure 2.1

The seabed at Loch Carron, Scotland, showing the great range of phyla that can be present even in shallow temperate waters. (Photograph courtesy of Paul Naylor.)

Phylum	Marine	Symbiotic/Parasitic	Freshwater	Terrestrial
Acanthocephala		√ *		
Annelida	√ B P *	√	√	√
Arthropoda	√ B P *	√	√	√
Brachiopoda	√ B *			
Bryozoa	√ B		√	
Chaetognatha	√ B P *			
Chordata	√ B P	√	√	√
Cnidaria	√ B P		√	
Ctenophora	√ P *			
Dicyemida	√ E *			
Echinodermata	√ B P *			
Echiura	√ B *			
Gastrotricha	√ B		√	
Gnathostomulida	√ B *			
Hemichordata	√ B *			
Kamptozoa	√ B	√	√	
Kinorhyncha	√ B *			
Loricifera	√ B *			
Mollusca	√ B P	√	√	√
Nematoda	√ B P	√	√	√
Nematomorpha	√ E	√		
Nemertea	√ B P	√	√	√
Onychophora				√*
Orthonectidae	√ E *			
Phoronida	√ B *			
Placozoa	√ B *			
Platyhelminthes	√ B P	√	√	√
Pogonophora	√ B *			
Porifera	√ B	√	√	
Priapula	√ B *			
Rotifera	√ B P	√	√	√
Sipunculida	√ B			√
Tardigrada	√ B		√	√
Totals	31	12	14	11
Endemic	15	1	0	1
	10 benthic			
	1 pelagic			
	2 both benthic & pelagic			
	2 endoparasitic in			
	marine animals			

Figure 2.2

The distribution of animal phyla by habitat. tick = present; B = benthic; P = pelagic; E = endosymbiotic or parasitic in marine animals; * = endemic to that habitat. (Modified from Angel 1993; reproduced with permission of Wiley-Blackwell.)

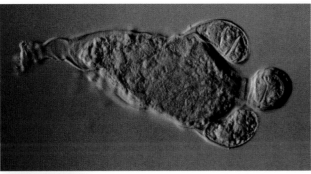

Figure 2.3

Symbion pandora (Phylum Cycliophora). (Photograph courtesy of Peter Funch.)

Figure 2.4

500 mya jellyfish fossil. (Photograph courtesy of Paulyn Cartwright & Bruce Lieberman, PlosOne.)

now been placed in the Annelida), and that established phylogenetic relationships are being modified, if not completely overturned (Bourlat et al 2008; Philippe et al 2009). Whatever the detail, terrestrial and freshwater habitats are comparatively phyla-impoverished with only around 11 and 14 respectively. Of the 15 endemic marine phyla, 10 are bottom living (benthic), 1 lives in the water column (pelagic), 2 are both benthic and pelagic, and 2 are endoparasites. At the level of the phylum, it is the seabed which is most diverse, possibly because of the wide variety of habitats available. Briggs (1994) claimed 34 phyla in the marine realm, compared with only 15 on land. This high number presumably includes protozoan phyla.

Marine species richness in geological time

Many marine organisms are essentially soft bodied (take for example many cnidarians, ctenophores, annelids, and tunicates), and have left few fossils over the half a billion years that metazoan life has existed in the sea. Nonetheless, traces of many ancient life forms can occasionally be found (Figure 2.4), as long as it proves possible to recognize them for what they are. Another problem is that ancient fossils may occur in deposits so far below the surface of the earth that even large geological upheavals may not expose them for study. Thus attempting to look back at the fossil history of marine organisms and commenting on changes in species richness is difficult. Nonetheless, scientists have looked back in time at marine diversity both for organisms in general, and more frequently, for specific groups.

Generally, marine biodiversity has increased with occasional mass extinctions (Sala & Knowlton 2006). This

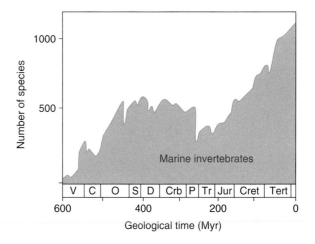

Figure 2.5

Patterns of diversification of families of marine invertebrates over the last 600 million years. (From Benton & Emerson 2007; reproduced with permission of Wiley-Blackwell.)

increase in the diversity of marine animals appears to be exponential, especially since Palaeozoic times (Stanley 2007), and this is likely to continue far into the future, providing this natural process is not disrupted. One of the largest threatened disruptions is climate change, in particular temperature, which Mayhew et al (2008) have shown to be directly related to biodiversity levels over the last 500 million years. Diversity has taken off in the last 250 million years (Figure 2.5; Benton & Emerson 2007), following the mass extinction event that occurred at the Permian/Triassic boundary. Over 90% of all marine species may have gone extinct at this time, but their recovery has been impressive and according to models we probably now have at least twice the number of marine invertebrates that we had 200 mya. According to Benton & Emerson, this pattern is not solely a marine phenomenon, and the shape of speciation graphs for vascular land plants,

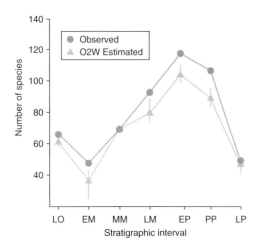

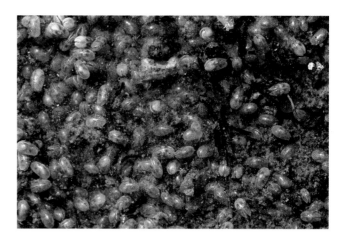

Figure 2.6

Species richness in Caribbean corals over various geological time periods. LO = Late Oligocene (lower boundary 28.45 mya); EM = Early Miocene (lower boundary 23.03 mya); MM = Middle Miocene (lower boundary 15.97 mya); LM = Late Miocene (lower boundary 11.61 mya); EP = Early Pliocene (lower boundary 5.33 mya); PP = Late Pliocene/Early Pleistocene (lower boundary 1.6 mya); LP = Late Pleistocene (lower boundary 1 mya). (From Johnson et al 2008a; reproduced with permission of *Science*.)

Figure 2.7

An aggregation of ostracods on the seabed photographed in a shallow upper estuarine habitat. (Photograph Peter Henderson.)

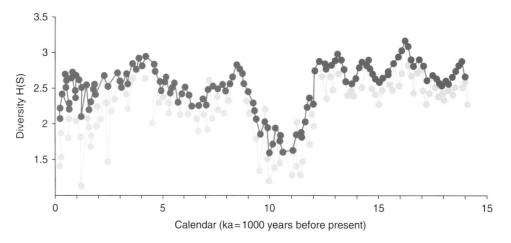

Figure 2.8

Ostracod species diversity (Shannon–Wiener index) over the last 20,000 years in northwest Atlantic deep-sea sediment cores. Light blue plots = calculations based on raw census data; dark blue plots based on 3-point moving sum dataset. (From Yasuhara et al 2008; reproduced with permission of Micropaleontology Press.)

nonmarine tetrapods, and insects follows the same trend as that of invertebrates in the sea.

Changes in diversity can also be observed over relatively small timescales. Coral species diversity, an extremely contentious issue in the present day as we shall see later in this chapter, appears to have been extremely variable over the last 30,000 years. Johnson et al (2008a) used fossil and stratigraphic data to plot the changes in coral species richness in the Caribbean over this time period (Figure 2.6), and it is clear that coral richness peaked in the Early Pliocene (around 5 mya), and has been declining ever since. In another example, ostracod species diversity was studied

from fossil deposits laid down over the last 20,000 years in the northwest Atlantic (Yasuhara et al 2007). Ostracods (Figure 2.7) or "seed shrimps" are small crustacea. They possess a pair of calcareous shells (valves) which make them look superficially like tiny bivalve molluscs or barnacle larvae. The results of this study are shown in Figure 2.8. (Yasuhara et al 2008). Note the considerable variations in diversity over the time period, with a particularly significant dip between 12,000 and 13,000 years ago. This event is likely to be linked to cold periods which caused changes in deep-water circulation patterns, and since ostracods are crucial links in marine planktonic food chains, it is tempting to predict the likely impacts of such declines on both their predators and especially their prey.

Present marine species richness

We are still far from having an adequate inventory of marine species, though great efforts are being made to obtain reliable measures of species richness. Several international initiatives are now sampling, identifying and cataloging marine organisms from all parts of the oceans. Leading the way in these endeavors are the World Register of Marine Species (WoRMS), the Census of Marine Life (COML), and the Ocean Biogeographic Information System (OBIS).COML, which was formed in 2000 and runs until 2010 (O'Dor 2003; O'Dor & Gallardo 2005), has a number of tasks: looking at the history of marine species diversity, assessing the present-day situation, and making predictions for the future of life in our seas. COML provides a wealth of online information about marine species richness and taxonomy which is constantly updated. By 2008, COML and WoRMS estimated that more than 120,000 species had been properly validated (i.e. they had had their identifications plus many pseudonyms confirmed), but there is still much more work to be done. Clearly, certain marine habitats are much easier to sample than others, because of their inaccessibility, or simply their vastness. It is estimated that we have found and identified about half of the species from near-shore and shallow seas, but many more than 50% remain unknown for ice oceans (both Arctic and Antarctic), hydrothermal vents, and the deep-sea. Some geographical locations are far better known than others. For example, Mora et al (2008) suggest that global species censuses are even now particularly incomplete for tropical areas, and that only a few developed countries have species inventories around their coasts that are reasonably complete.

Some groups of marine organisms are better known than others. As might be expected, fish are much better known than worms, which in turn are more fully described than microbes. The rate of discovery of new fish species has held fairly constant for the last 170 years, according to WoRMS. Flatworms (Platyhelminthes; Turbellaria), on the other hand, have only started to accumulate species rapidly in the last few decades. The smaller and more hidden the group of organisms, the less complete our knowledge of their species richness. Meiofauna (organisms defined loosely as those that are able to pass through a 1-mm mesh but will be retained by a 45-μm mesh – Figure 2.9) are rarely sampled even from such exciting sites as hydrothermal vents (Flint et al 2006). Smaller still, microbial species are extremely hard to identify with any confidence, even using sophisticated RNA and DNA techniques. As Pedros-Alio (2006) asks, "The total number of prokaryotic cells in the oceans

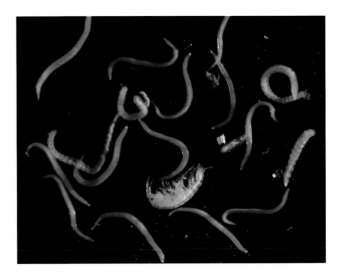

Figure 2.9

Meiofauna from sandy sediments in the Severn Estuary, England. The red coloration is not natural, the animals were stained with rose bengal to aid in their sorting from the inorganic matrix. There were amphipods, copepods, polychaetes, oligochaetes, and nematodes in the sample. (Photograph by Peter Henderson.)

has been estimated to be circa 10^{29}, but into how many species are these cells partitioned?". Sogin et al (2006) may provide at least part of the answer to this question. These authors suggest that bacteria, archaea, and protists account for the majority of the living biomass in the world's oceans, and say, "the accumulation of microbial mutations in the sea over the past 3.5 billion years should have led to very high levels of genetic and phenotypic variations".

So using the available data, the overwhelming dominance of invertebrate species compared with other major groups of plants and animals is shown in Figure 2.10. In comparison with terrestrial habitats, the invertebrate total is low. More than 950,000 species of terrestrial insect have been described, of which 300,000 are beetles. As J. B. S. Haldane noted "The Creator must have an inordinate fondness for beetles." As our knowledge increases, it is likely that he will also be shown to have a fondness for small benthic worms.

While it is certain that the total number of marine species is considerably greater than the number presently described, there is no consensus on even the approximate total. Table 2.1 gives estimates for marine habitats which range between about 10^5 and 10^7 species. Estimates above 10^6 are based on the estimated species richness of benthic invertebrates calculated from species acquisition curves. However, extrapolating from the species number in a few square meters of seabed to the entire floor of the world's oceans is clearly speculative and has been subjected to fierce criticism. Briggs (1991) referred to the estimate of Grassle and Maciolek of 10^7 as "an act of statistical legerdemain," but it certainly now seems that Brigg's own estimate of fewer than 200,000 is far too low. The difficulty of estimating

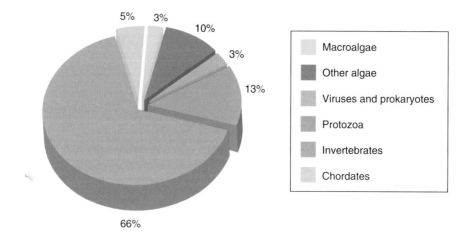

5% 3% 10% 3% 13% 66%

Macroalgae

Other algae

Viruses and prokaryotes

Protozoa

Invertebrates

Chordates

Figure 2.10

The relative abundance by taxonomic group of described marine species. (From Angel 1993; reproduced with permission of Wiley-Blackwell.)

Table 2.1 Predictions of marine species richness

REFERENCE	ESTIMATED NUMBER	HABITAT
Grassle & Maciolek 1992	10 million	Deep-sea benthos
May 1994	0.5 million	Deep-sea benthos
Poore & Wilson 1993	5 million	Benthos
Briggs 1994	0.2 million	All sea
Reaka-Kudla 1997	0.6–0.95 million	Coral reefs
Knowlton 2001	0.6–9 million	Coral reefs
Adrianov 2003	20–30 million 20–30 million 10 million	Macrobenthos Meiobenthos Nematodes
Malakoff 2003	1 million+	All taxa/sites

the total from species acquisition curves is illustrated by the fact that, using the same data on deep-sea benthic fauna, Grassle & Maciolek (1992) estimated 10 million undescribed species, while May (1993) estimated 0.5 million.

Geographical variation in marine diversity

As on the land, the greatest marine diversity is within the tropics. This is clearly illustrated by the distribution of bivalve molluscs (Clarke & Crane 1997). The contours of increasing species richness are generally aligned with latitude (see later in this chapter), but there are clear indications of bivalve species "hotspots" on the tropical Pacific coast of South America and in the Indo-West Pacific waters between the Pacific and Indian Oceans. Data for other animal groups reinforces the conclusion that the tropics are far from uniform in their species richness, and in fact there is considerable variation in sponge, coral, bivalve mollusc, echinoderm, and shore fish species richness between four main tropical regions, Indo-West Pacific, East Pacific, West Atlantic, and East Atlantic. Sponge and fish species number is particularly variable and it is clear that at least for these groups the Indo-West Pacific is particularly species rich (Pauley 1997).

Coral reef is the most visually spectacular tropical marine habitat and is notably species rich. As shown in Table 2.1, coral reefs may hold between 600,000 and 9 million species and are often thought of as the marine equivalents of tropical rain forests. The great number of animal species found on a reef is related to the structural complexity of the coral, which is in turn determined by coral diversity (see later in this chapter). The biodiversity of the coral is therefore the foundation upon which total reef diversity is supported. It is important at this stage though to note that just as forests and woodlands vary in structure and species composition in different parts of the world, so do coral reefs. Put simply, there is no such thing as a global standard coral reef as Figure 2.11a–d shows. With this premise in mind, Figure 2.12 shows the geographical variation in coral species number for the 13 most species rich coral families (Knowlton 2001). Again, the most diverse coral communities are found on the Great Barrier Reef and in the Maldives. This suggests that the generally higher species diversity of the Indo-West Pacific region may in part be related to the diversity of the coral reef. Not all corals occur in shallow water however, and in recent years there has been an upsurge in interest of cold, deep-water coral communities, both in terms of their evolutionary significance and biodiversity, and also their commercial worth as fishing grounds (Roberts et al 2006). Deep-water corals such as species in the genus *Lophelia*

Figure 2.11

Coral reefs in a) Sabah, b) Red Sea, c) Honduras, d) Sulawesi. (Photographs courtesey of: Martin Speight (a,c); Annalie Morris (b); Joanna Barker (d).)

(Figure 2.13) occur from 50 m to more than 2400 m (Mortensen et al 2008), and though slow growing (less than 1 cm per year), may cover large areas of seabed and provide extremely complex habitats for all manner of marine organisms (Armstrong & van den Hove 2008). They are also very important given their rarity in relation to the impact they are suffering, primarily from commercial fishing and hence are greatly in need of conservation (see Chapters 9 & 10). Though still in the Scleractinia, deep-water corals do not possess photosynthetic symbionts (Buhl-Mortensen & Mortensen 2004), for the obvious reason that they live at depths where there is negligible light and are known as azooxanthellates. Contour lines in Figure 2.14 from Turley et al (2007) show that though deep cold sites with any of these corals cluster in regions such as the Caribbean, Western Europe, and Scandinavia, species richness again peaks in Australasia. A final global scale example involves foraminifera (Figure 2.15), planktonic protists somewhat

similar to amoeba, which form the basis of many marine food chains. As Figure 2.16 shows, though not particularly diverse (but vastly abundant), the species richness of forams peaked both in the Pacific and Atlantic Oceans in middle latitudes, not at the Equator (Rutherford et al 1999). We shall discuss the possible reasons for global distributions in later sections of this chapter.

Differences in the distribution of biomass and species number

Before proceeding to consider why species diversity varies in marine ecosystems, it is useful to think about biomass instead of species richness and abundance. While species

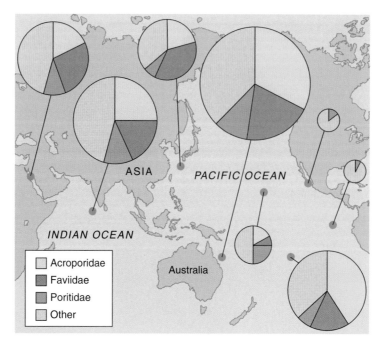

Figure 2.12

Coral biodiversity. Circle area is proportional to total number of coral species, which ranges from 13 associated with reefs of the Galapagos Islands to 321 associated with the Great Barrier Reef. (From Knowlton 2001; reproduced with permission of *Science*.)

Figure 2.13

Deep sea coral (*Lophelia pertusa*). (Photo courtesy Ed Bowlby, NOAA/ Olympic Coast NMS; NOAA/OAR/Office of Ocean Exploration.)

number is generally the focus of biodiversity studies, as we have said, it is important to note that species number hotspots do not always correspond to biomass hotspots. The results presented by Callaway et al (2002) for epibenthic animals and demersal fish in the North Sea well illustrate this point. Epibenthic fauna live on the surface of the seabed, whether hard or soft, and may be sessile (or at least sedentary) such as feather stars and mussels, or mobile such as crabs and snails. Demersal fish such as cod or haddock still associate with all manner of marine sub-

strates from rocky reefs to shipwrecks, mainly by loitering in the vicinity of structures. In both groups of animals, Callaway et al showed that the numbers of species and species biomass are not spatially correlated. For the epibenthos, species richness is higher towards the northern part of the North Sea while biomass is greatest in the south. Demersal fish numbers are also higher in the south, presumably because of the increased food resources on offer, although otter trawling caught the greatest number of species in the north. A distinct difference in lifestyle was noted, with the epibenthos in the southern region of the North Sea dominated by free-living species whereas the northern region was dominated by sessile forms. Reference to Chapter 1 may give some suggestions as to the physical or oceanographic mechanisms potentially involved in this demarcation.

Factors determining biodiversity and species richness

Having discussed the observed variations in marine biodiversity, we now introduce a variety of factors that appear to influence this variation. Apart from the first one of these (sampling effort), all factors have important ecological and evolutionary implications. Though presented separately for

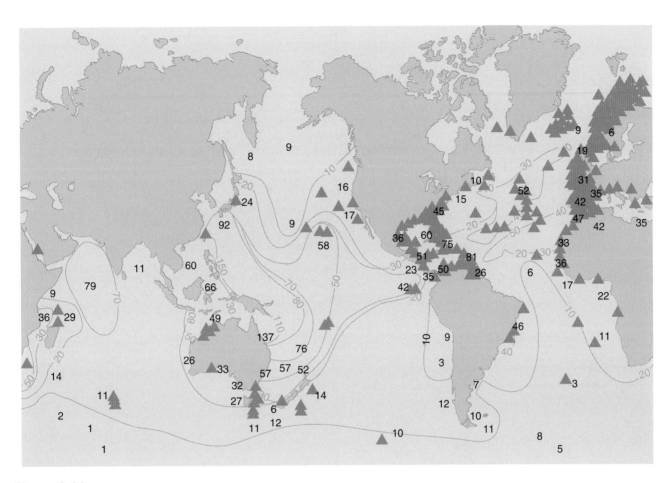

Figure 2.14

Global species richness of azooxanthellate (deep-water) corals. Contour lines delimit regions of similar richness; numbers not on contours indicate actual number of coral species at that location; green triangles indicate known sites. (Modified from Turley et al 2007.)

Figure 2.15

Drawing of a foraminiferan, Challenger Reports Vol IX.

convenience and clarity, it is clear that complex interactions between these factors occur, both in time and space.

Sample size and sampling effort

The total number of species observed depends on the sampling method and the effort expended. When sampling within a single limited area and habitat, it is generally found that the rate of new species observed declines with increasing sample number or effort (Figure 2.17; Smith & Whitman 1999). Thus many new species can be expected to be found in the first few samples, but as more and more samples are taken, fewer and fewer new species appear. In this example, there would appear to be little point in taking more than 10 samples, whereas if only 5 or fewer were collected, a serious underestimation of species richness would occur. This type of data is particularly important when sampling very difficult, time-consuming or expensive marine habitats, as for example the meiobenthos of mussel beds on hydrothermal vents 3000 m deep (Zekely et al 2006). This feature is also shown in Figure 2.18, which plots the total marine fish

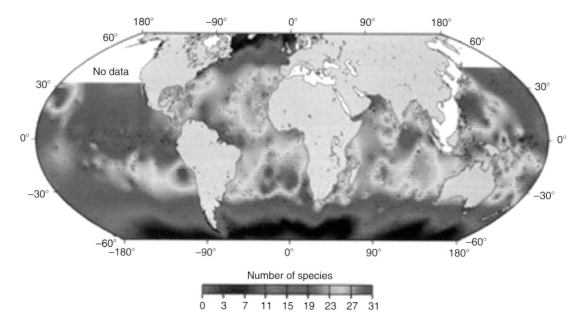

Figure 2.16

Global distribution of plantonic foramifera species. (From Rutherford et al 1999; reproduced with permission of *Nature*.)

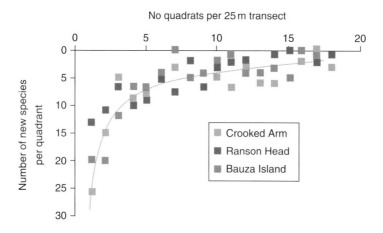

Figure 2.17

The rate of accumulation of new species from successive 0.25-m² quadrats in shallow waters off New Zealand. (From Smith & Whitman 1999; reproduced with permission of *Ecology – ESA*.)

species number accumulated through time by regular sampling in a bay in the lower Severn estuary (Henderson 2007). In addition, the graph shows that while the rate of accumulation decelerates with sampling effort (time), it does not stabilize to a constant number, but settles into a slow, approximately linearly, increase. This is a commonly observed pattern.

We can also measure sampling effort in terms of the number of individuals sampled, and as shown in Figure 2.19, plot species number against number of individuals

(Grassle 1989). This plot shows that for the benthic macrofauna species number initially increases approximately linearly with the number of individuals encountered. Again, this type of relationship is fairly widespread, and in a different example Fonseca et al (2007) found a log-linear relationship between the abundance of nematodes from sediments on continental slopes (50–2000 m) and the number of species in each sample. In this work, though, the log abundance/richness relationships were linear and with a positive gradient within a site; different sites in the Southern Ocean and western Indian Ocean showed significantly different slopes, with the latter having a much steeper relationship.

Time of sampling can also influence results, especially when marine habitats are influenced by environmental factors such as tides. Tropical marine lagoons (Figure 2.20) present demanding environmental conditions, second only perhaps to intertidal communities. They are frequently very warm, even in relation to normal coral reef waters, low in oxygen, and high in salinity. Any winds, waves or paddlers can stir up soft sediments, causing extreme difficulties for small sessile filter feeding inhabitants, and if the fringing reefs break down for natural or anthropogenic reasons, the entire lagoon ecosystem can easily be destroyed. As Figure 2.21 shows, species richness in larval fish assemblages varies with the tidal cycle, so that richness in samples tends to decline at low tide and increase at high water (Aceves-Medina et al 2007). Note also that richness can also vary with position in the water column – there are clear differences between the surface layers and near-bottom. The point to note is that significantly different estimates of species richness will be obtained depending on the state of the tide at the time of sampling.

Body size

Generally, there is a greater number of small than larger species in a habitat. This general pattern is illustrated for the fish families in Micronesia by Ormond & Roberts (1997). For the 94 fish families studied, there was a rapid decline in average species number with increasing median body length (Figure 2.22). Presumably it is easier to pack more species into a given habitat the smaller each species is, just as a tree can support many species of small insects but few large primates. However, a particular animal taxon tends to be most species rich at an intermediate position within the total size range for the taxon. For example, there is a general tendency for there to be more small than large mantis shrimp species (Figure 2.23). However, it is not the smallest size class that is most species rich (Reaka-Kudla 1997), possibly indicating that the very smallest are rather specialized to one or two specific niches, and there is not the variety of niches offered to larger bodied forms (Figure 2.24).

The species richness reported from a sample will be, in part, determined by the size range selected for study. All sampling and sorting methods are to some extent size-selective, making it difficult to compare species richness between samples collected using different methods.

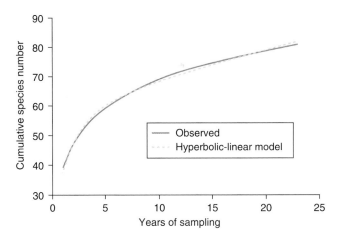

Figure 2.18

Species accumulation curve for an intensively sampled area. The curve shows the total number of marine fish species captured in Bridgwater Bay, England. The curve is fitted to data collected monthly over a 24-year period. Also shown is the fit to the hyperbolic-linear model of Henderson (2007). (Reproduced with permission of Cambridge University Press.)

Depth

As we explain in detail in Chapter 1, the sea has enormous depth ranges, with concomitant variations in easily measured

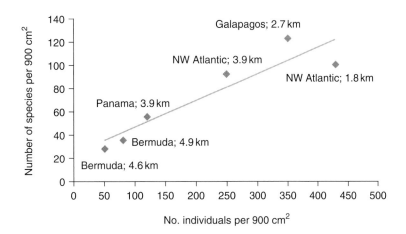

Figure 2.19

Total number of species of microfauna in relation to total number of individuals in 900-cm² samples in various locations and depths. (From Grassle 1989; reproduced with permission of Elsevier.)

Figure 2.20

Mixed motile and sessile fauna in a tropical lagoon, Indonesia. (Photograph Martin Speight.)

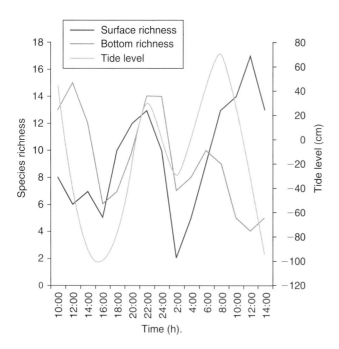

Figure 2.21

Tide levels and species richness of larval fish assemblages in a Mexican tropical lagoon. (From Aceves-Medina et al 2007; reproduced with permission of Elsevier.)

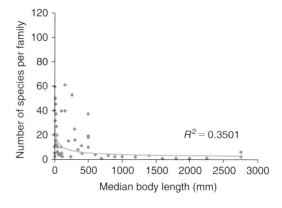

Figure 2.22

The relationship between median body size and the number of species in each of 94 fish families on reefs in Micronesia. (From Ormond & Roberts 1997; reproduced with permission of Cambridge University Press.)

physical variables such as light levels, temperature, salinity, pressure, and wave action. Therefore it is not surprising that for all marine taxa, species richness tends to vary with depth. The variation in species richness with depth can occur over

short depth gradients, a matter of a few tens of meters perhaps, or alternatively, over truly bathyal ranges of thousands of meters. In this section we give examples of both situations.

As a first example, Figure 2.25 shows the change in the coral species number with depth in the Chagos Atoll in the Indian Ocean (Huston 1985). It might be expected that with light-demanding corals, the highest species richness would occur at the surface where the daylight is brightest, but as can be seen, species richness increases rapidly and fairly linearly from around 1 to 10 m deep. Peak richness occurs between 10 and 20 m, followed by a more gradual

Figure 2.23

Mantis shrimp, *Odontodactylus scyallarus*. (Photograph courtesy of Jeff Jeffords.)

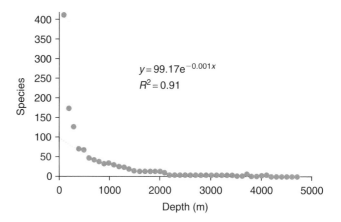

Figure 2.26

Bathymetric gradient of polychaete richness using data split into 100-m depth bands. (From Moreno et al 2008; reproduced with permission of Wiley Interscience.)

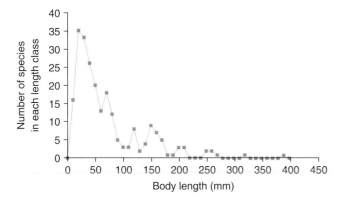

Figure 2.24

The relationship between body size and species number for mantis shrimps. (From Reaka-Kudla 1997; reproduced with permission of Cambridge University Press.)

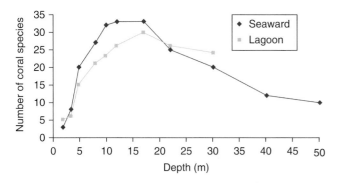

Figure 2.25

The relationship between depth and coral species richness in the Chagos Atoll. (From Huston 1985; reproduced with permission of Annual Reviews Inc.)

decline again. Notice that the maximum depth of the lagoon is around 30 m, and that maximum richness is a little higher on the exposed seaward side of the reef. Clearly a variety of factors are coming together to explain these observations. The general decline with depth can be related to the fact that corals generally require light because of their symbiotic relationship with photosynthetic algae, the zooxanthellae. Therefore, as the light declines (see later section), the species richness of the coral also declines. On the other hand the reduced species richness at the shallowest depths is probably linked to physical disturbance by wave action, and other factors that may include excess light, UV light in particular, and suboptimal warming of the water.

Moving on to consider deeper waters, we find conflicting results. There appears to be a general trend for both macro- and meiofaunal species for species richness to peak in the bathyal zone, between 2000 and 3000 m deep (e.g. Shimanaga et al 2008). Gray (2001) describes a basic paradigm in marine biodiversity, species richness in general increases with depth to a maximum at around 2000 m, thereafter it declines. According to Gray, these conclusions were reached using data from the 1960s, and the actual state of affairs is likely to be more complex. Olabarria (2006) studied prosobranch molluscs in the northeast Atlantic, and found evidence of a decrease in richness from 250 to 1500 m, an increase from there to 4000 m, and another decline to 5000 m. Alternatively, some authors suggest that biodiversity decreases significantly (and often exponentially) with depth (e.g. Danovaro et al 2008). In practice, it seems sensible not to generalize but to consider different taxonomic groups separately.

An example of exponential decline is given by Moreno et al (2008) who studied the bathymetric variations in species richness of benthic polychaete annelids (Figure 2.26) from the intertidal to the abyssal zone at nearly 5000m. As the figure shows, species richness in this case plummets

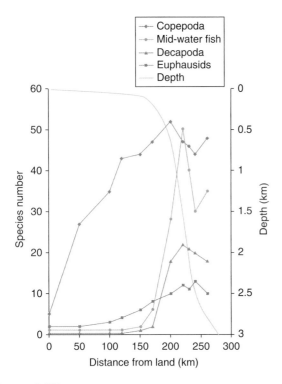

Figure 2.27

Changes in the numbers of pelagic species across the continental shelf, shelf break, and continental slope off Florida. (From Angel 1997; reproduced with permission of Wiley-Blackwell.)

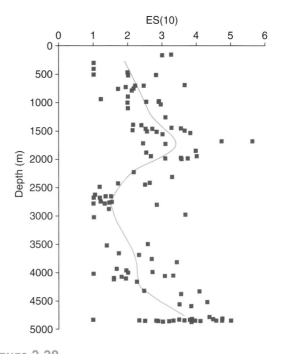

Figure 2.28

Change in diversity of asteroid (starfish) communities with depth in the Porcupine Seabight and Porcupine Abyssal Plain, northeast Atlantic. Diversity is estimated as ES(10) values, which are the expected number of species obtained from a sample of 10 individuals. Solid line indicates mean ES(10) per 500-m depth band. (From Howell et al 2002; reproduced with permission of Elsevier.)

within the first 500 m, and becomes much more constant over the next 4000 m. This basic relationship holds true for a range of latitudes, from 18°S to 55°S (see later). As a further example, Etter & Mullineaux (2001) studied the variation in biomass per unit area and animal abundance in numbers per unit area for the western North Atlantic. These samples were collected with an anchor dredge, a box corer, or a modified Ekman grab. For these samples, both biomass and abundance showed an approximately linear decline with depth on a logarithmic plot which indicates that both variables declined approximately exponentially with depth.

An example of peaks in mid-depth is shown in Figure 2.27. Angel (1997) reports on the number of species within three groups of crustacea and one of fish relative to the continental shelf break and the continental slope down to 3000 m. It can be seen that the numbers of species increase offshore, and tend to peak at or close to the shelf break and down the continental slope. Both benthic and planktonic species richness can vary similarly with depth. As mentioned earlier in this chapter, ostracods are common crustaceans found in all aquatic habitats. The species richness of planktonic ostracods in the North Atlantic achieves a maximum at around 1000 m, which is independent of latitude (Angel 1993). In a study of benthic communities in comparison, Rex et al (1997) plotted standardized species richness data obtained from box coring soft sedi-

ments on the Atlantic continental shelf rise of America. Species richness reached a maximum at an intermediate depth of about 1500 m. It is worth noting that while benthic communities may show maximum richness at intermediate depths, biomass and number of individuals may not behave in a similar fashion.

Finally we give examples of more complex patterns of variation with depth. Howell et al (2002) studied starfish (Asteroidea) distributions from 150 to 4950 m (Figure 2.28). The data appear somewhat scattered, but the authors concluded that distinct patterns of zonation with depth could be identified: (i) an upper continental slope zone from 150 to 700 m; (ii) an upper bathyal zone from 700 to 1100 m; (iii) a mid-bathyal zone from 1100 to 1700 m; (iv) a lower bathyal zone from 1700 to 2500 m; and finally (v) an abyssal zone starting beyond 2500 m.

Clearly there are no simple rules to cover all observations; Moreno et al (2008) propose three possible explanations for the observed depth gradients in diversity. These are: (i) Rapoport's rule, which suggests that zones in the sea with higher species richness should also contain species with smaller depth ranges; (ii) the Mid-Domain effect, which concerns itself with large overlaps of species ranges at mid-depth compared with the hard barrier to a species range at the surface and bottom of the sea; and (iii) the Source-Sink hypothesis, where abyssal zones are very species

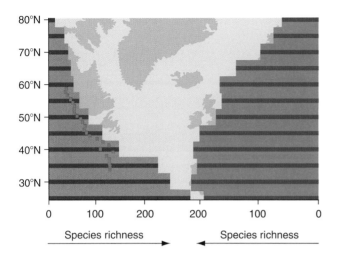

Figure 2.29

Species richness of fish in the north Atlantic, showing plots for both the west and east sides separately. A secondary dataset is also provided for the west side (squares). (From Frank et al 2007; reproduced with permission of Elsevier.)

poor, only maintained by influx of prologues from the much more productive shallow waters. Undoubtedly, no one theory will fit all situations.

Latitude and longitude

For most marine animal and plant groups biodiversity is greatest towards the equator and lowest close to the poles. This so-called latitudinal diversity gradient (LDG) holds true for many terrestrial systems from forest to grasslands, as well as in the sea (Jablonski et al 2006). So robust is this relationship that evidence for LDGs can be detected back in the Palaeozoic period. There are many examples of this trend, or cline as it is called, and we only provide a few here. According to Linse et al (2006), molluscs have been critical in detecting evidence of LDGs, exhibiting low diversity at the poles. Gastropod species richness on the continental shelf of the east Pacific coast shows a very clear peak at around 20°N and not, perhaps surprisingly, at the Equator (Rex et al 2005a). There is a smaller peak in the Antarctic which rather goes against the general prediction of LDG theory. Looking at a wider range of animal groups, Angel (1997) found that fish and various crustacea species also peaked in richness at around 20°N, exactly comparable to the gastropods discussed previously. Fish species richness in the north Atlantic (Figure 2.29), increases as predicted from temperate to tropical latitudes (Frank et al 2007), but there is also a slight difference in longitudinal patterns which we shall return to below. There appears to be a steeper gradient of increasing species richness on the western side of the ocean. Even marine bacteria show an increase in taxon richness with decreasing latitude (Fuhrman et al 2008).

The LDG rule is therefore robust over a wide range of taxa. Of course there are exceptions; not all marine species

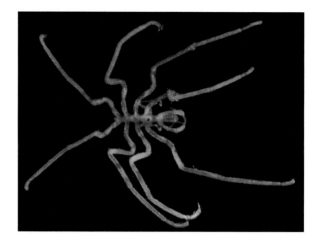

Figure 2.30

A typical pycnogonid or sea spider. Photograph taken from specimen caught in Southampton Water, England. (Photograph Peter Henderson.)

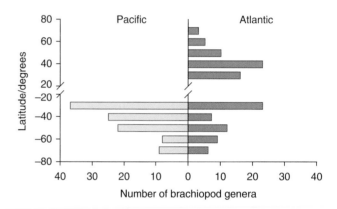

Figure 2.31

Generic richness of Pacific and Atlantic brachiopods according to latitude. (From Barnes & Griffiths 2008; reproduced with permission of Wiley InterScience.)

have diversity maxima on or near the equator, as we have already seen. Large oceanic predators such as sharks, tuna and swordfish have been shown by Worm et al (2003) to occur in distinct diversity hotspots, peaking consistently between 20 and 30° North and South These hotspots presumably offer the best hunting grounds for these predators. Another exception to the rule are the sea spiders or pycnogonids (Figure 2.30), which are particularly diverse in Antarctic waters which hold 38.75% and 21.5% of the 80 and 1164 described genera and species respectively (Leon 2001).

We can also combine latitudinal with longitudinal variations in species diversity. Take for example the number of brachiopod species found according to latitude in the Pacific versus the Atlantic oceans (Figure 2.31) (Barnes & Griffiths 2008). As can be seen, generic richness in the Atlantic shows a fairly typical LDG effect, whereas there seem to be no brachiopods north of the Equator in the Pacific.

The explanations for latitudinal, and to some extent at least, longitudinal events too, must lie, at least in part, in physical factors such as temperature, light energy, seasonality, or thermal stratification. Then there is the added complication of habitat size (area or volume as appropriate) coupled with habitat complexity. These last two factors will be discussed next.

Habitat size

While small islands of productivity such as hydrothermal vents are important for diversity, it is well established that the amount of biodiversity supported is proportional to the size of the habitat. Big rockpools support a higher number of species of organism than do small pools, for instance (Zhuang 2006). This is clearly shown in an analysis of the factors explaining the variability in coral and fish species number on coral reefs (Bellwood & Hughes 2001), where much more of the variability in species number was explained by the size of the reef than either the latitude or longitude of the reef. Figure 2.32 gives an example of the influence of reef size. De Vantier et al (2006) studied 599 sites on 135 reefs along the Great Barrier Reef in Australia. As the size of the coral habitat increased, assessed by percentage cover, coral species richness also increased eventually to a possible plateau with a large variance in the data.

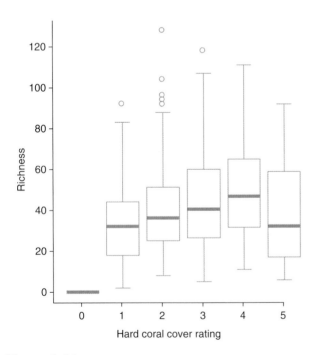

Figure 2.32

Box and whisker plots showing the relationship between species richness and hard coral cover. Ratings: 0=0%; 1=1–10%; 2=11–30%; 3=31–50%; 4=51–75%; 5=76–100%. Horizontal bars indicate median values, boxes cover the middle half of the data, lines mark 1.5 × the range of the boxed data, and circles denote extreme values. (From DeVantier et al 2006.)

Within a single area of reef, the abundance of coral will clearly be important in determining diversity; reefs with the greatest coral cover show the highest species richness and abundance of fish (Walter & Haynes 2006). A possible explanation is that the abundance of coral determines overall species richness which in turn determines fish species presence.

Clearly reefs are three-dimensional objects with nooks and crannies in abundance and where both habitat surface area and volume have been shown to correlate well with fish species richness (Balasubramanian & Foster 2007). A much simpler type of habitat is exemplified by glass sponges in the deep-sea. These are one of the few solid-substrate forming entities in the abyssal plains of the deep-sea, and provide excellent habitats for sessile epifauna to attach to (Beaulieu 2001). The bigger the sponge surface area, the more species of epifauna can attach, a simple and neat example of island biogeography theory in action.

In summary, the more habitat in terms of volume, biomass, and/or area covered, the more species will be present.

Habitat complexity

Even within an extensive habitat, species diversity is strongly linked to structural complexity. Large areas of a homogeneous habitat are unlikely to support large numbers of species. So, it is commonly observed that fish diversity and abundance tends to be higher around and within features rising vertically from the seabed such as rock reefs, coral reef, or sunken ships. Undoubtedly, the diversity of marine organisms is closely related to complex three-dimensional habitats such as coral reefs (Mumby et al 2007), as shown clearly by Mortenson et al (2007) who studied deep-water coral reefs between 800 m and 2400 m on the Mid-Atlantic Ridge. *Lophelia pertusa* (Figure 2.13) was the dominant species of coral, and observations showed clearly that the number of megafauna taxa (bivalves, asteroids, crinoids, and squat lobsters) was on average 1.6 times higher in areas where corals were present. Indeed, in many parts of the world, both temperate and tropical, artificial reefs are now constructed out of all manner of materials from cinder block to warships to enhance the diversity of local marine life (see Chapters 9 and 10). In fact, manipulating marine habitat complexity with natural or even artificial systems is an excellent method for investigating diversity/complexity relationships; a more detailed example will illustrate this crucial association.

Cabaitan et al (2008) used coral transplantation and giant clam restocking on degraded patch reefs in the Philippines to study the effects of habitat complexity on fish community structure. *Acropora* and *Pocillopora* corals were attached to concrete blocks and placed in the "restoration" patch reefs, along with specimens of the clam, *Tridacna gigas*, singly or together, to produce different types of "restoration" reef. The fish communities within these restoration reefs were compared with reefs consisting of mainly dead and damaged corals, the "controls." It is clear that the more complex restoration reefs supported significantly greater species richness than the degraded ones.

At sea, habitat complexity may be difficult to assess quantitatively, but relative methods work well. Gratwicke & Speight (2005) developed a habitat assessment score (HAS) system to estimate by eye alone habitat complexity indices (HCI). Divers, snorkelers or photo-interpreters used a check list of three-dimensional structure, holes, rugosity, and so on to give each site or habitat visited an HCI rating, which could then be related to factors such as fish species richness (Figure 2.33). As shown in the figure, there was a progressive increase in fish abundance and species number with the complexity of the habitat structure, from sand, the least complex, through algae and seagrass to coral reef, the most complex.

Complexity of course is likely to be linked to the age of the habitat; a young or new marine habitat will have to grow to extend its complexity. Copley et al (2007) studied mussel beds on hydrothermal vents on the East Pacific Rise, and found that the youngest mussel beds hosted the lowest species richness of nematode assemblages. A newly introduced, but complex from the start, habitat will need time to acquire diversity. Artificial reefs made from various arrangements of cinder block placed in shallow water seagrass beds in the British Virgin Islands took between 4 and 5 weeks to accumulate all their fish species (Gratwicke & Speight 2005).

Disturbance

The sea is rarely a quiet, calm place. Waves, winds, tides, and storms constantly batter the shallow oceans, whilst currents and avalanches beset parts of the deep-sea. Disturbance of some sort is an integral feature of marine ecology, and of course a whole variety of events can be considered to be disturbing. Take for instance the coral communities of Lizard Island in the Great Barrier Reef. Wakeford et al (2008) studied the impacts of three distinct disturbance events on coral cover (Figure 2.34), namely bleaching,

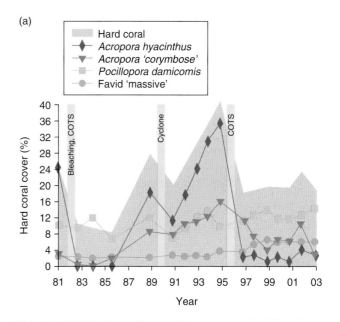

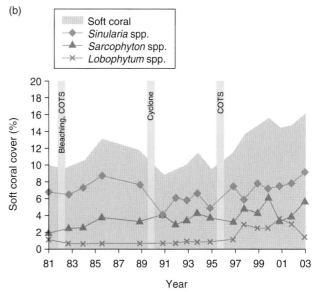

Figure 2.34

Changes in the percent cover of hard and soft corals on Lizard Island, GBR, after three distinct disturbance events. COTS=crown of thorns starfish. (From Wakeford et al 2008.)

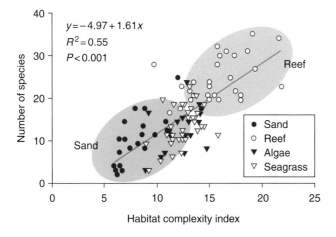

Figure 2.33

The relationship between habitat complexity and fish species richness in the British Virgin Islands. Shaded clusters indicate less complex (brown) and more complex (blue) habitats. (From Gratwicke & Speight 2005; reproduced with permission of Wiley-Blackwell.)

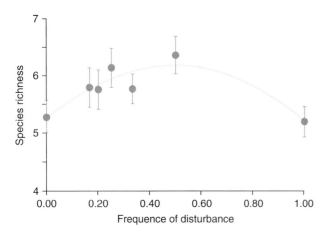

Figure 2.35

Species richness (± s.e.) of sessile invertebrates and macroalgae on experimental PVC rings after 24 weeks in Swedish waters. (From Svensson et al 2007; reproduced with permission of *Ecology, ESA*.)

Figure 2.36

Tubificid worms, with close up of head. (Photograph Peter Henderson.)

storm (cyclone) damage, and attacks by crown of thorns starfish create disturbances that are biotic in origin (see Chapter 9 for information on crown of thorns population explosions and also for a discussion of bleaching and climate change). We have argued above that percent coral cover is a good predictor for coral reef species richness, and it can be seen that after all three events, coral cover is reduced significantly, and only recovers over a period of years. The authors point out that a shorter interval between disturbances reduced the amount of dominant hard corals.

Ecologists have long suggested that extreme high and low levels of disturbance may result in low levels of species diversity. The former may wipe out all but the very hardy or specialized species, whilst the latter allows the most dominant few to outcompete all the others (e.g. Kimbro & Grosholz 2006). Hence habitats and communities exposed to neither very low nor very high disturbance are likely to support the highest diversity – this is the Intermediate Disturbance Hypothesis (IDH). Figure 2.35 illustrates evidence in favor of the IDH. Svensson et al (2007) conducted field experiments on the west coast of Sweden. PVC rings were suspended from buoys and allowed to be colonized by sessile invertebrates and macroalgae such as sponges, hydroids, anemones, fan worms, barnacles, seasquirts, and seaweeds. Benthic grazers and predators were excluded. Disturbance at different levels of intensity was applied using the simple expediency of scraping settled organisms off some of the rings periodically. The figure shows species richness data after 24 weeks, and it is clear that the highest species richness is found at intermediate levels of scraping.

Not all research in marine communities has found evidence of the IDH. For example, neither McClintock et al (2007) working on invertebrates in a pebble beach in New Zealand, nor Sugden et al (2008) studying subtidal benthic communities in the UK, found evidence in support of the IDH, both concluding that species richness or diversity was unaffected by levels of disturbance.

So far, we have primarily considered natural disturbance events; however, in Chapter 9 we discuss in detail the impacts of anthropogenic disturbance on the world's oceans (see Knowlton & Jackson 2008 for an overview). Human disturbance is not universally deleterious. Probably the most disturbing event that can happen to a marine ecosystem is a nuclear explosion. Such events occurred at Bikini Atoll in the Pacific Ocean in the early 1950s. Both surface and sub-surface thermonuclear tests were undertaken. Of the many disturbances to the reefs of the Atoll, the biggest left a crater 73 m deep. Richards et al (2008) went back to Bikini nearly 50 years after these tests and surveyed the scleractinian coral communities. Very surprisingly, they concluded that overall coral species richness has remained much the same as it was before the tests, the only differences related to the loss and gain of a few coral species.

Productivity

In many cases, increasing the productivity of a marine system selects for a few dominant species to the detriment of others (see Chapter 4). Certainly, reef corals become less diverse in water with the highest nutrient concentrations (Taylor 1997). When nutrient levels become artificially high, as is frequently the case in estuaries supporting large cities, diversity can be greatly reduced. Highly organically enriched sediments in the Thames Estuary in the UK, for example, may hold only a single tubificid worm species (Figure 2.36) at densities approaching a million per square meter. However, the higher productivity of waters with higher natural levels can increase the diversity of some animal groups. Taylor (1997) showed that bivalve molluscs, which actively and selectively filter-feed plankton, are more species rich in nutrient-enriched (eutrophic) waters than in low-productivity coral reefs.

Grazing and predation

The interactions of herbivores and predators in marine ecosystems are covered in detail in Chapters 4 and 5 respectively. Here we look briefly at how the interactions of members of higher trophic levels may influence diversity, via top-down mechanisms. Many studies have shown that feeding by predators such as starfish, and grazers such as sea urchins, maintains the diversity of ecosystems beneath them in food chain, by removing large numbers of individuals from species that would otherwise dominate and eliminate less competitive species, thus reducing diversity. The classic example involves the starfish *Pisaster ochraceus* feeding on mussels in North America (Paine 1974). The predator kills mussels, leaving patches which can be colonized by barnacles that are in turn kept in check by the predatory snail, *Nucella*. Removal of the top predator, *Pisaster*, allows mussels to outcompete all other sessile species. Predators can also influence the diversity of trophic levels several stages below them, via trophic cascades. Bruno & O'Connor (2005) set up experimental 30-liter aquaria (so-called mesocosms) within which densities of macroalgae, herbivores, and predators could be manipulated and their interaction assessed. The major generalist carnivores in the experiments were killifish, blennies, and shrimps; the major herbivores were amphipods and isopods, and the algae a mixture of red, brown and green seaweeds. Figure 2.37 shows that after 22 days mesocosms containing the three major predators showed the highest species richness, diversity and evenness of the algal communities.

Genetics and dispersal

Biodiversity encompasses genetic variability, and there has been considerable concern that the activities of man have reduced the genetic variability of many populations, resulting in a potentially reduced ability to respond to future environmental challenges. There has therefore been a growing interest in measuring genetic variability and understanding how it is maintained in natural populations.

Spatial genetic variability is linked to the geographic distances between populations or assemblages, and the ability of a species to disperse. This in turn is associated with the duration of the dispersal phase. According to Angly et al (2006), viruses are the most common biological entities in the sea, and despite their ease of dispersal in currents, a significant positive correlation was found between geographic distance and genetic difference. The further apart

they were geographically, the further apart they were genetically. Another example, this time from the metazoa, is the work by Goldson et al (2001) on the population genetics and larval dispersal in sessile bryozoa (Figure 2.38). Bryozoa,

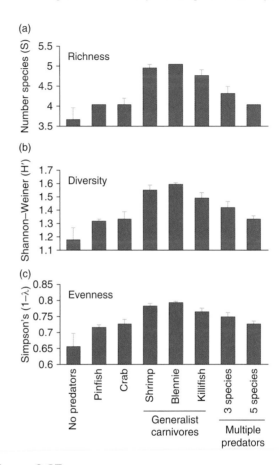

Figure 2.37

Effects of predator type and number on species richness, diversity and evenness on macroalgal assemblages in experimental aquaria. (From Bruno & O'Connor 2005; reproduced with permission of Wiley-Blackwell.)

Figure 2.38

Bryozoa. (Photograph courtesy of Alex Hayward.)

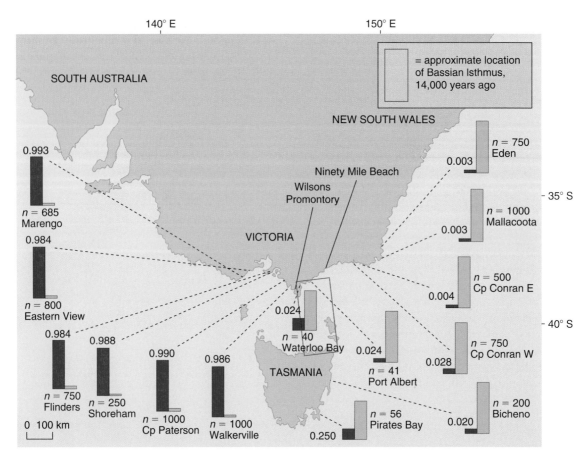

Figure 2.39

Relative abundance of two species of *Nerita* in southeast Australia. Dark green bars= *N. atramentosa*; light green bars= *N. melanotragus*. (From Waters 2008; reproduced with permission of Elsevier.)

or sea mats as they are called, are colonial filter feeders that construct encrusting or erect skeletons, in some ways analogous to corals, and they employ a special set of filter feeding tentacles, the lophophore. In the Menai Straights, North Wales, the bryozoan *Electra pilosa* releases planktotrophic larvae that swim for several weeks before they are ready to settle; in contrast, *Celleporella hyalina* broods lecithotropic larvae (see Chapter 7) which, once released from the maternal colony, swim for up to 4 hours, but settle within 1 hour in the presence of suitable substrate. Random Amplification of Polymorphic DNA (RAPD) analysis of *C. hyalina* samples showed genetic differentiation over distances as small as 10 m, as would be predicted by the short larval stage and the very limited current movements in the study locality. *E. pilosa* samples on the other hand exhibited high levels of genetic heterogeneity only over much larger distances (70 km), in accordance with expectations for a long-lived larval dispersal in coastal water currents. Similarly, work on seamounts in the Southwest Pacific by Samadi et al (2006) compared the genetics of populations of gastropods on mounts in the order of 300 km apart. The only species, *Nasasaria problematica*, to show significantly different genetic structures between the two mounts was also the

Figure 2.40

Scalloped hammerhead shark, *Sphyrma lewinii,* aggregation at the Galapagos Islands. (Photograph courtesy of Scott Henderson.)

only species amongst those studied to have a nonplanktotrophic larval stage. All three of these examples point to a simple mechanism for promoting marine species diversity

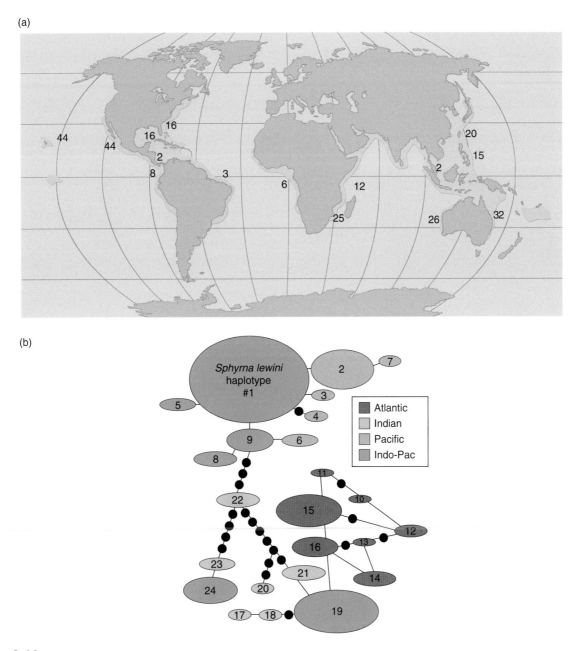

Figure 2.41

(a) Global oceanic range of the scalloped hammerhead shark, *Sphyrna lewinii*, showing sample sites for genetic material. (b) Haplotype network for *Sphyrna lewinii* constructed by statistical parsimony. (From Duncan et al 2006; reproduced with permission of Wiley-Blackwell.)

and marine endemism; this is geographic isolation over large spatial scales.

As Rocha & Bowen (2008) point out, there are very few actual geographic barriers in the oceans, but some examples point to some type of reduction in propagule dispersal ability, over and above the huge scale of the oceans. These barriers may not necessarily be obvious physical structures; differences in salinity, temperature, and current flow can also act to reduce or prevent gene flow in the sea. In many cases, it may not be obvious that barriers exist. The Baltic Sea, for instance, though appear-

ing to be fairly homogeneous from the surface, can be split into at least three major basins based on changes in temperature and salinity. These discontinuities are particularly distinct around the time of year when herring, *Clupea harengus*, spawn (Jørgensen et al 2005). Herring populations in the Baltic can be separated into three major zones, each genetically isolated from one another, not so much by distance but by barriers to fish migration and dispersal. Other seas of the world show similar nonsolid barriers to gene flow. Wallace & Muir (2005) describe unique faunas in the Indian Ocean. Because of the direction of a

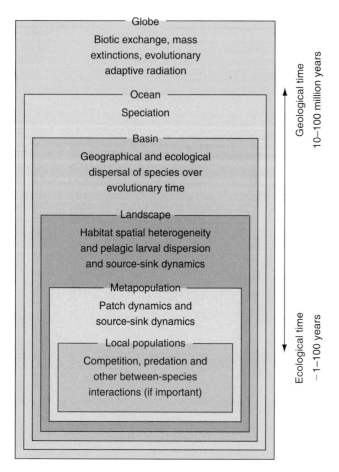

Figure 2.42

Interactions in time and space to produce observed biodiversity in marine ecosystems. (From Gage 2004; reproduced with permission of Elsevier.)

major ocean current, the Indonesian through flow, Pacific Ocean species can invade from the east, whereas there is little possibility of Indian Ocean species migrating into the Pacific. This results in the maintenance of genetic isolation of these species, especially in the far North West around the Red Sea and the Arabian Gulf.

Physical barriers do of course occur frequently in the sea, and have done so over geological time. Just as land bridges come and go between, for example, the UK and Continental Europe, or Indonesia and Australia, allowing the movement of flora and fauna between the two from time to time, similar geological features prevent the movement of marine organisms, hence increasing reproductive isolation and eventually species richness. 14,000 years ago, Tasmania was connected to the rest of Australia by a land-bridge, the Bassian Isthmus, which disappeared as the sea level rose at the end of the last ice age. Waters (2008) mapped the present day locations of two species of *Nerita*, an intertidal gastropod. As Figure 2.39 shows, it is clear that the ranges of *N. atramentosa* and *N. melanotragus*

hardly overlap at all, which seems to correlate well with the historical barrier promoting isolation and hence speciation. It is interesting that even though the barrier has gone, the two species remain distinct and nonoverlapping.

Physical coastal features can also act as genetic isolating mechanisms. For example, a study of mitochondrial DNA length variants in four common coastal species by Levinton (2001) showed discontinuities in the vicinity of South Florida. Studying the relative frequencies of two of the most common monophyletic groups of length variants showed clearly that South Florida represents a major biogeographic discontinuity between the southwest Atlantic and the Gulf of Mexico.

Two final examples will combine all the various aspects of genetic isolation, dispersal and speciation discussed so far, on a global scale. As a first example, consider the global genetic structure of hammerhead sharks shown in Figure 2.40. *Sphryna lewinii* population structure suggests that the species originated in the Indo-West Pacific with late Pleistocene radiations into the central Pacific (Hawaii) and eastern Pacific (central America), and a more recent interchange between oceans via southern Africa (Figure 2.41a,b). Genetic discontinuities are primarily associated with oceanic barriers, though having distinct and ancestral nursery areas may add to the genetic distinctiveness of different regions of the seas.

As a second example Lessios et al (2001) studied speciation in the tropical sea urchin genus, *Diadema*. Sequences of mitochondrial DNA taken from 482 individual urchins from around the world showed that the genus began to speciate (split into extant clades) between 9 and 14 million years ago. Today, closely related clades are always allopatric, and only distantly related ones overlap geographically. The root split into the Indo-Pacific and the eastern Pacific plus Atlantic forms is thought to be derived from geographic isolation when seawater levels were 80 m lower than today. This extreme low-sea-level event occurred approximately 10 million years ago.

Conclusions

There are a large number of factors that can influence diversity and richness in the world's oceans, many of which will act together to produce the observed assemblages. The vast scale of the oceans is important in determining species isolation and offers large units of habitat. Part of the key to understanding marine diversity is to appreciate the huge scale and the great age of marine ecosystems (Gage 2004). These vast regions then hold a hierarchy of structures at global, regional and local scales (Figure 2.42) which create habitat heterogeneity and allow high species richness to be maintained.

Chapter 3

Primary production and chemosynthesis

Primary production is the production of organic molecules from carbon dioxide; this is achieved through both photosynthesis and chemosynthesis. Organisms capable of primary production are termed primary producers or autotrophs. All organisms either practice primary production or are reliant on primary producers at the base of food chains for their energy. In marine ecosystems algae are the dominant photosynthetic group. Chemosynthesis in deep-water habitats such as hydrothermal vents is undertaken by bacteria.

In photosynthesis, the energy used to construct the organic molecules is derived from the sun. A simplified equation for the production of carbohydrate by photosynthesis is:

$$6CO_2 + 12H_2O + \text{photons} \rightarrow C_6H_{12}O_6 + 6O_2 + 6H_2O$$

or in words:

carbon dioxide + water + light energy $\rightarrow$ glucose
+ oxygen + water

The reliance on sunlight restricts photosynthesis to surface waters and the intertidal and shallow subtidal zones of oceans down to a depth of about 200 m. The reduction in light level and spectrum with depth is described in Chapter 1 and illustrated in Figure 1.11.

Chemosynthetic organisms use the oxidation of inorganic molecules such as hydrogen gas, hydrogen sulfide, or methane as a source of energy. In the deep, dark, waters of the oceans it is common for chemosynthesizers to use the energy from the oxidation of hydrogen sulfide. A simplified equation for the synthesis of carbohydrate using this pathway is:

$$6CO_2 + 6H_2O + 3H_2S \rightarrow C_6H_{12}O_6 + 3H_2SO_4$$

The photosynthetic communities of marine environments

In terms of their distribution and physical structure, the photosynthetic organisms of marine habitats can be classified into five main groups, attached large seaweeds (Figure 3.1; see Figures 2.21b & 9.4), seagrass beds (Figure 3.2), mangrove (Figure 3.4), phytoplankton (Figure 3.5), and photosynthetic symbioses between algae and animals. Attached plants can only live in the shallow coastal zone. Gattuso et al (2006) estimated that the difference between primary production and respiration is

Marine Ecology: Concepts and Applications, 1st edition. By M. Speight and P. Henderson. Published 2010 by Blackwell Publishing Ltd.

positive over about 33% of the global shelf area. Over the rest of the shelf area light availability so limits photosynthesis that there is no net production.

Seaweeds (Figure 3.1)

These are multicellular, marine algae which can reach considerable sizes. In fact, the giant kelp, genus *Macrocystis*, is the largest benthic organism in the world (Graham et al 2007). Seaweeds are classified into red algae, green algae, and brown algae which do not comprise a monophyletic group. While red and brown algae are almost exclusively marine, green algae are also common in freshwater and on land. Because they are attached to the substrate, they are restricted to the intertidal zone and shallow coastal waters. The need for a stable anchorage restricts the development of large seaweed beds to rocky substrates. Where good community development is possible, seaweed beds can be extremely productive per unit area and are believed to be amongst the most productive autotrophs in the world. For example, in the Indian Ocean, kelp forests may fix $2000\,\mathrm{g}$ $\mathrm{C\ m^{-2}\ yr^{-1}}$.

Compared with many terrestrial plants seaweeds experience relatively little herbivory so that about 90% of production decomposes. However, as discussed in Chapter 4, the germlings or sporelings of macro-algae are often heavily grazed and this frequency limits seaweed cover in the intertidal zone.

Seagrasses (Figure 3.2)

Seagrasses are flowering plants from the families Posidoniaceae, Zosteraceae, Hydrocharitaceae, and Cymodoceaceae. They are so named because they frequently form large "meadows" in shallow, calm, marine waters which look like underwater grassland. The leaves are long and narrow and usually green and appear superficially similar to grass. As they are rooted plants that can derive their nutrients from the substrate they are not limited in their growth by the nutrient concentrations in the water column. Under ideal conditions, they form the most productive regions of the ocean per unit area and may fix $4000\,\mathrm{g}$ $\mathrm{C\ m^{-2}\ yr^{-1}}$. However, the abundance, growth and distribution of seagrass are extremely sensitive to the availability of light. Increased turbidity caused by eutrophication is a major cause of seagrass decline. As would be expected for any other angiosperm, light reduction can have serious consequences for seagrass production and survival, as shown diagrammatically in Figure 3.3 (Leoni et al 2008). As light intensity and duration declines, growth, biomass and shoot density also declines, whilst nitrate and phosphate levels increase due to increased physiological imbalances as well as breakdown of dead material.

Figure 3.1
A mixture of red, green and brown macroalgae (seaweeds) in the UK. (Photograph Martin Speight.)

Figure 3.2
Seagrass community in Honduras. (Photograph Martin Speight.)

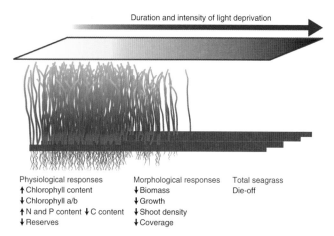

Duration and intensity of light deprivation

Physiological responses	Morphological responses	Total seagrass
↑ Chlorophyll content	↓ Biomass	Die-off
↓ Chlorophyll a/b	↓ Growth	
↑ N and P content ↓ C content	↓ Shoot density	
↓ Reserves	↓ Coverage	

Figure 3.3
Responses of seagrass meadows to light reduction. (Modified from Longstaff & Dennison 1999, from Leoni et al 2008; reproduced with permission of Wylie Interscience.)

The importance of seagrass habitats to shallow-water marine ecosystems is discussed in Chapter 4. As with any terrestrial community of green plants, seagrass meadows support a host of herbivores, from the smallest protozoan micrograzers to large turtles, dugong, and manatee (Nakaoka 2005).

Seagrass meadows are easily damaged by the acitivies of man and are vulnerable in particular to increases in suspended solids which reduce the available light and under extreme conditions can be buried under the sediment. This is discussed in Chapter 9.

Mangrove forest

Mangrove forest comprises trees and shrubs that grow in saline coastal habitats in the tropics and subtropics. The most important plants in the mangrove forest belong to the Rhizophoraceae. Mangrove develops in habitats sheltered from strong wave action where soft, muddy sediments normally rich in organic matter accumulate. Mangrove covers a total area of between 100,000 and 230,000 km^2. They form a highly productive community which fixes an average of 900 g C m^{-2} yr^{-1} and are important sources of organic material to adjacent coastal waters via the export of detritus and living organisms. They account for about 11% of the total terrestrial carbon input into the oceans.

We discuss the vital and multivarious importance of mangroves in several parts of this book, and as mentioned elsewhere, probably the biggest threat to them and their associated species and ecosystem services is their conversion to aquaculture, in particular for shrimps and prawns as well as various species of brackish water fish. Huge areas of mangroves are being destroyed all over the world to turn them into lakes and ponds for the intensive production of

Figure 3.4

Mangroves in Indonesia. (Photograph Martin Speight.)

shrimp (see Figure 9.42). This topic is discussed in further detail in Chapter 9. The conservation and rehabilitation of mangrove is discussed in Chapter 10.

Microphytobenthic communities

Shallow-water sediments and rock surfaces receiving sufficient light develop microphytobenthic communities composed of diatoms, chlorophytes, dinoflagellates, and other microscopic algae. This community is subdivided into two components. The epipsammic component comprises the algae attached to sediment particles and the epipelic component comprises free-living forms that move through the sediment. Linares (2006) presents evidence that shows that the high primary productivity of microphytobenthic communities in sandy sediments low in dissolved inorganic nitrogen is maintained by the use of dissolved free amino acids. The subtidal microphytobenthic community in the Bay of Brest, France, was found by Longphuirt et al (2007) to have an average seasonal production ranging from 57 in winter to 111 mg C m^{-2} in summer. This was estimated to range between 12 and 20% of total primary production, demonstrating the importance of this easily overlooked benthic component.

Epiphytic communities composed of algae, fungi, and bacteria form on seagrasses and have been found by Vizzini and Mazzola (2008) to be grazed on by amphipods, and other crustaceans in Mediterranean coastal lagoons. The high nutritional value and central trophic role of the epiphytes of seagrass beds has been reported by Kitting et al (1984), Klumpp et al (1992), Connolly et al (2005), and others.

On many coral reefs the algal turf community is the dominant autotrophic component (Carpenter and Williams 2007). This community comprises free-living algae of high species diversity which form a thin layer over dead coral. It is likely that the algal turf community makes an important contribution to the high productivity of coral reef systems.

Many animals feed on these organisms see Chapter 4 for a description of the grazing community.

Phytoplankton

Planktonic or free floating organisms which live by photosynthesis are termed phytoplankton. Almost all phytoplankton are small algae or bacteria. The main phytoplankton groups are diatoms, dinoflagellates, coccolithophores, and prochlorophytes. Phytoplankton are present in all surface ocean waters although densities can vary greatly between localities and with the seasons. Photosynthesis by marine

Table 3.1 Estimates of total net primary production (NPP) by different primary producers and the amount of this production that is consumed by herbivores, decomposed, or stored in the sediments. Values in parentheses are the percentage of the total herbivory, decomposition or storage in the ocean. (From Duarte & Cebrian 1996; reproduced with permission of American Society of Limnology and Oceanography Inc.)

PRIMARY PRODUCER	AREA COVERED (10^6 KM²)	TOTAL NPP	HERBIVORY (PG C YR^{-1})	DECOMPOSITION	STORAGE
Oceanic phytoplankton	332	43^5	24.4(88)	14.7(77.5)	0.17(26.5)
Coastal phytoplankton	27	4.5	1.8(6.5)	1.8(9.8)	0.18(27.0)
Microphytobenthos	6.8	0.34	0.15(0.5)	0.09(0.4)	0.02(3.1)
Coral reef algae	0.6	0.61	0.18(0.6)	0.45(2.0)	0(0.7)
Macroalgae	6.8	2.55	0.86(3.1)	0.95(4.2)	0.01(1.6)
Seagrasses	0.6	0.49	0.09(0.3)	0.25(1.1)	0.08(12.0)
Marsh plants	0.4	0.44	0.14(0.5)	0.23(1.0)	0.07(11.3)
Mangroves	1.1	1.1	0.10(0.3)	0.44(1.9)	0.11(17.6)
Total	–	53.0	27.8(52)	19.0(36)	0.65(1.2)

phytoplankton contributes roughly half of total global primary production (Behrenfeld et al 2006a). Table 3.1 shows the dominant role of oceanic phytoplankton in the supply of organic matter (carbon) to marine ecosystems. In comparison, the contribution from the seaweeds (macroalgae) and microphytobenthos is two orders of magnitude less and that from seagrasses and mangrove almost insignificant on a global scale.

Diatoms, dinoflagellates, and coccolithophores (Figure 3.5)

Diatoms are unicellular algae usually placed in the family Bacillariophyceae. Each cell is protected by a silica shell made of two valves. In some species the cells are linked together to form chains. Diatoms are extremely abundant both in the plankton and sediments of marine and freshwater ecosystems. There are over 10,000 living species of diatoms and they are responsible for up to 40% of the total primary production in the oceans (Roberts et al 2007).

Because diatoms sink relatively rapidly they dominate the flocs of "marine snow" that rain down onto the seabed and are therefore important exporters of production from the photic zone to the ocean depths (Sarthou et al 2005).

Dinoflagellates are single-celled protists that use two flagella for propulsion. About half of all dinoflagellates are photosynthetic, and these make up the largest group of eukaryotic algae after the diatoms. Because they can swim towards the light or nutrients they prosper under different conditions from those that favor diatoms which gradually sink in still water lacking circulation within the photic zone. Some species, called zooxanthellae, are endosymbionts of anemones and corals.

Some species of dinoflagellate occasionally concentrate in surface waters in sufficient density to produce a visible discoloration at the water surface. Densities of more than a million cells per milliliter have been recorded. In some cases the water appears red and these red tide events are linked to the release of a neurotoxin which can result in large scale poisoning of marine and even terrestrial littoral life. People have been killed by a condition called paralytic shellfish poisoning following consumption of shellfish which had ingested neurotoxins produced by dinoflagellates. Dinoflagellates produce benthic resting spores which can generate regular seasonal bursts of abundance. Some species bioluminescence producing showers of light in the wake of ships and breaking waves.

Coccolithophores or coccolithophorids are single-celled algae belonging to the haptophyte division. They possess calcite plates called coccoliths. Coccolithophores are found in large numbers throughout the surface euphotic zone of the oceans. They are an important component of the carbon cycle – as they sink they transport carbon to the sediments on the ocean floor. Many limestone rocks are composed of coccoliths. Coccolithophore blooms cause water to appear white. They flourish in cold waters where other phytoplankton species are rare and produce most of the dimethyl sulfide (DMS) derived from the ocean. DMS is linked to the generation of rainfall. This group may be of

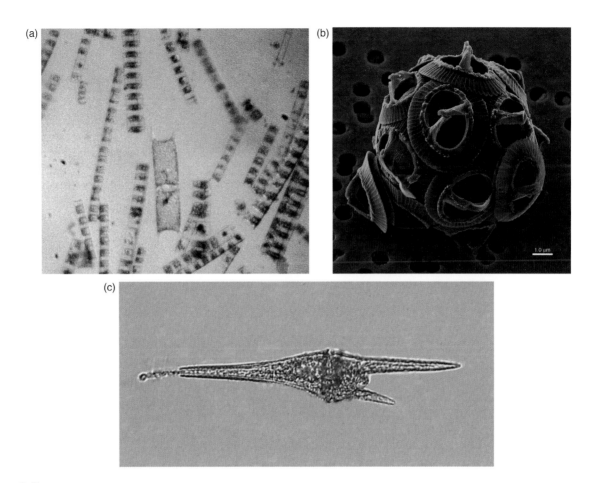

Figure 3.5

Members of the phytoplankton community. (a) Diatoms during the early summer bloom in Southampton Water, England. (Photograph Peter Henderson.) (b) The coccolithophore *Gephyrocapsa oceanic*. (Photograph courtesy of NEON ja, coloured by Richard Bartz; from http://en.wikipedia.org/wiki/File:Gephyrocapsa_oceanica_color.jpg#filelinks.) (c) The dinoflagellate *Ceratium furca*. (Photograph courtesy of Minami Himemiya; from http://en.wikipedia.org/wiki/File:Ceratium_furca.jpg.)

great future importance. Iglesias-Rodriguez et al (2008) showed in laboratory experiments that the net primary production of the coccolithophore *Emiliania huxleyi* increased with increasing CO_2 partial pressure and that biomass has increased by an average of 40% over the last 220 years.

Prochlorophytes

Prochlorophytes are photosynthetic bacteria closely related to cyanobacteria. Because of their extremely small size they are termed picoplankton and were only recently discovered. They are now believed to be responsible for up to 50% of total oceanic primary productivity. They are the most abundant organisms in the oceans and are particularly abundant in tropical and subtropical oceans where they can live as deep as 200 m from the surface. The deep chlorophyll maximum layer, which is mostly contributed by picoplankton, extends, on average, over depths from 67 to 126 m and is a consistent feature of tropical and subtropical waters (Perez et al 2006).

Phytoplankton diversity and the efficiency of resource utilization

Recent experimental studies have suggested that a high diversity of phytoplankton species within a single community results in a greater efficiency of light and nutrient utilisation and a higher standing crop of algal biomass. However, at present, direct experimental evidence that phytoplankton resource use and carbon fixation is directly linked to the diversity of phytoplankton communities is mainly supported from studies of freshwater and brackish water communities (Ptacnik et al 2008).

Photosynthetic symbioses

Symbiotic associations between animals and algae and cyanobacteria are widespread amongst sponges (Porifera)

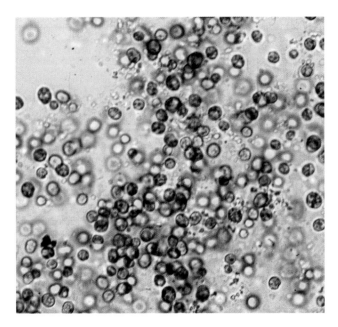

Figure 3.6

Symbiotic *Symbiodinium* in cells of coral polyp. (Photograph courtesy of Scott Santos.)

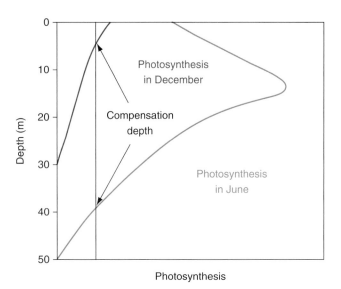

Figure 3.7

Seasonal variation in the amount of photosynthesis with depth showing the differing vertical position of the compensation depth in the North Sea.

and corals and sea anemones (Cnidaria), and have recently been reviewed by Venn et al (2008). In terms of habitat formation, the most important symbiotic relationship is that between dinoflagellates of the genus *Symbiodinium* (Figure 3.6) and reef-building corals. In most cases of algal symbiosis the algal cells are held within the animal cells, one notable exception is the giant clams where *Symbiodinium* is held extracellularly, but is still capable of meeting a large part of the clam's energy requirements. Jantzen et al (2008) in a study of two species of giant clam in the Red Sea found that *Tridacna maxima* was limited to water depths at which light was sufficient for photosynthesis to meet respiratory needs, while *T. squamosa* could live at greater depths, where it needed to be a mixotroph. Cnidaria supporting *Symbiodinium* have cellular transportation systems to deliver bicarbonate ions to the cells supporting *Symbiodinium* to ensure there is sufficient CO_2 for photosynthesis to occur. Experiments using incubated symbiotic *Symbiodinium* suggest that they may release to the host a range of low molecular weight compounds including glycerol, organic acids, glucose, and amino acids. However, there is uncertainty if the same behavior occurs within the host cell. A study by Whitehead and Douglas (2003) of the photosynthetic snakelocks anemone *Anemonia viridis* suggested that within the host the candidate compounds released by *Symbiodinium* were glucose, succinate, and fumarate.

It has generally been assumed that symbiotic algae are important for the well being of to the animal host. There is certainly evidence that coral reef growth is fastest in regions where the rate of photosynthesis is highest. Nakamura and Nakamori (2007) produced a model that explained the characteristic morphology of a coral reef and its division into a shallow lagoon, the reef crest, and the outer reef slope in terms of differential growth in relation to photosynthesis. They argue that the increase in carbonate species moving offshore results in higher reef growth further offshore creating the initial differentiation between the slower growth lagoon and the faster growing crest. Further, the crest also grows faster because it receives more light.

The availability of light and the spring bloom

The availability of light depends on wavelength, angle of incidence of sunlight, and the transparency of the water. The rate of attenuation of light with depth was described in Chapter 1. Because light penetration varies with the angle of incidence, there is a latitudinal and seasonal variation in the amount of light entering the water column. Similarly, because turbidity and hence transparency varies seasonally, this also changes the annual cycle of light penetration.

The compensation depth is defined as the depth at which light intensity reaches a level at which oxygen evolved from a photosynthesizing organism equals that consumed by its respiration. It is not possible for phytoplankton living below the compensation depth to grow. The compensation depth varies greatly between localities, in the North Sea it is typically at 35 m from the surface, while in the clear waters of the mid Pacific it is at 150 m. The region from the surface to the compensation depth, in which 95% of all marine primary production occurs, is termed the photic zone and comprises less than 4% of the volume of the oceans.

Figure 3.7 shows how photosynthesis varies seasonally with depth in the northern latitudes of the North Sea. Under

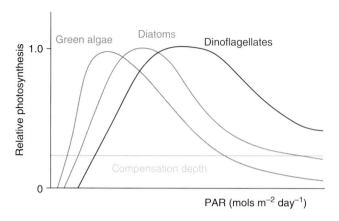

Figure 3.8

The general effect of available light on the relative photosynthesis of different phytoplankton groups showing the differences in their responses to the available photosynthetically active radiation (PAR) that is used by algae for photosynthesis.

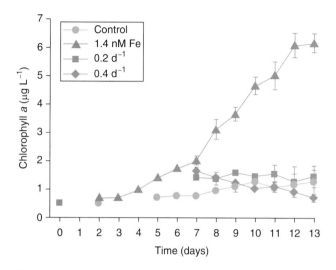

Figure 3.9

Daily total chlorophyll *a* (Chl-*a*) concentrations (μg L^{-1}) in bottle cultures under two conditions of growth medium (violet square and blue diamond) and enhanced Fe levels (green triangle) compared with a control (pink circle) ($\pm$1 s.d.). (From Hare et al 2007; reproduced with permission of Elsevier.)

the winter conditions of December, the maximum rate of photosynthesis is at the surface and the compensation depth is at about 5 m depth. In contrast, at the height of the summer light in June, the photosynthetic maxima occur below 10 m and the compensation depth is around 40 m.

The existence of the compensatory depth results in a powerful regulatory limit on the upper abundance of phytoplankton. As cell numbers increase, turbidity increases and the penetration of light decreases, raising the compensatory depth and reducing phytoplankton growth. In contrast, when grazers or disease reduce phytoplankton density the turbidity decreases, light penetration decreases, and the compensation depth increases, resulting in a larger phytoplankton habitat and more scope for growth.

As shown diagrammatically in Figure 3.8, phytoplankton groups differ in their relative levels of photosynthesis in response to the level of photosynthetically active radiation (PAR), so that the compensation depth differs substantially between groups. These differences can, in part, explain seasonal and spatial differences in the dominant phytoplankton group.

Primary production towards the poles can be light-limited because of the formation of surface ice. In the Beaufort and Chukchi Seas intense spring phytoplankton blooms are observed in open waters close to the edge of the retreating ice (Wang et al 2005). In these northern waters the spring bloom is sufficient to appreciably reduce nutrient levels in surface waters, and Hill and Cota (2005) found that during the summer the highest rates of primary production occurred at 25–30 m depth at the top of the nutricline.

The intensity of the spring bloom and annual primary production can vary with climatic conditions. Skogen et al (2007) report that mean annual production in the Nordic seas is 73 g C m^{-2} yr^{-1} but varies by 20% between the years with highest and lowest production. This considerable

between-year variability which must impact the growth and reproduction of higher trophic levels and the reproductive success of fish in particular was linked to the North Atlantic Oscillation, sea ice abundance, and water movement into the Nordic seas.

The availability of nutrients and large-scale differences in productivity

In comparison with the soil, seawater offers very low levels of nutrients to autotrophs and ocean primary production is generally limited by nutrient availability. Phytoplankton increases are often associated with human derived nutrient enrichment of coastal waters from sources such as fertilizer runoff and wastewater discharges (Cloern et al 2007). This is why when light and waves conditions are good, seagrass beds can be so comparatively productive – unlike other marine plants, they have roots that can obtain nutrients from the substrate. While the essential nutrients nitrogen and phosphorus are normally in low concentration in the sea, and are often limiting (Elser et al 2007), the most striking difference from terrestrial systems is the low availability of iron in the oceans. This is because iron is rapidly oxidized in seawater and becomes insoluble, so that the only significant source of soluble iron is wind-blown, terrestrially derived dust. Iron deficiency limits phytoplankton growth in Antarctic waters as it does in large parts of the Pacific and Southern Oceans which are far from land. As Figure 3.9 shows, phytoplankton primary productivity in the Antarctic waters of the Southern Ocean, measured by chlorophyll *a* concentrations in shipboard culture bottles, increased very markedly as iron concentrations were raised (Hare et al 2007).

Moore et al (2006) have shown that primary production during spring bloom of the central North Atlantic is also iron limited. While enrichment experiments undertaken by Coale et al (1996), Boyd et al (2000), and Tsuda et al (2003) have clearly shown that iron availability in the oceans can limit diatom growth, analysis of nutrient stress using phytoplankton fluorescence reported by Behrenfeld et al (2006a) in the Pacific Ocean indicates a more complex situation. They used diel differences in the normalized variable fluorescence to identify four physiological regimes shown diagrammatically in Figure 3.10. Behrenfeld et al concluded that within the tropical Pacific there were three physiological regimes present, each characterized by different nutrient availabilities. Only two of these are characterized by iron limitation.

Iron enrichment leads to enhanced phytoplanktonic carbon fixation. This observation explains the assertion by John Martin in 1991: "Give me half a tanker full of iron and I'll give you an ice age." The idea is that increased primary production would remove the atmospheric carbon dioxide that warms the planet. There has been considerable recent research on the ability of iron fertilization to reduce atmospheric CO_2 levels by increasing net primary production. Jin et al (2008) conclude that following iron fertilization most of the dissolved inorganic carbon removed from the euphotic zone by increased biological export is replaced by the uptake of CO_2 from the atmosphere. However, they conclude that over a decadal time scale, the fertilization of patches of medium size with an area of 3.7×10^5 km^2 in the Pacific Ocean will remove little net CO_2 from the atmosphere.

The most common macronutrient limiting plant growth in marine environments is nitrogen. This is present in inorganic (NH_4^+, NO_3, and NO_2) and organic forms (urea, peptides, and amino acids). Nitrates (NO_3) are transported from deep waters below the photic zone by turbulence and

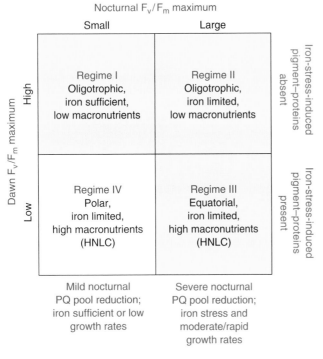

Figure 3.10

The environmental conditions corresponding to the four physiological regimes distinguished by Behrenfeld et al (2006a). The regimes are distinguished by the dawn normalized fluorescence values and the decrease in fluorescence at night. The three regimes observed in the tropical Pacific are shaded yellow. (From Behrenfeld et al 2006b; reproduced with permission of Nature.)

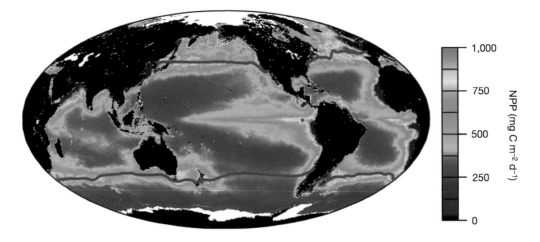

Figure 3.11

Annual average global ocean phytoplankton productivity (NPP). Low latitude permanently stratified waters with annual average surface temperatures above 15°C are delimited with a black contour line. (From Behrenfeld et al 2006b; reproduced with permission of *Nature*.)

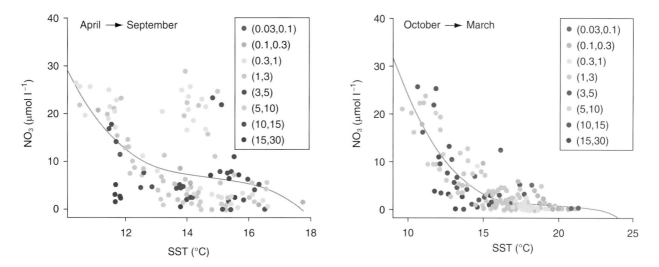

Figure 3.12

Relationship between nitrate concentrations and sea surface temperatures in the Benguela system for two periods of the year. The color scale refers to chlorophyll *a* concentration ranges in mg m^{-3}. (From Silio-Calzada et al 2008; reproduced with permission of Elsevier.)

upwellings, and regions where there are large upwellings of nutrient-rich water are noted for their high productivity. Ammonia (NH_4^+) is excreted by animals and is efficiently recycled within the photic zone. Organic nitrogen is utilized by heterotrophic bacteria. The availability of nutrients explains the general pattern of ocean productivity shown in Figure 3.11 (Behrenfeld et al 2006b). Productivity is highest in northern latitudes and in southern waters in which there are coastal upwellings. The southern ocean is limited by iron availability and much of the warm open oceans are thermally stratified, limiting the movement of nutrients from deeper, richer waters into the photic zone. Indeed, there appears to be a fairly close relationship between seawater temperature and nutrient (in this case nitrate) concentrations (Figure 3.12). In this study off the southwest Africa coast, Silio-Calzada et al (2008) showed that warmer water contained much less nitrate than colder water, with a resulting reduction in chlorophyll levels.

Seasonal changes in the phytoplankton community and the availability of resources

There is a seasonal rhythm to primary production which is particularly notable at extreme northern and southern latitudes. Lutz et al (2007) illustrated this difference by mapping the global distribution in the seasonal variation index

(Figure 3.13), which is simply the coefficient of variation in primary production over the year. The near constant low level of production in equatorial waters and the high seasonality of temperate waters are both clearly represented on this figure. There are some low latitude regions showing clear seasonality; these tend to be waters where the climatic conditions generate seasonal upwelling.

In high and low latitude waters there is a regular seasonal succession in the phytoplankton community. This is illustrated in Figure 3.14 for the diatom and dinoflagellate community of the North Atlantic. The seasonal succession in this region is described in greater detail by Taylor et al (1993). The annual growth season starts with the spring diatom bloom. This is supported by the increase in light (PAR), the relatively high nutrient levels, and relatively low levels of predation. This is then followed by a summer collapse in diatom abundance caused by a depletion of inorganic nutrients in the photic zone and increased zooplankton grazing. This depletion is linked to the establishment of the thermocline which stabilizes the water column and reduces the transport of nutrients from waters below the photic zone. There is also an increased settlement of diatoms in the generally calmer conditions. However, dinoflagellates do not generally collapse with the diatoms and continue to grow into the summer. This is believed to be because they actively swim to stay in the photic zone, suffer lower rates of grazing because they are small, and also are auxotrophic and able to use organic nutrients.

Mesocosms are enclosed bodies of water kept under controlled experimental conditions. Using mesocosms, Sommer and Lengfellner (2008) argue that climate change resulting in spring and winter warming may lead to substantial effect

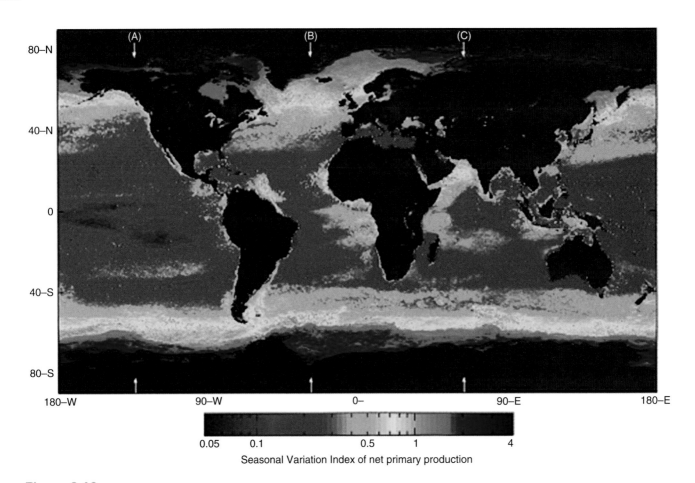

Figure 3.13

The geographic distribution of the seasonal variation index (annual standard deviation divided by average) of net primary production (NPP). (From Lutz et al 2007; reproduced with permission of AGU.)

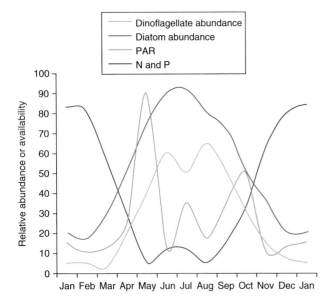

Figure 3.14

The seasonal succession of diatom and dinoflagellate abundance in North Atlantic waters. The figure also shows the change in availability of nitrogen (N), phosphorus (P), and light (PAR). The figure is a generalization of the actual pattern which tends to be more variable as it reflects local conditions and variation in the climate.

on the magnitude and composition of the phytoplankton spring bloom in the Baltic Sea. They found that higher temperatures resulted in a reduced peak biomass and a decrease in mean cell size. They concluded that these changes would disadvantage copepods which feed on the phytoplankton and result in lower energy transfer to higher trophic levels and fish in particular.

Spatial variation in net production and the influence of temperature

The net production of a phytoplankton community is given by the difference between the amount of CO_2 fixed by photosynthesis and the amount released by respiration. It has been argued that there are large expanses of the epipelagic ocean where the respiration of the planktonic community exceeds the gross primary production, so that the long-term existence of these communities must be sustained by the importation of organic carbon (delGiorgio et al 1997;

Duarte et al 2001). This potential deficit in the overall oceanic budget of organic matter could only be made up from terrestrial, freshwater and estuarine ecosystems. As these inputs occur at the periphery of the oceans, their contribution is largely restricted to continental-shelf waters. As Williams (1998) pointed out, it is difficult to envisage that such imbalances in the deep ocean areas could be sustained by organic-matter import from the peripheries. He concluded following an analysis of depth-integrated measures of production and respiration from five open-ocean regions that, in the upper 100 m of the water column, biological production generally exceeds consumption so that the open oceans as a whole are not substantially out of organic carbon balance.

The delicate balance between photosynthesis and respiration is still an active topic of research. Lopez-Urrutia et al (2006) point out that the rates of photosynthesis and respiration show different responses to temperature. They predict that global warming has the potential to increase respiration at a faster rate than photosynthesis, so that in the future the oceans may capture less CO_2 than at present. How phytoplankton will respond to increased atmospheric CO_2 and temperature is still far from clear. There are indications that some organisms will respond by increasing their productivity. Iglesias-Rodriguez et al (2008) showed in laboratory experiments that the net primary production of the coccolithophore *Emiliania huxleyi* increased with increasing CO_2 partial pressure. They also noted that over the last 220 years there has been a 40% increase in average ocean coccolith mass, suggesting that coccolithophores have already responded to enhanced CO_2 levels.

The effects of excessive nutrient loads

The addition of excessive levels to nitrogen and phosphates to surface water is termed eutrophication and is often fueled by nutrient-rich waters draining from agricultural areas. The increased availability of nutrients can generate massive algal blooms (Figure 3.15) in coastal waters. When these dead algae and zooplanktonic feces are decomposed by oxygen-consuming bacteria and when large levels of decomposition occur in sheltered waters which mix little with surrounding waters, the oxygen concentration in water close to the seabed can become low or even negligible. These hypoxic conditions can lead to the death of all aerobic organisms that cannot escape.

Diaz & Rosenberg (2008) reviewed the occurrence of dead zones in coastal waters bottom waters caused by the depletion of oxygen following eutrophication. They noted that dead zones have been increasing at an exponential rate since the 1960s, have now been reported from 400 systems, and now affect a total area greater than 245,000 km².

Figure 3.15
Red algal bloom at Leigh, near Cape Rodney. (Photograph courtesy of Miriam Godfrey.)

Chemosynthetic ocean communities

A surprisingly large number of different of marine communities have chemosynthetic microbes as the basis of their food chains, either in the form of free-living bacterial mats which accumulate in regions of high methane, sulfur and petroleum seeps in the sea (Gilhooly et al 2007), or more commonly as symbionts with a multitude of higher animals. Table 3.2 summarizes this diversity of symbiotic relationship (Dubilier et al 2008). The first thing to notice is that at least seven animal phyla are known to host chemoautotrophic symbionts, occurring from the shallowest marshes and lagoons to the deepest oceans. The microbes involved tend to be of two main types of bacteria, the gammaproteobacteria commonly found as endosymbionts in clams, tube worms and snails, and the epsilonproteobacteria found for example on the gills of deepwater shrimps such as *Alvinocaris* (Nakagawa & Takai 2008). Those microbes that use sulfides are known as thiotrophs, whilst those that use methane are not surprisingly labeled methanotrophs. Not only are the animal hosts of these symbionts diverse, but the chemosynthetic bacteria themselves have been found to be genetically variable, even within one individual host clam for example (Duperron et al 2007). In terms of evolution, it would appear that free-living chemosynthetic bacteria developed first, to be taken up by their hosts as endosymbionts later. Undoubtedly, these relationships have evolved many times in separate animal phyla, providing excellent examples of convergent evolution (Dubilier et al 2008).

Table 3.2 A summary of marine chemosynthetic symbioses compiled from various original authors. (From Dubilier et al 2008; reproduced with permission of *Nature Reviews – Nature Publishing Group*, MacMillan Publishers)

PHYLUM OR MAJOR GROUP	SUBGROUPS*	HOST‡	COMMON NAME	SYMBIONT LOCATION	HABITAT§	SYMBIONT TYPE¶
Ciliophora	Oligohymenophora Peritrichida	*Zoothamnium*	Colonial ciliate	Epibiotic; cell surface	Shallow water	Sulfur-oxidizing symbiont
Ciliophora	Polyhymenophora Heterotrichida	*Folliculinopsis*	Blue-mat ciliate	Epibiotic and endobiotic; cell surface and cytoplasm	Vents	Unknown
Ciliophora	Karyorelictea Kentrophoridae	*Kentrophoros*	Free-living ciliate	Epibiotic and endobiotic; cell surface and cytoplasm	Shallow water	Unknown
Porifera	Demospongiae Cladorhizidae	*Cladorhiza*	Sponge	Intracellular and extracellular	Seeps	Methane-oxidizing symbiont
Platyhelminthes	Catenulida Retronectidae	*Paracatenula*	Mouthless flat worm	Intracellular; trophosome	Shallow water	Sulfur-oxidizing symbiont
Nematoda	Desmodorida Stilbonematinae	*Stilbonema Laxus*	Nematode	Epibiotic	Shallow water	Chemoautotrophic symbiont
Nematoda	Monhysterida Siphonolaimidae	*Astomonema*	Mouthless nematode	Endosymbiont; gut lumen	Shallow water	Sulfur-oxidizing symbiont
Mollusca	Aplacophora Simrothiellidae	*Helicoradomenia*	Worm mollusc	Epibiotic and endocuticular; sclerites and mantle cavity	Vents	Unknown
Mollusca	Bivalvia Solemyidae	*Solemya Acharax*	Awning clam	Intracellular; gill	Vents, seeps, wood falls and shallow water	Sulfur-oxidizing symbiont
Mollusca	Bivalvia Lucinidae	*Lucina Codakia*	Clam	Intracellular; gill	Vents, seeps and shallow water	Sulfur-oxidizing symbiont
Mollusca	Bivalvia Thyasiridae	*Thyasira Maorithyas*	Clam	Extracellular; gill Intracellular; gill	Vents, seeps, whale falls and shallow water	Sulfur-oxidizing symbiont
Mollusca	Bivalvia Vesicomyidae	*Calyptogena Vesicomya*	Clam	Intracellular; gill	Vents, seeps and whale falls	Sulfur-oxidizing symbiont
Mollusca	Bivalvia Mytilidae	*Bathymodiolus Idas*	Mussel	Intracellular and extracellular; gill	Vents, seeps, whale falls and wood falls	Sulfur-oxidizing and methane-oxidizing symbionts
Mollusca	Gastropoda Provannidae	*Alviniconcha Ifremeria*	Snail	Intracellular; gill	Vents	Sulfur-oxidizing and methane-oxidizing symbionts

Taxon	Genus†	Common name	Symbiont location	Habitat§	Symbiont metabolism#
Mollusca — Gastropoda Lepetodrilinae	*Lepetodrilus*	Limpet	Epibiotic; gill	Vents	Chemoautotrophic symbiont§,∥,¶
Mollusca — Gastropoda Peltospiridae	Not named yet	Scaly foot snail	Intracellular; oesophageal gland	Vents	Unknown
Annelida — Polychaeta Terebellida	*Alvinella*	Pompeii worm	Epibiont; integument	Vents	Chemoautotrophic symbiont¶
Annelida — Polychaeta Vestimentifera	*Riftia*, *Lamellibrachia*, *Escarpia*	Tube worm	Intracellular; trophosome	Vents, seeps, whale falls and wood falls	Sulfur-oxidizing symbiont
Annelida — Polychaeta Monilifera	*Sclerolinum*	Tube worm	Intracellular; trophosome	Vents, seeps and wood falls	Sulfur-oxidizing symbiont
Annelida — Polychaeta Frenulata	*Siboglinum*, *Oligobrachia*	Beard worm	Intracellular; trophosome	Vents, seeps, wood falls and shallow water	Sulfur-oxidizing and methane-oxidizing symbionts
Annelida — Polychaeta incertae sedis	*Osedax***	Bone-eating worm	Intracellular; root (ovisac)	Whale falls	Heterotroph
Annelida — Clitellata Phallodrilinae	*Inanidrilus*, *Olavius*	Gutless oligochaete	Extracellular; subcuticular	Shallow water	Sulfur-oxidizing and sulphate-reducing symbionts
Annelida — Clitellata Tubificinae	*Tubificoides*	Sludge worm	Epibiotic	Shallow water	Sulfur-oxidizing symbiont
Arthropoda — Decapoda Alvinocarididae	*Rimicaris*	Hydrothermal vent shrimp	Epibiotic; gill chamber	Vents	Chemoautotrophic symbiont¶
Arthropoda — Decapoda Galatheoidea	*Kiwa*	Yeti crab	Epibiotic; setae	Vents	Unknown

* The orders and families to which chemosynthetic hosts belong are still under debate, and therefore the nontaxonomic term subgroup is used here.

† An example of one or more genera is listed.

§ Shallow water includes all marine habitats less than 200 meters deep.

∥ Based on enzymatic, molecular, or stable-isotope data.

¶ Function inferred from phylogenetic data.

If possible, recent literature was chosen.

** Osedax hosts are included here, even though they have heterotrophic symbionts, because they are closely related to tube worms with chemosynthetic symbionts.

Figure 3.16

Deep-sea shrimp, *Alvinocaris*. (Photograph copyright C.R. Fisher/Ridge 2000.)

Figure 3.17

Hydrothermal vent system and black smoker This black smoker chimney in the Main Endeavour Hydrothermal Field is called Sully. The chimney has undergone dramatic changes since it was perturbed by a series of earthquakes in 1999–2000. Prior to 2002, tubeworm communities were absent.

Figure 3.18

Members of the deep ocean cold seep community. Tubeworms, soft corals and chemosynthetic mussels at a seep located 3,000 metres down on the Florida Escarpment. Eelpouts, a galatheid crab and an alvinocarid shrimp feed on mussels damaged during a sampling exercise. (Image from NOAA Ocean Explorer, courtesy of the National Oceanic and Atmospheric Administration.)

Deep-sea marine communities are probably the best-known systems that rely entirely on sulfide or methane oxidation by these bacteria for their primary production (though note that oxygen is still required for these aerobic reactions which may still have to be derived from the surface of the sea). Such communities are cold seeps and hydrothermal vents, both of which frequently harbor enormous densities of animals with high species richnesses (Figures 3.16 & 3.17). Cold seeps occur where chemical-rich fluids flow out of the seabed at cool or lukewarm temperatures, and they are usually located at the margins of the continental plates and are may contain crude oil products and methane. Some seeps may be over 100 million years old (Chen Zhong et al 2007). Dominant animals at cold seeps which harbor chemosynthetic autotrophs include bivalve clams and mussels in the genera *Bathymodiolus* and *Idas*. *Bathymodiolus* in particular seems fairly ubiquitous in deep-sea chemosynthetic habitats, with various species in deep waters from Florida and Louisiana to the eastern Pacific for example (Duperron et al 2008). The primary-producing bacteria occur in specialized cells at the surface of the gills, and receive their sulfides or methane from the surrounding waters.

Hydrothermal vent communities are the best-known ecosystem supported by chemosynthesis. They develop where geothermally heated water rich in minerals and hydrogen sulfide erupts from the seabed where tectonic plates are moving apart, forming so-called black smokers. The earliest fully identifiable black smoker chimney fossils with recognizable microbial associations are an astonishing 1.4 or more billion years old (Li & Kusky 2007), and there are strong suggestions that life on earth may indeed have originated at vents such as these. Specially adapted animals including polychaetes, molluscs and crustaceans are often members of these communities and these can form symbiotic relationships with the bacteria. For example, tubeworms such as *Riftia pachyptila* (see Figure 6.33) have no mouth or gut (Robidart et al 2008), and living close to the vents depend entirely on their symbiotic relationship with bacteria hosted within the trophosome, their bright-red haemoglobin-rich anterior structure, for energy and food. *Riftia* larvae (see Chapter 7) disperse on the deep ocean currents like many of their distant fan-worm relatives, and the endosymbiont bacteria must be acquired by juvenile worms from the surrounding waters, since these larvae don't carry bacteria with them.

As well as sessile worms and clams, mobile species also carry chemosynthetic symbionts. Schmidt et al (2008) describe the energy sources of dense aggregations of the shrimp *Rimicaris exoculata* at hydrothermal vents on the Mid-Atlantic Ridge. The branchial cavity of the shrimps is colonized by chemosynthetic bacteria which may use different metabolic pathways at different sites. In addition to sulfide oxidation, the authors also suggest that at one site hydrogen oxidation may be the primary energy source. There are also active undersea volcanoes offering far larger islands of geothermal activity.

In general, which species occur where on vent chimneys depends a lot on temperature, concentrations of sulfides, and even pH, so that clear zonations of biogeochemical events have been identified (Brazelton et al 2006). As chimneys age and cool down, some anaerobic microbes dominate which are still able to produce methanotrophically, a last-ditch effort before the whole chimney dies and all life on it becomes locally extinct. Finally, chemosynthetic microbes live on the islands of food offered by the bodies of dead whales (see Chapter 6) and tree trunks that sink to the sea floor. Most excitingly, we must try to imagine the global extent of cold seeps (Figure 3.18) and hydrothermal vents. As we have said many times in this book, the exploration of the deep-sea is in its infancy. The very first hydrothermal vent wasn't observed until the 1970s, and so much is out their waiting to be discovered.

Primary consumption: marine herbivores and detritivores

In this chapter, we discuss the ecology of primary consumption. Though slightly inaccurate, we use the term herbivore to cover the consumption of living primary producers, and detritivore to cover the consumption of dead organic material itself mostly derived from primary producers. Primary consumers feed on a wide range of primary producers including phytoplankton, macroalgae, and bacterial mats, and are divided into herbivores, planktivores, algivores, and bacterial feeders respectively. The remains of dead organisms derived from both land and sea form deposits covering benthic habitats or incorporated within marine sediments, or detritus suspended in the water column. This material and its associated bacteria are consumed by detritivores. Both detritivores and herbivores are major components of a myriad of marine food webs.

We are familiar with animals feeding on multicellular terrestrial plants such as grass and trees. Similarly, in the oceans, there are also herbivores feeding on large, multicellular seaweeds; however, the great majority of marine herbivory is undertaken by animals feeding on single-celled algae within the plankton. Detritivores, while consuming dead, decomposing, organic particles, probably often derive an important part of their nutrition from the consumption of bacteria. Shallow-water surface sediment feeders consume both plant and detrital food sources. As we go deeper down into the sea, the types of primary consumer and their food changes, so that deposit feeders tend to dominate in the depths (Jayaraj et al 2008). Similarly, habitats where little or no current flow occurs will have very few suspension or filter feeders (see Chapter 1). Clearly, any interference with primary consumer activity will have consequences further up food chains (Cerrano & Bavestrello 2008).

As this is the first chapter that discusses organisms defined by their feeding behavior, we start by introducing the concept of trophic guilds.

Trophic guilds

In ecology, a trophic or feeding guild can be thought of as a group of organisms that share the same method of feeding. For example, a guild might comprise a group of fish that feed by scraping algae off rocks and other hard surfaces. Members of a guild overlap in their niche requirements and compete with one another for resources; members of different guilds are not generally in direct competition for food. The division of the species present into trophic or feeding guilds is used in the analysis of community structure and food webs. They can give a general idea of the relative frequency and importance of different life styles within a community. For example, Figure 4.1 shows the proportion of the fish community allocated to different trophic guilds on a tropical reef (Williams & Hatcher 1983). The figure shows the importance of benthic invertebrates as a food source and the relatively small number of species

Marine Ecology: Concepts and Applications, 1st edition. By M. Speight and P. Henderson. Published 2010 by Blackwell Publishing Ltd.

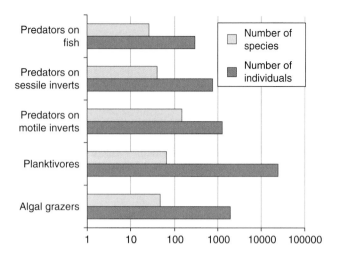

Figure 4.1

The proportion of tropical reef fish species in different trophic guilds. (Data from Williams & Hatcher 1983.)

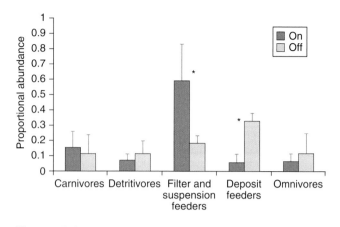

Figure 4.3

Trophic guild composition on and off cold-water coral mounds (bars = ± 1 s.d.). (From Henry & Roberts 2007; reproduced with permission of Elsevier.)

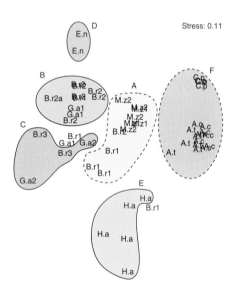

Figure 4.2

An nMDS ordination plot showing clusters of species indicating different trophic guilds in tropical rock pools. A = omnivore guild; B = small-prey carnivore guild; C = large-prey carnivore guild; F = herbivore guild. Numbers refer to size classes of fish. B.r. = *Bathygobius ramosus* (Panamanic frillfin); A.c. = *Abudefduf concolor* (dusky sergeant); A.t. = *Abudefduf troschelii* (Panamian sergeant major); H.a. = *Halichoeres aestuaricola* (Mangrove wrasse); C.p. = *Chaenomugil proboscideus* (snouted mullet); M.z. = *Malacoctenus zonifer* (glossy blenny); G.a. = *Gobiesox adustus* (Panamic clingfish); E.n = *Echidna nocturna* (freckled moray). Note that D & E = two individual fish species which do not conform to any particular guild. (From Castellanos-Galindo & Giraldo 2008; reproduced with permission of Springer.)

that feed on algae or plankton. Planktivores, however, dominate in terms of numbers of individuals. In a similar study, Castellanos-Galindo & Giraldo (2008) investigated guild structures amongst fish living in tropical Pacific intertidal rock pools. In Figure 4.2, nMDS (nonlinear multi-

dimensional scaling) analysis is used to represent four possible trophic guilds by looking at the similarities in the types of food eaten by various fish species. nMDS is a multivariate statistical method which creates an ordination of the species by examining the similarity in the chosen attributes, in this case their diet (Henderson 2003). The key idea behind MDS as used here is to find the best arrangement within a two-dimensional space that places the species with the most similar diet closest together and the least similar furthest apart. The stress value on the figure gives a measure of the ability to represent the relationships between the species in a two-dimensional space, the greater the stress, the poorer is the representation of the species similarity in the two-dimensional space. In this example from the west coast of Columbia, the authors suggest that guild formation is a consequence of fish species converging in their feeding to utilize a small number of highly abundant, patchy, food resources. Notice that moray eels, *Echidna nocturna*, and mangrove wrasse, *Halichoeres aestuaricola*, have their own single species clusters (D and E), indicating their assignment to their own unique feeding guilds.

Different guilds may occur in separate, but adjacent, locations, which may result in reduced niche overlap and interspecific competition. Figure 4.3 provides an example from deep-sea macrobenthos communities on or adjacent to cold-water coral (*Lophelia pertusa*) mounds in the Porcupine Sea Bight in the northeast Atlantic (Henry & Roberts 2007). At depths of between 850 m and 1000 m, different guilds occupy different sites, so that filter and suspension feeders were more common on the coral mounds, whereas deposit feeders were more abundant off the mounds on more mobile, muddy substrates. Such differences in guild locations have also been found for planktonic species. In the Fram Straight in the Arctic Ocean, depth variations in the proportional presence of predator, herbivore and omnivore zooplankton guilds were recognized

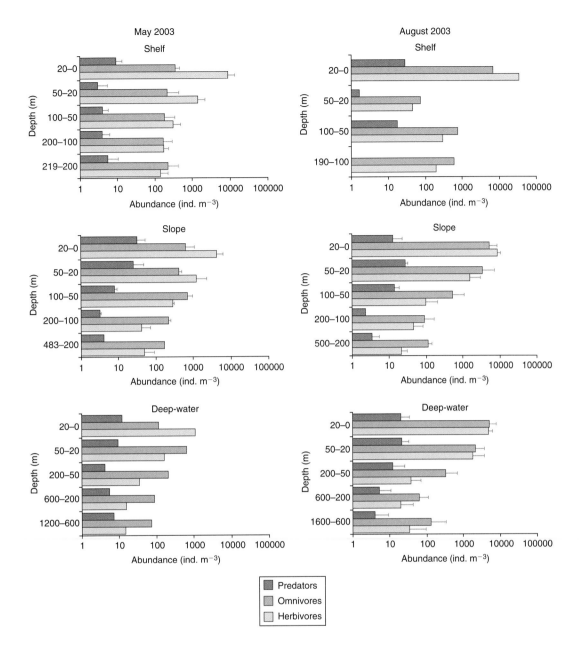

Figure 4.4

Vertical distribution of different trophic levels in the zooplankton of the Fram Straight in three depth categories at two different times of year (spring and autumn). (From Blachowiak-Samolyk et al 2007; reproduced with permission of Elsevier.)

according to depth and season (Figure 4.4.) (Blachowiak-Samolyk et al 2007). Herbivores (such as barnacle and copepod larvae) tended to dominate the shallow water whilst omnivores (such as small adult copepods) were more abundant in deeper water. Predators (such as chaetognaths) varied seasonally and were common in the surface layers (0–20 m) in spring and in deeper layers (20–50 m) in autumn.

Figure 4.5 shows the distribution of invertebrate biomass between the main functional feeding groups according to geographic region and sediment type. The macroinverte-

brate biomass (usually defined as animals retained by a 1 mm² mesh sieve) in temperate regions is 14 times higher on rocky shores than on sedimentary shores, owing to dense populations of suspension feeders such as mussels and barnacles, and grazers such as limpets (Figure 4.6), that account for an average 55% and 35% respectively of total biomass (Ricciardi & Bourget 1999). In all regions suspension feeders are either the dominant or subdominant group in terms of biomass. This high relative abundance can be explained by their exploitation of diverse particulate food resources

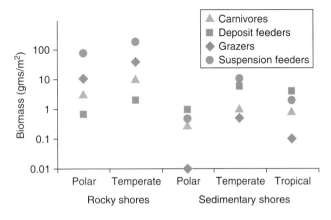

Figure 4.5

Distribution of macroinveretbrate biomass among functional feeding guilds in polar (>60 degrees N), temperate (25–60 degrees N & S), and tropical (0–25 degrees N & S) shores. (From Ricciardi & Bourget 1999; reproduced with permission of InterResearch.)

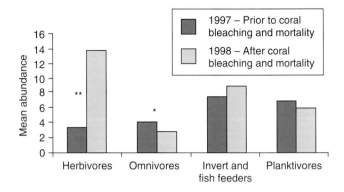

Figure 4.7

Changes in the proportions of the fish trophic guilds before and after an extensive coral dieback in Tanzania. * = Significant difference at $P<0.05$. (From Lindahl et al 2001; reproduced with permission of Elsevier.)

Figure 4.6

Dense populations of suspension-feeding barnacles and young grazing winkles on a temperate rocky shore. (Photograph Martin Speight.)

produced within a large volume of water and made available and replenished by wave action and tidal currents. In addition, sessile or sedentary suspension feeders have relatively low energy demands for food acquisition.

Changes in the relative proportions of the various trophic guilds present can also reflect major changes or damage to an ecosystem. For example, Figure 4.7 shows the changes in the guild structure of a coral reef after a large coral dieback event (Lindahl et al 2001). The main trophic change after bleaching was the increase in herbivorous parrot fish and surgeon fish, because of the increased food in the form of filamentous algae which was overgrowing the dead coral. These observations suggest that the herbivores in this system are usually constrained by a shortage of algae for food.

Finally, some organisms can change their guild or at least feeding behaviors to reflect changes in food supply or abundance (both quantity and quality of the resource). Barnacles are well known to be able to switch between active filter feeding (where they beat their cirri regularly to collect suspended particles) and passive filter feeding (holding their cirri still in a current), depending on the amount of suitable food in the water. Other sessile invertebrates are also able to change feeding methods to suit the habitat and food availability. For example, *Mesochaetopterus taylori* is a polychaete worm that lives in a sand tube and uses a pair of tentacles to collect suspended material in the water column when covered by the tide (Busby & Plante 2007). However, at certain stages of the tide, especially when current velocity is low, the worm sweeps the substrate with its tentacles, thus becoming a deposit feeder. This is not thought to be a change in diet, since the same types of food are ingested. Similarly, the amphipod *Corophium volutator* switches between filter feeding and deposit feeding, depending on the concentration of algae in the surrounding water (Riisgard & Schotge 2007). Both these examples show that feeding behaviors in marine organisms are frequently variable, and it is difficult to assign fixed habits to many species. With that in mind, the next sections describe the major types of primary consumption, with particular reference to the influences of feeding on the food supply (top-down factors), and food supply on the feeders (bottom-up factors).

Sediment and deposit feeders

The type of sediment determines the animals found living and feeding within it (these are collectively termed infauna). The size of sediment particle dictates the types of

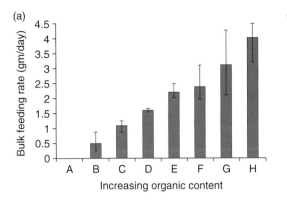
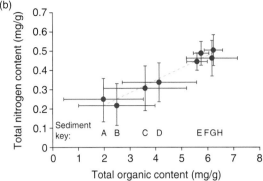

Figure 4.8

Feeding rate of the lugworm *Abarenicola pacifica*, in relation to the organic content of the sediment. (a) Bulk feeding rate in eight sediments if increasing organic content. (b) The organic and nitrogen content of the eight sediments used in the experiment. Capital letters denote different sediment types. (Redrawn from Linton & Taghon 2000; reproduced with permission of InterResearch.)

organisms present and their method of feeding. Coarse sediments such as sands tend to have the highest species richness of shrimps and polychaete annelids, and these tend to be suspension feeders. Fine sands and muds support a majority of the deposit feeders (Lourido et al 2008). The movements and feeding activities of these sediment dwellers are themselves important to the ecology of the substrates, since many organisms, including worms, crabs, bivalves, and sea urchins, create vertical mixing of sediments and the organic matter contained within them. Sediment and deposit feeders are detritivores, which digest the organic matter and bacteria found on the surface of and within seabed sediments.

Familiar examples of detritivores are the large worms dug for fishing bait on muddy and sandy beaches throughout the world. A typical example are intertidal lugworms. The commonest European species is *Arenicola marina* (see Chapter 1), whilst *Abarenicola pacifica* is found intertidally on the North American Pacific coast. These deposit-feeding polychaetes live in U-shaped burrows, which may extend 20–25 cm below the sediment surface. They feed upon sediment, and pass rope-like faecal strands to the sediment surface. It has been argued by Hylleberg (1975) that this species actively stimulates the bacterial growth upon which it feeds.

As shown by Linton & Taghon (2000), the feeding rate of *A. pacifica* increases with the organic content of the sediment (Figure 4.8.). This increasing feeding rate is also reflected in increased growth and increased fecundity (Figure 4.9a,b). Notice that whilst growth rates appear to be fairly linearly related to increasing organic content of the sediments, fecundity seems to reach a plateau after a while. Even small worms with limited food supplies are able to reproduce maximally, an important trait when living in unpredictable or ephemeral habitats. These observations together with many others indicate that sediment-feeding animals are often limited in their growth and reproduction by the availability of organic matter. It is believed that organic matter

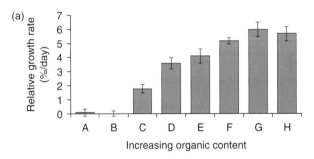
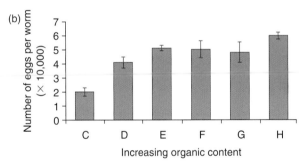

Figure 4.9

(a) Growth rate as a proportion of the final weight and (b) variation in fecundity, measured as the number of eggs produced per worm, of *Abarenicola pacifica* feeding on sediments with increasing organic contents. Means ±1s.d. are shown. (Redrawn from Linton & Taghon 2000; reproduced with permission of InterResearch.)

content, especially organic nitrogen, is a limiting resource in many soft sediment benthic communities.

As well as being vital in the natural ecology of benthic communities, some deposit feeders may be of assistance in intensive aquaculture. One of the big problems in the intensive culture of bivalves such as oysters and mussels is the build-up of their waste products that can rapidly pollute and degrade coastal habitats where this type of aquaculture is practised. Paltzat et al (2008) looked at the growth rates of deposit-feeding sea cucumbers, *Parastichopus californicus*, under intensive oyster beds in British Columbia (Figure 4.10). Clearly, the sea cucumbers grew, suggesting that not only could they be used to reduce the

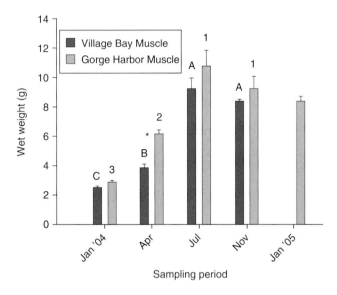

Figure 4.10

Mean (±s.e.) muscle wet weights of sea cucumbers grown in trays under oyster culture rafts at two sites (VB & GH) in British Columbia. Different letters and numbers indicate significant differences (*P*<0.05). (From Paltzat et al 2008; reproduced with permission of Elsevier.)

polluting effects of intense oyster culture, but also provide a harvestable cash crop in their own right.

Filter and suspension feeders

We consider filter feeding and suspension feeding together because both systems attempt to collect and consume the same types of food. Both collect living and dead particulate matter from water columns using a variety of selective or nonselective, active or passive techniques. Filter feeding may be considered a subset of suspension feeding that is more selective about the suspended material removed and consumed. Filter feeders can be active or passive.

Active filter feeders have mechanisms which create small, but locally effective, water currents to bring food particles to them. Examples of active filter feeders include sponges (Porifera), fan worms (polychaete Annelids), barnacles (Crustacea), bivalve molluscs, sea mats (Byrozoa), and seasquirts (Urochordata) (Figure 4.11). Note that active filter feeding has evolved in two of the most phylogenetically separate marine animal phyla, the sponges and the seasquirts, illustrating the success of the system and the number of times it has independently evolved.

Passive filter feeders use a static device such as an arm or tentacle held in a water current to collect whatever may

Figure 4.11

Various active filter feeders from a temperate rocky shore. From left to right, seamats (Bryozoa), colonial seasquirts (Chordata-Urochordata:Tunicata), sponges (Porifera), with scattered fanworms (Annelida:Polychaeta). Note three tentacles of a sea-anemone, not a filter feeder but a "sit and wait" predator. (Photograph Martin Speight.)

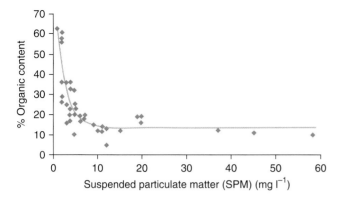

Figure 4.12

The organic content of seawater in relation to the amount of suspended particulate matter (SPM). (From Yukihira et al 1999; reproduced with permission of InterResearch.)

pass by. Examples include hydroids and corals (Cnidaria), and brittle stars (Echinodermata).

Filter and suspension feeders can occur together in great numbers, and can then have a profound impact on the ecosystem, reducing the abundance of phytoplankton and suspended solids to such an extent that the water is visibly clearer. This in turn can promote plant growth and change the entire trophic structure. The loss of huge beds of filter feeders, as occurred in the Hudson Estuary with the loss of the oyster beds (see Chapter 9), is likely to have had a profound impact on water quality and the main energy and nutrient pathways within the ecosystem.

We must first consider what materials are usually suspended in seawater which suspension or filter feeders may eat. Suspended particulate matter (SPM), also termed seston, consists of a variety of items, which can be listed under: (a) living material from the smallest viruses and bacteria to relatively large zoo- and phytoplankton; (b) organic material produced by the death and breakdown of living organisms, large and small (known as particulate organic matter, or POM); and (c) inorganic matter, ranging in size from fine mud to stones, depending on the current velocity (see Chapter 1). Clearly, only categories (a) and (b) can be used by suspension feeders; inorganic material may well inhibit feeding. In fact, as Figure 4.12 shows, as the total amount of SPM increases, the proportion which is organic decreases rapidly (Yukihira et al 1999). In extreme habitats, for example macrotidal estuaries which have very large tidal ranges such as the Severn Estuary, England, suspended solid levels can reach 3000 mg l⁻¹ or more, and filter feeders such as mussels cannot grow, or grow very slowly.

Turbidity can be very deleterious to marine suspension feeders, not just because it contains little useful food, but also because it clogs and damages the delicate and sensitive feeding apparatus. Take for example the slipper limpet, *Crepidula fornicata* (Figure 4.13). Though it is called a limpet, this gastropod does not graze on algae like a typical

Figure 4.13

A stack of slipper limpets, *Crepidula fornicata*. (Photograph courtesy of Paul Naylor.)

limpet (see below), but is in fact an active filter feeder. Figure 4.14a illustrates how filtration rate decreases exponentially as sediment concentration in the surrounding water increases, presumably because of clogging (Johnson 1972). Figure 4.14b relates this to the growth rate of *Crepidula*, showing that shell growth is significantly faster in low turbidity water. Note also that there is a slight increase in growth rate as the height above the seabed increases; faster currents even a few centimeters above the substrate bring more food to the limpet. Species such as *Crepidula* are still fairly tolerant of high levels of turbidity, which they accommodate by increasing the rate of production of pseudo-faeces which removes unwanted particulate matter from the filtration mechanisms (Chaparro et al 2002).

In general, the growth of filter feeders is directly related to the food available in the surrounding water. As shown in Figure 4.15, the growth rate of mussels increases with the density of their algal food, although as algal density increases, growth reaches a maximum determined by the upper limit on the rate of filtration and digestion (Clausen & Riisgard 1996). As mentioned above, a potential problem for filter feeders is that their filtering systems can become overloaded with nondigestible particles. It is notable that in this experiment the presence of 0.5 mg l⁻¹ of silt suspended in the water did not affect growth. Mussels are adapted to dealing with a certain amount of inert material

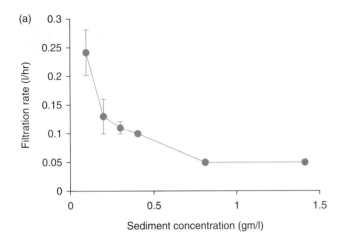

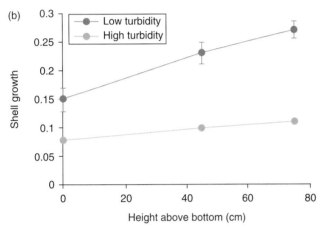

Figure 4.14

(a) The changes in slipper limpet filtration rate in relation to sediment load.
(b) Growth rate of slipper limpet in two levels of turbidity. ((a,b) From Johnson 1972; reproduced with permission of Veliger, Bangor University.)

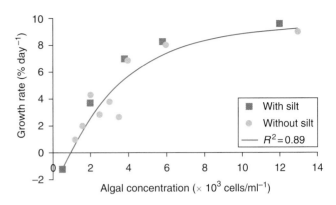

Figure 4.15

Growth rate of mussels as a function of algal density. (Redrawn from Clausen & Riisgard 1996; reproduced with permission of InterResearch.)

of no food value suspended in the water, which they are able to handle at no appreciable cost to their growth.

There is a huge particle size range in the seston, and most filter feeders select specific size ranges according to their body size and the design of their filter apparatus. As is listed in Table 4.1, particle sizes range from the ultra-small (less than 1 μm) to nearly 100 μm for the various sessile invertebrates listed.

Active filter feeders have "pumps" of various forms, and in some cases, the power of this pump has actually been measured (Table 4.2) (Petersen 2007). The pump power output is measured in micro-watts (μW), and it can be seen that some filter feeders (mussels) can generate a lot more power than others (seasquirts) (Figure 4.16). Petersen suggests that the higher power in the mussels is an adaptation to life in turbid water.

Table 4.1 Particle sizes used by various marine filter feeders

SPECIES OR GROUP	TAXON	PARTICLE SIZES	REFERENCE
Corophium volutator	Amphipod crustacean	>= 7 μm	Moller & Riisgard 2006
Halocynthia papillosa	Ascidian seasquirt	0.6 to 70 μm	Ribes et al 1998
Sabellid worms	Polychaete annelid	3 to 30 μm	Shields et al 1998
Dreissena polymorpha	Bivalve mussel	1 to 4 μm	Silverman et al 1996
Aplysina lacunosa	Tubular sponge	0.75 to 18 μm	Duckworth et al 2006
Rhabdocalyptus dawsoni	Glass sponge	0.4 to 5 μm	Yahel et al 2006
Crassostrea gigas	Oyster	4 to 72 μm	Dupuy et al 1999
Electra crustulenta	Bryozoan	5 to 30 μm	Lisbjerg & Petersen 2001

Table 4.2 A comparison of the ascidian and bivalve filter pump in terms of clearance rates and pump characteristics. Ascidian data from *Ciona intestinalis* and *Styela clava;* bivalve data from *Mytilus edulis*. (From Petersen 2007)

	ASCIDIAN	BIVALVE
Clearance rate DW$_{body}$, (1 h^{-1} g^{-1})	11	9.5
Clearance rate gill area, (1 h^{-1} cm^{-2})	0.28	0.34
Pump power output, (μW)	0.5–2.3	10
Water processing potential, (1 H$_2$O ml^{-1} O$_2$)	82	15–50
Pump efficiency, %	0.1–0.3	1.5

Clearance rate is a measure of filter feeder efficiency, and can be thought of as the volume of water emptied of food cells (algae, etc.) per unit time. It can be determined using the formula $CR = Mnt^{-1}(\log_e C_o C_t^{-1})$, where CR is clearance rate, M volume of water at the start, n number of individual feeders in the experimental tank, t time elapsed, C_o the number of particles at the start, and C_t the number of particles at time t (Shumway et al 2003). These figures can be compared to a blank control where particles are counted over time t without any feeders present. Lisbjerg & Petersen (2001) studied the clearance rate and ingestion efficiency of food particles for a model system comprising the marine protist, *Rhodomonas* sp. and the bryozoan *Electra crustulenta*. Bryozoans such as *Electra* are colonial invertebrates which normally encrust rocks and seaweed fronds, actively filter feeding with their special crown of tentacles known as a lophophore. Clearance rate decreased fairly linearly as the concentration of food particles increased, but ingestion rate reached a plateau, suggesting a maximum rate or saturation was achieved. This is not linked to turbidity or clogging, merely that the lophophore mechanism and associated feeding devices have an upper limit to the amount of material handled per unit time. Also note that this example utilizes only one size of food particle; bryozoans, like many active filter feeders, change their feeding behaviors and efficiencies with the size of particles available (Okamura 1990). Undoubtedly, filter feeders can exert a significant impact on their food supply. In Figure 4.17, the rapid and extremely efficient filtration of a single oyster can be seen (Dupuy et al 1999). While this is a small and simple experimental system, multiply this observation up to an entire oyster bed in a shallow-water marine system and a significant impact on marine planktonic primary producers can be envisaged.

So far we have mainly concentrated on active filter feeders, but as mentioned earlier, some phyla such as the Cnidaria comprise only passive feeders. There is also an intermediate area between filter (or suspension) feeding and other forms of heterotrophy. Scleractinian corals vary enormously in the sizes of their polyps. Small species have polyps that use their tentacles as a filtering apparatus (though without any ability to generate their own feeding currents), whilst others have polyps so large that they can

(a)   (b)

Figure 4.16

(a) Sea squirts, *Ciona intestinalis*. (Photograph courtesy of Paul Naylor.) (b) Mussels, *Mytilus edulis*. (Photograph Peter Henderson.)

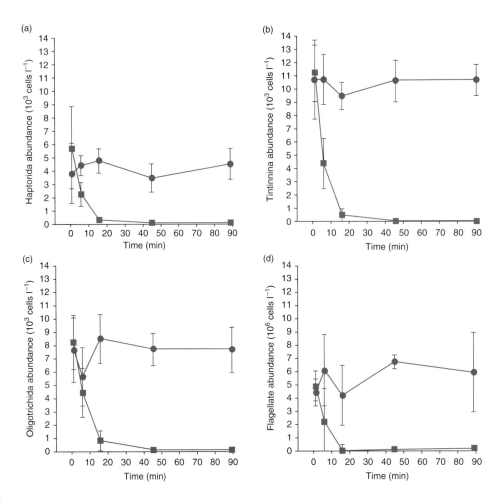

Figure 4.17

Retention of various protist taxa by the oyster *Crassostrea gigas*. Mean protist abundance (± s.d.) was measured in 400 ml of water. Purple diamonds = with a filtering oyster; blue squares = without a filtering oyster. (From Dupuy et al 1999; reproduced with permission of InterResearch.)

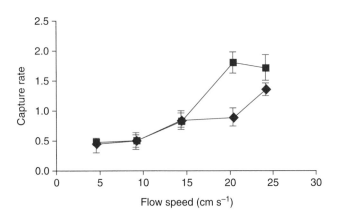

Figure 4.18

Mean (± s.e.) captures rates of brine shrimp cysts by symbiotic (diamonds) and asymbiotic (squares) *Oculina* corals related to water current velocity. (From Piniak 2002; reproduced with permission of Springer.)

catch live prey, and can be classified as sit and wait predators (see Chapter 6). Since they cannot generate their own, even very local, water currents, passive filter feeders depend entirely on the sea to bring suspended material to them, so as current flow speeds up, the rate of capture of suspended material is expected to increase. That this is indeed the case is shown in Figure 4.18 for the temperate coral, *Oculina arbuscula*. The particle capture rate (defined as the number of food particles, in this case brine shrimp cysts, divided by the concentration of cysts) increases with current velocity (Piniak 2002). This species of coral occurs naturally with or without symbiotic zooxanthellae, and it can be seen from the figure that the capture rate of food particles is significantly lower at high flow speeds for corals with symbionts. Presumably, the asymbiotic individuals have to rely solely on heterotrophy, so mechanisms to improve the efficiency of filter feeding are more critical.

Passive filter feeding can be efficient in clearing suspended material. Sea fans (Figure 4.19) belong to a group of corals called gorgonians or octocorals, and have tiny polyps, only

Figure 4.19

Sea fans in Plymouth Sound. (Photograph courtesy of Paul Naylor.)

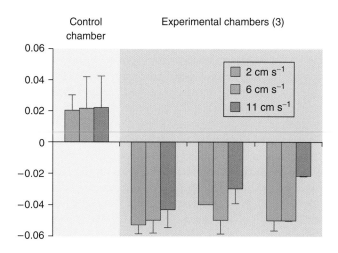

Figure 4.20

Mean (±s.d.) growth rates of heterotrophic bacteria per hour under different water flow rates, and the presence (experimental) and absence (control) of red coral, *Corallium rubrum*. (From Picciano & Ferrier-Pages 2007; reproduced with permission of Springer.)

a millimeter or two in diameter. The Mediterranean red fan coral *Corallium rubrum* is shown in Figure 4.20 feeding on pico- and nanoplankton cells of marine heterotrophic bacteria. *Corallium* is extremely efficient at removing bacteria from seawater, as shown by the clear and consistent negative bacterial growth rates in tanks containing corals (Picciano & Ferrier-Pages 2007).

Once food particles have been ingested, the efficiency at which filter feeders can assimilate varies from species to species, but can be high. Tatian et al (2008) report mean assimilation efficiencies of macro-algal particles to be 26–51% in a clam, and 26–72% in sea squirts.

How important are filter/suspension feeders to marine ecosystems? Taking as an example coral reef, even the casual observer will realize that the majority of fixed organisms on a reef are filter feeders. Sponges, fan worms, giant clams, and of course the multitude of hard and soft corals, make up the vast bulk of biomass and diversity, and as such might be expected to have a significant impact on the primary producers present. Fabricius & Dommisse (2000) studied reefs on the Great Barrier Reef in Australia, and concluded that chlorophyll and particulate organic carbon were depleted downstream of a reef, indicating that they had been removed from the water by the reef filter feeders. In this study, the depletion of chlorophyll averaged 35% of the standing crop, whilst the net depletion of organic carbon was 15%. These figures may not seem extremely high, but note that this is just one reef at one time, in a vast and dynamic marine landscape. A further question concerns whether filter feeders such as oysters are regulated by their food supply – whether phytoplankton dictates the behavior and dynamics of consumers in a bottom-up fashion – or alternatively, filter feeders regulate primary producers in a top-down way. We shall return to this topic when we consider grazers later in this chapter, but for now it seems that incontrovertible evidence is hard to come by, and what evidence there is, is contradictory (Bruschetti et al 2008; Helson et al 2007).

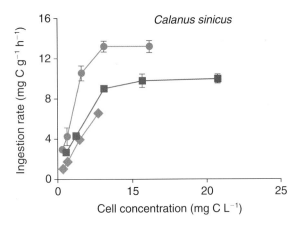

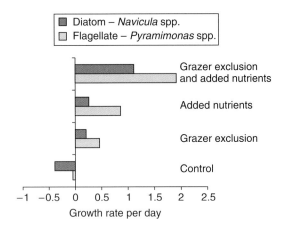

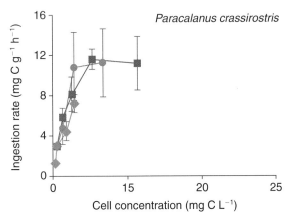

Figure 4.21

The ingestion rates of two species of copepod on three algal diets at different algal carbon concentrations. Circles = toxic dinoflagellate *Alexandrium tamarense*; squares = nontoxic dinoflagellate *A. tamarense*; diamonds = diatom *Thalassiosira weissflogii*. Mean ± s.d. (From Liu & Wang 2002; reproduced with permission of Springer.)

Figure 4.22

Growth rates per day of two types of phytoplankton in the Red Sea, with either zooplanktonic grazers removed, and/or nutrients added. (From Sommer 2000; reproduced with permission of InterResearch.)

Phytoplankton grazers

The phytoplankton of open waters is grazed by zooplankton, small animals many of which are crustaceans, of which the most important, in terms of biomass, are the copepods. Copepods like many other zooplanktonic species are, strictly speaking, more omnivores than herbivores, eating a whole range of other zooplankton and organic particulate matter, as well as phytoplankton (Schnetzer & Steinberg 2002). However, studies on copepod "grazers" have shown that the feeding rate increases with phytoplankton cell density until a saturation density is reached. For example, the ingestion rate of dinoflagellates and diatoms by copepods increases almost linearly with food concentration up to a plateau (Figure 4.21) (Liu & Wang 2002). Above this density ingestion remains constant no matter how much more food is available. Note that toxic dinoflagellates are still ingested at much the same rate as nontoxic ones, but feeding on the toxic species reduces copepod fecundity considerably.

Experimental manipulation has found evidence for both top-down and bottom-up regulation of medium-sized phytoplankton. Sommer (2000) undertook an experiment in which the growth of the diatom *Navicula* and flagellate *Pyramimonas* were studied following the exclusion of grazers and addition of nutrients. The grazers were dominated by crustacean nauplii and ciliate protozoa in the size range 20–100 μm. Growth rates of the phytoplankton showed a strong positive response to nutrient enrichment in the water, but a weaker, though significant, response to grazer exclusion (Figure 4.22). From this example, top-down regulation of phytoplankton by herbivores has an influence, but bottom-up control seems to be more important.

Within the oceans, both the density and the feeding rate of grazers such as copepods increases with the biomass of phytoplankton. Figure 4.23 illustrates an example from Arctic seas, where the highest rates of herbivory occurred when the highest densities of herbivores interacted with the highest abundances of phytoplankton (Saunders et al 2003). Identifying cause and effect (top-down versus bottom-up regulation) here is again difficult, but it is possible that herbivore impacts on phytoplankton are kept at low levels because the herbivores themselves are adversely influenced by their own predators (see Chapter 5).

Finally in this section we consider some of the bigger, mobile marine filter feeders. This group or guild of animals includes basking sharks, and blue whales, amongst others. Large baleen whales are unlikely to be feeding on phytoplankton, whereas their own food, krill, is certainly a microalgal filter or suspension feeder (Figure 4.24). This is an example of a trophic cascade (Sommer 2008) which we shall return to in later chapters. Krill are small (mostly 1 to 2 cm long) prawn-like decapod Crustacea in the order Euphausiacea which are enormously important in some marine food chains and are also becoming increasingly

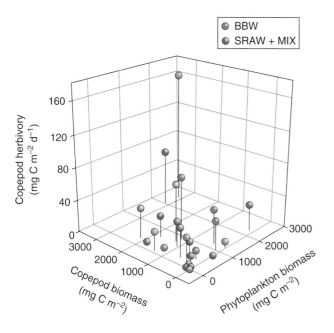

Figure 4.23

Total herbivory rate of copepods in relation to copepod abundance (biomass) and phytoplankton density (biomass). BBW = Baffin Bay water; SRAW + MIX = silica-rich Arctic water plus transitional water. (From Saunders et al 2003; reproduced with permission of InterResearch.)

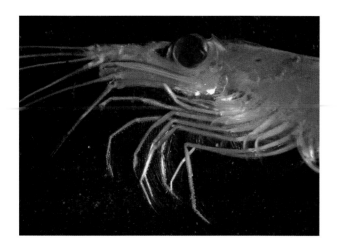

Figure 4.24

Photograph of krill (Euphausia) showing the head and feeding limbs. The illustrated species is *Meganyctiphanes norvegica*, a common species in North Atlantic temperate waters. (Photograph Peter Henderson.)

exploited by Man. There are many species in both the northern and southern hemispheres. They filter feed on plankton using modified front appendages. Many of the common species appear to feed mainly as herbivores on diatoms and other members of the phytoplankton, but others have a more mixed diet, also eating other Crustacea such as copepods (Dalpadado et al 2008). The impact of krill on phytoplankton can be variable. Tanimura et al (2008),

Figure 4.25

Limpet, *Patella vulgata*. (Photograph courtesy of Paul Naylor.)

for example, report that the dominant krill species, *Euphausia superba*, south of the Antarctic Circumpolar Current, removed less than 6% of the total phytoplankton standing stock. Nonetheless, Marrari et al (2008) suggest that unusually high krill reproduction in 2000/2001 was coincident with above-average chlorophyll *a* concentrations throughout most of their study area in the Southern Ocean. The phytoplankton peak suggested by the chlorophyll maximum resulted in the largest juvenile *E.superba* recruitment observed since 1981.

Grazing on surface algae

Large and mobile marine herbivores include gastropod molluscs such as limpets and periwinkles, echinoderms such as cushion stars and sea urchins, many species of temperate and tropical fish, and at the top of the size scale, dugongs. Most eat algae in some form, most frequently as recently settled biofilms of microalgae such as diatoms, or propagules of macroalgae (seaweeds) rather than mature fronds, though a diverse and important set of herbivores also eat seagrass, one of the very few marine angiosperms (see Chapter 3).

Little & Kitching (1996) describe a simple example of how limpets (Figure 4.25) reduce the covering of microalgae in rocks. As the density and biomass of grazing limpets increases, the cover of diatoms on rocks declines dramatically. Most grazers of this type are fairly indiscriminate about what they remove from hard substrates, so as well as single-celled biofilms, as in the previous example, the colonization and recruitment stages of macro-algae will also be removed. Note that limpets and winkles do not eat mature fronds of seaweeds, only the just-settled propagules (called germlings or sporelings), but clearly, if the settling

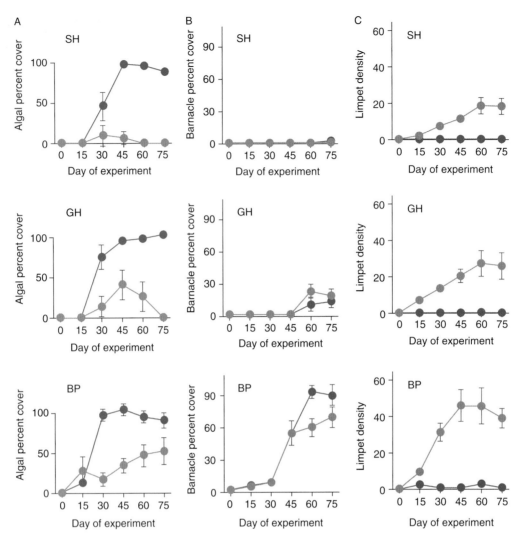

Figure 4.26

Results of a limpet grazing experiment at three different sites (SH, GH, BP) in Oregon. A = effect of limpets on the recolonization of bare surfaces by algae; B = effect of limpets on the relonization of bare space by barnacles; C = density of limpets over 75 days. Blue spots = treatments with no limpets; green spots = treatments with limpets present. (From Freidenberg et al 2007; reproduced with permission of Elsevier.)

stages are grazed away, mature seaweed can never develop. This effect is shown in Figure 4.26, from work in Oregon (Freidenberg et al 2007). The algae in this example were a mixture of diatoms and green seaweeds (*Ulothrix*, *Urospora* and *Enteromorpha* spp.). It is quite clear that in the absence of grazing pressure from limpets, algal cover on the rocks reached nearly 100% after only 30 days, whereas in the presence of the herbivore, algal cover varied from 10% to 50%. The real-world effect of such grazing pressure is admirably demonstrated by the photographs in Figure 4.27. The oil tanker, *Sea Empress*, ran aground on the shores of West Wales releasing huge quantities of crude oil (see Chapter 9), resulting in the deaths of most of the limpets on the local rocky shores. With this grazing pressure removed, large quantities of green and later brown seaweeds grew

luxuriantly until the limpets managed to re-colonize and restore grazing pressure.

Fish can also act as surface grazers of diatoms and other microalgae on both hard and soft surfaces. An interesting example is the feeding of grey mullet, *Liza* spp., on the surface algae of intertidal mudflats. The fish scrape the algae off the surface at high tide and leave characteristic marks on the surface. Ojeda & Munoz (1999) excluded fish such as the herbivorous blenny *Scartichthys viridis* from intertidal rocky shores in Chile (whilst allowing invertebrate grazers such as limpets continued access). Excluding fish had a huge impact on the volume of macroalgae (note the log scale on the *y*-axis in Figure 4.28). Similar conclusions were reached by Fox & Bellwood (2007) working on the Great Barrier Reef. They quantified

Figure 4.27.

Effects of removal of limpet grazing pressure due to oil pollution in Pembrokeshire, West Wales. Bare patches of rock occur where some limpets survived. (Photograph courtesy of Robin Crump.)

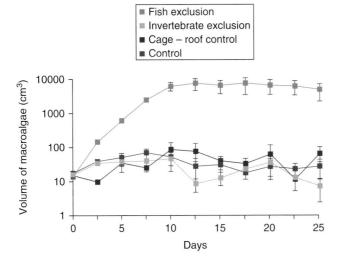

Figure 4.28

Abundance (volume cm³) of foliose macroalgae (red, green, and brown combined) through time on rocky surfaces in a mid-tidal area in Chile. (From Ojeda & Munoz 1999; reproduced with permission of InterResearch.)

the grazing impact of three reef fish, the parrotfish *Scarus rivulatus* and *Chlorus microrhinos* and the rabbitfish, *Siganus doliatus*. The feeding impacts of these grazers varied with depth, but results showed that parrotfish grazing per month covered between 40% and 100% of a square meter area of reef crest, whilst rabbit fish removed 26 cm³ of algae in the same time and place. Fox & Bellwood concluded that the structure of coral reef algal communities is shaped by fish herbivory.

So far, we have considered grazing effects on algae, which are the dominant photosynthetic primary producers in the oceans. However, seagrass deserves a separate mention. Seagrasses are flowering plants (angiosperms) with underwater and underground roots and rhizomes (Short et al 2007) (see also Chapter 3). There are only about 60 species worldwide, both tropical and temperate, with, as usual, the highest diversity in the Indo-Pacific. Seagrass habitats are unique but crucial to the ecology of shallow marine systems both temperate and tropical, as they home a nursery and feeding ground for a multitude of marine species from lobsters to moray eels, conch to jellyfish. They are also extremely sensitive to anthropogenic damage (see Chapter 9). As with any terrestrial community of green plants, seagrass "meadows" support a host of herbivores, from the smallest protozoan micrograzers to large turtles, dugong and manatee (Nakaoka 2005). Heck & Valentine (2006) summarize herbivory in seagrass under three headings: (a) grazing on live seagrass leaves; (b) consumption of algae growing on seagrass leaves; and (c) consumption of planktonic algae in the waters surrounding the seagrass meadows. In the past, herbivore levels were thought to be low in seagrass communities, simply because nothing obviously seemed to eat it. These days however research shows that somewhere between 3% and 100% of seagrass net primary production enters marine food webs via grazing (Heck & Valentine 2006). In a specific study in the Mediterranean, Tomas et al (2005) reported 50% of seagrass biomass removed in one summer season by grazing fish.

Undoubtedly, where they occur, dugongs are the biggest grazers of seagrass. These large mammals make distinctive tracks through seagrass meadows, and like equivalent large grazer systems on land, this type of herbivory has a large impact on the plant community. Figure 4.29 shows that dugongs in Indonesia remove most seagrass biomass in the depth layer 0–4 cm (de Iongh et al 2007), which is where the rhizomes occur. This grazing appears to be linked to the carbohydrate content of these rhizomes, and as these authors point out, seagrass meadows are vital for dugong survival.

Seagrass supports epiphytes. Just like any other surface in the sea, seagrass leaves rapidly become colonized by all manner of sessile organisms (Figure 4.30). Prado et al (2007) recorded 129 taxa growing on seagrass leaves in the Mediterranean, including a large range of primary producers such as cyanobacteria and red, green and brown seaweeds. Epifauna such as hydroids, byrozoans, and seasquirts

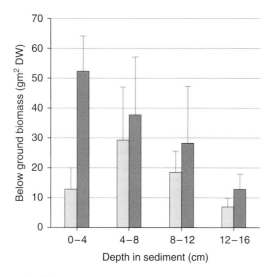

Figure 4.29

Distribution of the seagrass *Halodule universis* in Indonesia according to depth in the sediment (± s.d.). Dark blue bars = undisturbed seagrass beds; light blue bars = within dugong feeding tracks. (From de Iongh et al 2007; reproduced with permission of Springer.)

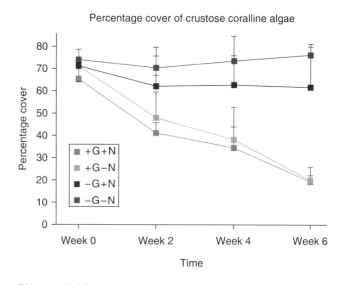

Figure 4.31

Mean (± s.d.) variation in crustose coralline algal cover in experimental treatments exposed to different levels of urchin grazing (G) and nitrogen enrichment (N) over a 6-week period. (From Wai and Williams 2005; reproduced with permission of Elsevier.)

Figure 4.30

Epifauna and flora on the leaves of seagrass. (Photograph Martin Speight.)

were also common, but less so than the algae, the latter providing a rich and abundant food source for grazers such as fish, urchins, crustacea, and molluscs. Prado et al report around 25% removal of all seagrass epiphytes by direct herbivore grazing. Clearly, dugong feeding is likely to have a deleterious effect on the epiphytes simply by removing their substrate; the seagrass leaves (Nakaoka 2005).

Finally in this section, we briefly introduce the special case of sea urchins. The grazing effects of urchins have received great attention in the marine ecological literature over many years, mainly because of what happens when they disappear. The best example of this involves the spiny urchin *Diadema antillarum* which underwent a mass mortality in the Caribbean in 1983 (Macia et al 2007), and is now still recovering (see Chapter 5). As with many other marine grazers discussed earlier in this chapter, urchins can have a significant effect on macroalgae, as shown in Figure 4.31 (Wai and Williams 2005). In this case, it is clear that urchin grazing (top-down control) is more important to encrusting coralline algae than nutrients (bottom-up control). Part of the grazing impact of urchins is of course to remove germlings of potentially ecologically dominant seaweeds and other algae, maintaining stability and diversity in both temperate and tropical marine ecosystems. One classic consequence of a change in urchin abundance is the so-called "phase shift" observed in many parts of the tropical world, especially, where coral reefs, normally dominated by hard corals, shift to a high percentage of algal cover instead (see later in this chapter).

An added complication with some sea urchins is their role in bio-erosion. The mineral skeleton of coral reef is gradually eroded away by algal film grazers such as urchins, and can result in the eventual destruction of large hard coral structures. In order to feed on algae growing over dead corals (see below), some urchins (and fish such as parrots and triggers) ingest coral material as well as surface algae, digesting the algae and leaving coral "dust" in their wakes. The analysis of the guts of sea urchins grazing the algae from hard surfaces in Tanzanian coral reef showed that

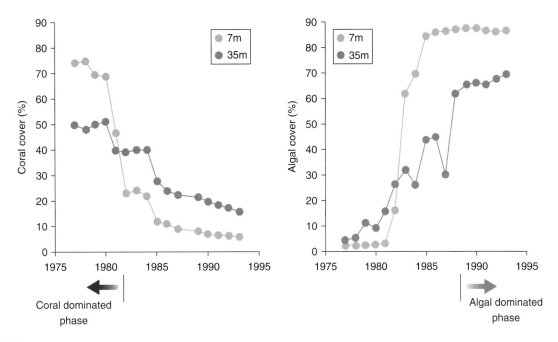

Figure 4.32

Degradation of Jamaican coral reefs over two decades through a combination of over-fishing, disease in grazing urchins, and hurricane damage. (From Hughes 1994; reproduced with permission of Science AAAS.)

approximately 80% of the contents were calcium carbonate derived both directly from dead coral and also from eroded sediment trapped in the algal turf (Carreiro-Silva & McClanahan 2001). A comparison with the gut contents of sea urchins grazing the surface of sea grass indicated that approximately 15% of this calcium carbonate was derived from direct erosion of the dead coral surface.

The role of algal grazers in shaping communities – phase shifts

Herbivorous fish and invertebrates clearly have significant impacts on primary producer communities in the sea, and as we shall discuss in Chapter 10, management tactics such as no-fishing zones or marine protected areas can have extremely important consequences for increasing herbivory and reducing algal dominance (Hughes et al 2007). The removal of fish and urchins for whatever reason can result in dramatic increases in algae. Figure 4.32 illustrates such a phase shift on Jamaican reefs (Hughes 1994). The pattern of coral decline is almost a mirror image of the increase in algal cover. Most significantly, this type of phase shift is essentially irreversible (Ledlie et al 2007) (though see Chapter 10), in that hard substrates covered in algae cannot

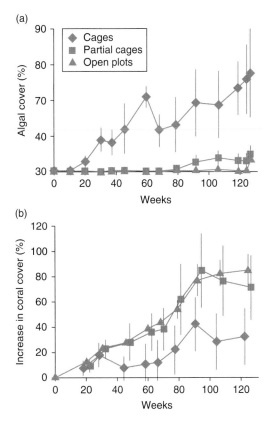

Figure 4.33

Contrasting "trajectories" of macroalgae and corals after the exclusion of herbivorous fish. (From Hughes et al 2007; reproduced with permission of Elsevier.)

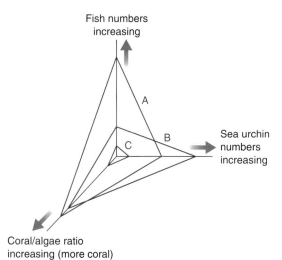

Fish numbers
increasing

A

C   B

Sea urchin
numbers
increasing

Coral/algae ratio
increasing (more coral)

Plane A – Pristine situation prior to fishing, with
grazing of algae shared equally between diverse
fish and inverts, sea urchins in particular

Plane B – Increased abundance of urchins and
other invert algal grazers compensate for loss of
grazing by fish due to over-fishing, so that ratio
of corals to algae remains high

Plane C – Mass mortality of sea urchins on top of
over-fishing allows explosive growth of algae which
overgrow coral

Figure 4.34

Graphical model showing the responses of coral, algal grazers, and fish to
changes in fishing pressure. (From Knowlton & Jackson 2001; reproduced
with permission of Marine Community Ecology, Sinauer.)

be colonized by new coral recruits which need a hard, clean
surface on which to settle (see Chapter 7). Phase shifts from
coral- to algal-dominated ecosystems because of the removal
of herbivory can be demonstrated in field experiments.
Hughes et al (2007) set up cages on the inner Great Barrier
Reef to exclude large herbivorous fish such as parrotfish
(Scaridae) and surgeonfish (Acanthuridae). In this way, the
biomass of these fish was reduced to levels between seven
and ten times lower than in adjacent partial cages (wherein
fish could access coral and algae) and open plots. Note that
in this experiment, live coral cover was very low at the start,
to simulate the ability of corals to recover from mortality
caused by bleaching. A clear phase shift from coral domination to algal domination in the fish-exclusion cages was
demonstrated over 2 years (Figure 4.33), showing that high
herbivore pressure is essential to enable coral reefs to
recover from serious perturbations such as bleaching.

In conclusion, studies on overfished and damaged coral
reef have shown that algal grazers have a key role in the
maintenance of a healthy reef. If grazers are removed, algae
can dominate a reef, causing a collapse in hard coral
abundance. A graphical model of change of relative abundance of herbivores, corals and algae in response to fishing
and other human interference in the Caribbean has been
produced by Knowlton & Jackson (2001). The model,
shown in Figure 4.34, comprises three planes plotted in the
three-dimensional space defined by the abundance of sea
urchins, fish, and coral/algae abundance ratio. With
increased human disturbance, the system moves to Plane C;
there is mass mortality of sea urchins on top of over-fishing,
allowing explosive growth of algae that overgrow coral.

Predators, parasites, and pathogens

CHAPTER CONTENTS

Predation
Parasitism

In this chapter, we discuss the occurrence and importance of secondary consumers in marine ecosystems. A secondary consumer is defined as an organism that eats all or part of a primary consumer. Carnivores, for example, eat other animals, whilst predators are a subset of carnivores that kill their prey before eating it. Parasites are also essentially carnivorous, though many may not kill their host outright, if ever. Pathogens in their turn are a subset of parasites. A carnivore that eats other animals that are already dead may earn the title of scavenger, but of course many animals we encounter in the oceans are catholic in their tastes. Many species of shark, for example, are classic predators, killing and eating all manner of unlucky prey, but they also commonly feed on dead animals such as whale carcasses, the latter being intrinsically easier to deal with. Note of course that one species of secondary consumer can eat another species of secondary consumer, the terms secondary and tertiary becoming rather blurred as we proceed up trophic levels in food webs. Further, it is common for fish to move up the trophic levels as they grow so we cannot define their trophic position without reference to their size or state of maturity. It is common practice to define the trophic position of a species by the food of the adults, although the great majority of individuals comprising the population may be young juveniles feeding at a far lower trophic level. The relationships between carnivores and their food, predators and prey, and parasites or pathogens and hosts, all follow similar principles and mechanisms, and, in the end, all may have significant impacts on tropic levels beneath them.

Predation

From the biggest (killer whales and giant squid) to the smallest (zooplantonic copepods and nauplii), the world's oceans are filled with predators. They may be sessile, such as tube-worms and some deep-sea sponges, or mobile such as arrow-worms and mantis-shrimps. They may be benthic, such as whelks and flatfish, or pelagic such as ctenophores or tuna. They may be active hunters such as sea-snakes and cone shells, or sit-and-wait predators such as sea anemones and octopuses. They may change their feeding tactics between day and night; a moray eel or lobster will hide in a hole or crevice during the day grabbing food that happens to walk or swim past, whilst venturing out at night to actively seek out prey. Some predators only catch and eat live prey occasionally, such as certain coral species who use their polyps at night to catch small living and dead organisms, whereas during the day, they retract their polyps and rely on their symbiotic zooxanthellae to provide them with the products of photosynthesis. For all but the largest animals the sea is a dangerous place and the majority of individuals of many species end their lives early in the jaws or attached to the tentacles of a predator.

The impact of predators on prey

One of the most basic questions in ecology concerns what effects predators have on their prey (Leitao et al 2008). Whelks are predatory gastropod molluscs (Figure 5.1)

Marine Ecology: Concepts and Applications, 1st edition. By M. Speight and P. Henderson. Published 2010 by Blackwell Publishing Ltd.

Figure 5.1

The dog whelk, *Nucella* (=*Thais*) *lapillus* with its barnacle prey. (Photograph courtesy of Paul Naylor.)

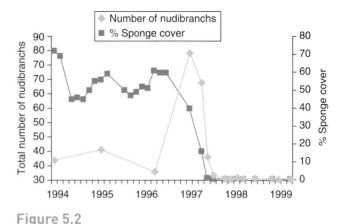

Figure 5.2

Mean percent cover of the breadcrumb spong, *Halichondria panacea*, and its predator the nudibranch *Archidoris montereyensis*, in 10 permanent quadrats in Alaska. (From Knowlton & Highsmith, 2000; reproduced with permission of InterResearch.)

which feed on a variety of prey in shallow waters, from limpets to periwinkles, and in this example, barnacles. Whelks have a modified radula which they use to rasp or drill holes in the shells of their prey, through which they are able to reach the flesh of the otherwise protected prey item. In 1970, Connell carried out an experiment in the USA to investigate the impact of whelk predation on barnacle populations. He discovered that the number of barnacles surviving inside cages which excluded whelks was much higher than when barnacles are exposed to predation. Note that this experiment does not take into account the recruitment of new stocks of barnacles, and that even when predators are excluded barnacle numbers still decline, though significantly more slowly, due to other factors such as old age.

It is possible for a predator to over-exploit its prey, as shown in Figure 5.2. Some species of nudibranch sea-slugs are one of the few marine animals that eat sponges, and the figure illustrates that they are capable of rapid expansions in numbers when the food supply is abundant (Knowlton & Highsmith 2000). The secret in many predator–prey relationships is to get the balance right – eat enough to make sure that your own species is successful, but don't eat so much that you destroy your own food supply. In this example, both the prey species (sponge) and the predator (nudibranch) are dispersed by planktonic larvae. However, the predator has a greater potential for local increase when the prey is abundant, resulting in heavy predation pressure on the sponge, virtually wiping it out within 2 years. Unfortunately for the nudibranch, sponge populations are unable to recover because their encrusting "niche" on open rock is taken over by heavy annual recruitment of macroalgae, and so the predator has to move on. To quote Knowlton & Highsmith verbatim, "The predator–prey relationship of *A. montereyensis* and *H. panicea* is an example

of a chase through space and time, with convergence resulting in extreme population fluctuations and an unstable community."

Predators not only seem to have large impacts on their prey, but they also tend to congregate in areas of high prey density, often called predator "hotspots," thus amplifying their impacts on prey populations. Predator hotspots can be identified on regional and global scales. Temming et al (2007) studied the spatial distributions of predator and prey fish species in the North Sea using fishing trawl hauls. Populations of predatory whiting, *Merlangius merlangus*, did not appear to be evenly distributed across wide spatial scales. At the highest local densities, the hauls amounted to more than 1000 kg of fish in a 30-minute trawl, whereas only 10 or 15 km away, the hauls were ten times less. Juvenile cod, *Gadus morhua*, are preyed upon by whiting, and it was found that in predator hotspots, several million cod were eaten per day. In fact, so intense was the predation pressure that Temming et al estimated that whiting could entirely wipe out an aggregation of 50 million young cod in just 5 days. A much larger scale example can be seen in Figure 5.3. Here, Worm et al (2005) map the distribution of tuna (family Thunnini) and billfish (family Istiophoridae) caught using long-line baited hook fishing over the last 50 years. There are clear density hotspots, in the northwest Atlantic, around Hawaii in the Pacific Ocean, and in northeast Australia. These are thought to be associated with thermal fronts, dissolved oxygen, and sea surface temperatures (SSTs). It appears these predators aggregate in response to physical, chemical and oceanographic conditions and not in a direct response to prey density.

These simple predator–prey relationships are the result of co-evolutionary processes in which predator and prey have a balanced set of advantages and disadvantages depending upon the conditions. However, we now frequently

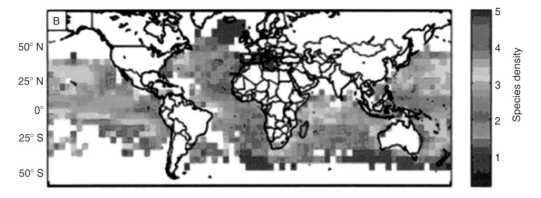

Figure 5.3

Global hotspots in the density of predatory billfish and tuna, measured using standard sample of 1000 hooks. (From Worm et al 2005; reproduced with permission of Wiley-Interscience.)

observe that man's activities can upset this balance (a topic we return to in depth in Chapter 9). Figure 5.4 illustrates an example of the effect of introducing a predator into a new region where it is able to upset a natural, co-evolved predator–prey system (Walton et al 2002). The European green crab, *Carcinus maenas*, was accidentally introduced into Tasmania in the 1990s, where it is now extremely abundant. The invader predates on the native Australian clam, *Katelysia scalarina*, with serious consequences for the local clam fishery. According to Walton et al, the Tasmanian catch of the clam peaked in 1995 at around 45,000 kg, worth approximately US$185,000. However, in recent years, the catch has fallen to around 10,000 kg per year. The figure shows that juvenile clams are most abundant low down the shore, possibly to avoid bird predation and to feed maximally. Adults move further up the tide line, thus avoiding competition with juveniles. Crab predation very significantly reduces overall densities of clams, especially of juveniles, and appears to redistribute adult populations back down the shore. It is important to note that indigenous Tasmanian crabs such as *Paragrapsus gaimardii* also eat clams, but do not significantly affect their population density.

Density-dependence

We have seen that predators eating prey can, unsurprisingly, have a big impact on the numbers of prey. The big question is whether or not it is the predator populations which "drive" those of the prey, as it often superficially seems to be, or in fact the other way round. We need to establish whether or not predators are able to respond to increasing numbers of prey in a density-dependent manner. Density-dependence is defined as a proportional increase in a factor such as mortality or feeding rate by a predator as the density of the population on which it is acting (the prey) increases.

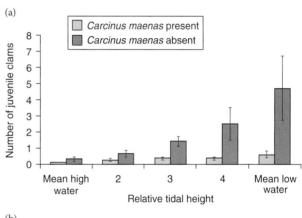

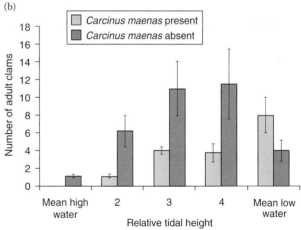

Figure 5.4

Carcinus maenas, on interstitial clam, *Katelysia scalarina*, populations in Tasmania (±s.e.). (a) Juvenile clams (<13 mm shell length). (b) Adult clams (>13 mm shell length). (From Walton et al 2002; reproduced with permission of Elsevier.)

Take the example of the mortality imposed on the clam *Mya arenaria* (Figure 5.5a) by the predatory blue crab *Callinectes sapidus* (Figure 5.5b) (Seitz et al 2001). It is clear

(a)

(b)

Figure 5.5

(a) Sand gaper clam, *Mya arenaria*. (Photograph courtesy of Julian Cremona.) (b) Blue crab, *Callinectes sapidus*. (Photograph courtesy of NOAA.)

(a)

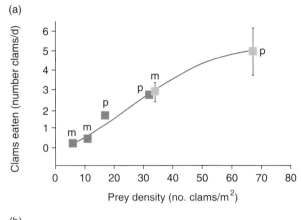

(b)

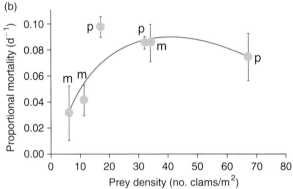

Figure 5.6

(a) Functional response of predatory blue crabs feeding on clams and (b) proportional mortality data of clams derived from the functional response. m and p denote two experiments (main and pilot) (means ±s.e.). (From Seitz et al 2001; reproduced with permission of *Ecology* – Ecological Society of America.)

from Figure 5.6 that the number of clams eaten per day by an individual crab increases as the density of the prey available also increases, but the relationship is asymptotic. This relationship is known as a "functional response". If we transform the data to look at proportional mortality (even though statisticians tell us we shouldn't plot one axis against another that is partially derived from the first), there is still a fairly linear increase with prey density at low or medium prey densities, followed by a drop in proportional mortality at high prey densities. Remembering the definition of density-dependence, the data from Seitz et al show a density-dependent relation between crabs and clams at low clam densities, which become inversely density dependent at higher clam densities. In a similar example, Johnson (2006) manipulated the densities of predatory juvenile bocaccio rockfish, *Sebastes paucispinis*, and their prey, juvenile kelp rockfish, *Sebastes atrovirens*, in shallow subtidal kelp habitats on the coast of central California. The proportion of prey killed (the per-capita mortality) by one predatory fish over a 48-hour period showed a linear

decline with increasing prey density. In other words, density-dependence was not in evidence even at low prey densities.

Generally, we can identify two types of functional responses, a Type II which is asymptotic and not density-dependent, and a Type III which is sigmoid, which can be density-dependent at low to medium prey densities (Figure 5.7) (Seitz et al 2001). There is in fact a Type I functional response, but this is a theoretical relationship only. So, why do both responses plateau out, and what features of predator–prey interactions influence the distributions? There are two factors which we recognize from predator behavior that have significant influences on the shape of functional responses; they are "attack or encounter rate," and "handling time." Attack or encounter rate involves the ability of a predator to find prey, say in a vast ocean or buried in sand, whilst handling time describes the ability of the predator to kill, eat and digest its prey once it has found it. Let us return to the example

Functional response

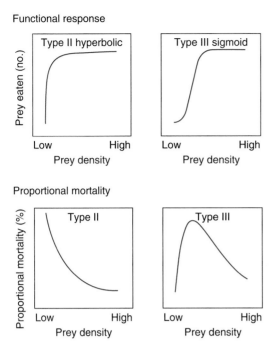

Figure 5.7

Two types of predator–prey functional responses. Conceptually, a Type II cannot be density dependent, whereas a Type III may be at low to medium prey densities, but not at high ones. (From Seitz et al 2001; reproduced with permission of *Ecology* – Ecological Society of America.)

Components of predation

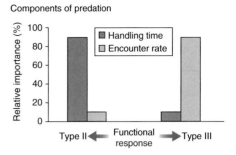

Figure 5.8

Handling time and attack/encounter rate as a function of prey armor or avoidance, and the types of functional response generated in their predator. (From Seitz et al 2001; reproduced with permission of *Ecology* – Ecological Society of America.)

of blue crabs eating clams (Seitz et al 2001). *Callinectes sapidus* mentioned above in fact feeds on a wide variety of prey, including several types of bivalve mollusc. Mussels (*Mytilus sp*) and oysters (*Crossostrea sp*) are attached to rocks in open view, whereas clams such as *Macoma* and *Mya* bury in the sand and mud. Because of their differing habitats and behaviors, mussels and oysters are heavily armored and are able to close their shell valves completely. Clams on the other hand rely on hiding to avoid being eaten by predators. So mussels are easy to find, but hard for a crab to deal with (handling time is the important factor), whereas clams are difficult to find but fairly easy to deal with once the crab has dug them up (attack or encounter rate are the important factors). As Figure 5.8 indicates, handling time tends to generate Type II functional responses, and attack rate, Type III responses. Thus in the case of blue crabs, there is more likely to be a density-dependent relationship between clams and crabs than between mussels and crabs. One way in which predators exhibit Type III functional responses rather than Type II is via a mechanism called switching. Assuming that there is a fairly wide range of prey available for a predator to choose (e.g. various species, multiple sizes, seasonal variations in abundance, etc.), then predators may concentrate on a particular sort

of prey, even when this prey is becoming less abundant. Godiksen et al (2006), for example, report that lesser sand-eel, *Ammodytes marinus*, and herring, *Clupea harengus*, in the Barents Sea fed copiously on larvae of the capelin, *Mallotus villosus*. It was thought that the predators retained a specific search image for this prey species, and fed exclusively on it for relatively long periods despite a high abundance of equally suitable alternative prey such as krill and copepods. In general, predator–prey switching can lead to increased density-dependent interactions, the reduction in predation pressure as a prey population collapses, and hence increased prey population stability. We shall return to the issue of density-dependence later in this chapter when we consider population regulation.

Multiple predators

In the sea, predators rarely occur singly. As we have described earlier in this chapter, predators tend to aggregate in areas or patches of high prey density. Sardines, *Sardinaps sagas*, mass along the coast of Kwa Zulu Natal in South Africa in vast numbers in sardine "runs" (Aitken 2008), and attract large numbers of predators such as common dolphins, bronze whaler sharks, and seabirds such as gannets. One special example concerns filter feeding whale sharks, *Rhinocodon typus*, who have been reported aggregating in large numbers (for whale sharks) at set times of year and in localities where the mass spawning of cubera snappers, *Lutjanus cyanopterus*, and dog snappers, *L. jocu*, takes place of the coast of Belize (Heyman et al 2001). When groups of predators aggregate together, their individual functional responses combine to form "numerical or aggregative responses." Higher densities of predators are likely to have increased impacts on their prey. Returning to Johnson's

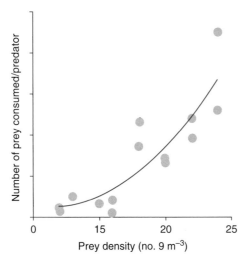

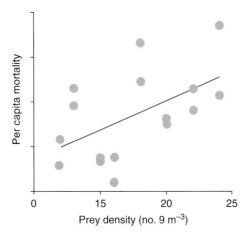

Figure 5.9

Response of a relatively high density of predatory rockfish to an increase in prey density. (From Johnson 2006; reproduced with permission of *Ecology* – Ecological Society of America.)

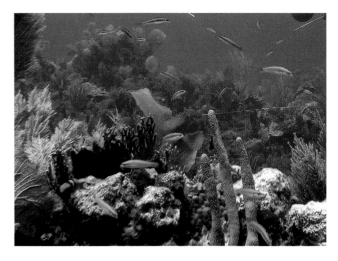

Figure 5.10

Bluehead wrasse, *Thalassoma bifasciatum* (bottom left). (Photograph courtesy of Jonathan Shrives.)

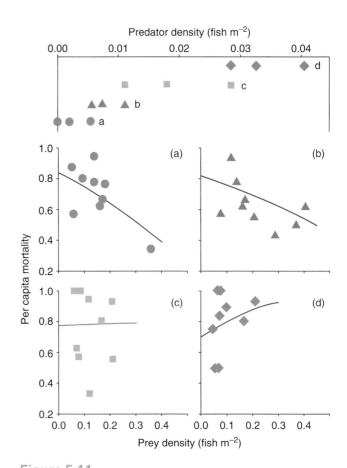

Figure 5.11

Plots of relationships between density and mortality of a prey species, settlers of the bluehead wrasse (*Thalassoma bifasciatum*), according to the density of its predator, the grouper (*Cephalopholis fulva*). Graphs (a) to (d) show per-capita (proportional) prey mortality, at four different predator density levels, as indicated in the top graph ((a) = lowest density; (d) = highest density). (From White 2007; reproduced with permission of Wiley Interscience.)

work on rockfish described above, as shown in Figure 5.9, high predatory rockfish densities (5 per m³ of kelp habitat in California) resulted in a combined functional/numerical response which was exponential, resulting in a linear and fully density-dependent response to prey density (Johnson 2006). White (2007) studied the predation on recently settled bluehead wrasse, *Thalassoma bifasciatum* (Figure 5.10), on patch reefs in the US Virgin Islands by the grouper, *Cephalopholis fulva*. As can be seen in Figure 5.11, the efficiency of density-dependent mortality is once again related to predator density, so that as the density of predators increased, wrasse mortality became positively density-dependent (positive gradient on per capita/prey density plot). Predation is inversely density-dependent (negative gradient on per capita/prey density plot) at low predator densities. Put simply, as might be predicted, the higher the density of predators, the greater the impact on the prey population.

Too many predators in one place may result in competition for prey and decreased hunting efficiency, a process known as "interference." There are various mechanisms whereby predators reduce or avoid interference. These include prey size and/or species selectivity, a form of niche partitioning, switching to other types of prey, foraging for prey in different localities or depths, or simply emigrating from a high density prey patch into one with fewer prey, but also fewer competing predators. One effect of predator competition is illustrated in Figure 5.12. Griffin et al (2008) studied resource partitioning in three species of intertidal predatory crabs in the UK, and concluded that the influence of predator diversity was only apparent at high predator densities, since the competitive interactions were intensified. The solution to the problem, as shown in the figure, is for each species of crab to occupy a slightly different "functional niche" and eat different species of prey so as to reduce the overlap in prey requirements. As can be seen, the effect of this resource partitioning is most clear at high predator densities. The velvet swimming crab, *Necora puber*, for instance shows a significant reduction in prey species selection at high densities compared with low ones. Competition in the sea is the subject of Chapter 6.

The effects of prey on predators

Predators in the sea do not by any means get everything their own way. If we return to the topic of hotspots for a moment, it can be seen that some prey species have hotspots of their own to which predators can only respond. Indeed, the example of sardine runs above could be considered an example of prey hotspots influencing the distribution of predators (to the eventual detriment of the prey). Houghton et al (2006) studied the distribution of leatherback turtles, *Dermochelys coriacea*, compared with that of its prey, the barrel jellyfish *Rhizostoma octopus* (Figure 5.13) in the Irish (Cambrian) Sea between Wales and Ireland (Figure 5.14). Though the distribution of turtles varies spatially and by decade, it is clear that the majority of occurrences of these predators are around the southeast coast of Ireland, the west coast of northern England, and the south coast of Wales, all three areas where jellyfish are known to be most abundant. In this case, the predators appear to be drawn to the hotspots of the prey.

Not all types of prey are equally easy or difficult for a predator to utilize. As we saw earlier in this chapter, certain types of bivalve are easier for crabs to feed on than others, depending on where they live and how armored they are. Another example involves starfish. Predatory species either consume prey directly like a normal predator, or some are able to evert their stomachs through their mouths and feed on their prey "extra-orally." Which method is employed depends to some extent on the type of prey in question;

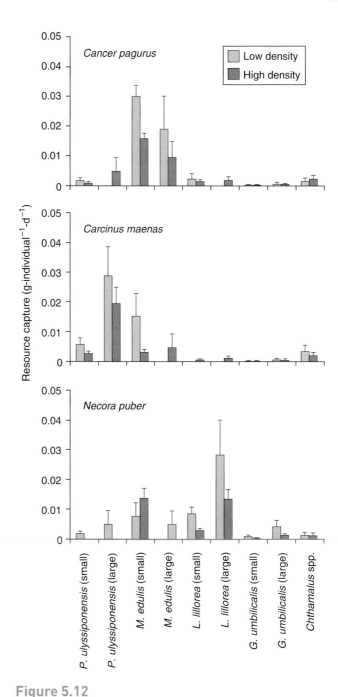

Figure 5.12

Predation rate (±s.e.) on various intertidal prey species by three species of predatory crab in the UK, at low (light blue bars) and high (dark blue bars) predator densities. Prey are limpets (*Patella ulyssiponensis*), mussels (*Mytilus edulis*), periwinkles (*Littorina littorea*), topshells (*Gibbula umbilicalis*), and barnacles (*Chthamalus* spp.). (From Griffin et al 2008; reproduced with permission of *Ecology* – Ecological Society of America.)

those with hard shells such as gastropods and bivalves can be particularly troublesome. Large genera such as *Asterias* and *Pisaster* are able to grasp the two valves (shells) of a mussel with their arms, and slowly but very surely pull them apart, against all the resistance of the mussel to avoid being

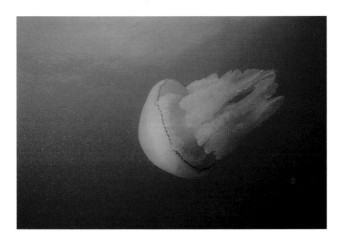

Figure 5.13

Barrel jellyfish, *Rhizostoma octopus*. (Photograph courtesy of Paul Naylor.)

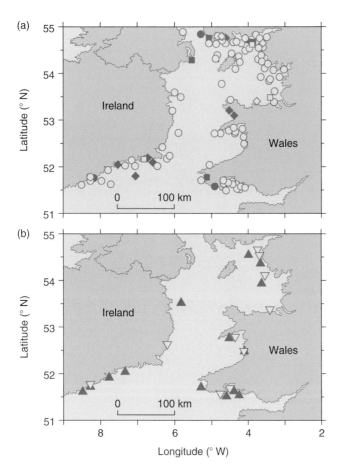

Figure 5.14

Sightings of leatherback turtle, *Dermocheles coriacea*, in the Irish Sea (upper graph), compared with the occurrence of their food jellyfish, *Rhizostoma octopus*, prey (lower graph). Upper graph data are plotted by decade: 2000–2005 = open circles; 1990s = solid circles; 1980s = open squares; 1970s = solid squares; 1960s = open diamonds; 1950s = solid diamonds. Lower graph data show sightings of turtles associated with jellyfish = solid triangles; confirmed foraging activity of turtles on jellyfish = open triangles. (From Houghton et al 2006; reproduced with permission of *Ecology* – Ecological Society of America.)

so eaten. In order to do this of course, the starfish has to get a firm grip on the mussel shell with its tube feet, and to maintain this grip perhaps for hours on end. Hence, the surface covering of the mussel can significantly influence the likelihood of the prey being eaten by the predator. *Asterias rubens* is a common species of starfish found all around the northeast Atlantic coasts, sometimes aggregating in very large numbers to spawn. Small starfish eat fewer mussels in a given time than larger ones, as might be expected, but more significantly, the predation rate is much greater on clean mussels found below the low tide mark than on similar prey situated in the intertidal region where they are frequently overgrown by barnacles (Saier 2001). The barnacles make it more difficult for the starfish to get a firm grip on the mussel shells, so that living intertidally may be thought of as a habitat semi-protected from predation (though living intertidally has many other difficulties). There are many features that can render prey less attractive to predators. In a similar intertidal habitat to the previous example, this time off the coast of California, Thornber (2007) found that the gastropod snail, *Tegula brunnea*, was much less likely to be preyed upon by starfish if its shell had been covered by crustose or coralline algae (epibionts). Bare snails tended to be eaten up to four times more readily than encrusted shells. Quite how this works is a matter for conjecture. Camouflage is unlikely to be a factor, since starfish do not have image-forming eyes. There may be a physical or chemical deterrent effect imparted by the algae which repel the starfish, or otherwise reduce their capture efficiency and increase their prey handling time. Whatever the mechanism, this so-called associational resistance of a prey species against its predator certainly works. Prey individuals have many other ways in which they can reduce the likelihood of being eaten by a predator, which include behavioral (shoaling, hiding), physical (camouflage, warning coloration, spines, and spikes) and chemical (nonpalatability, poisonous) defenses. Whatever mechanisms evolve, we can be sure that some predator somewhere in the sea will find a way around these defenses.

Predator–prey population cycles

The scientific literature has long discussed the cycles of predator–prey dynamics, and terrestrial ecologists have pondered the relationships between arctic foxes and arctic hares, or winter moths and their parasitoids, for many years. Such cycles are harder to detect in the sea, perhaps because of the enormous spatial scales involved, but also the numerous potential actors in the play. On a microlevel, if we examine Figure 5.15, it can be seen that the population densities of a prey population (in this case the copepod *Eurytemora affinis*) fluctuate rather wildly, as do those of the predator, a mysid shrimp, *Neomysis integer* (David et al 2006). Clearly both graphs oscillate through time, but whether this is linked to biotic factors such as predator–prey dynamics, or perhaps

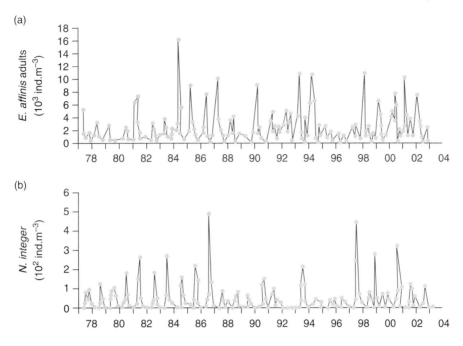

Figure 5.15

Long-term series of abundance of (a) adult copepods, *Eurytoma affinis*, and (b) predatory mysid, *Neomysis integer*. (From David et al 2006; reproduced with permission of Elsevier.)

physical fluctuations such as water temperate or salinity, is rather hard to disentangle. In this example, the authors conclude that the predator is likely to be responding to seasonal fluctuations in prey abundance.

Leitao et al (2008) studied fish populations around artificial reefs in Portugal, and were able to detect some degree of predator–prey oscillations between the major predator, bass, *Dicentrarchus labrax*, and various prey species including mackerel, *Scomber japonicus*, and seabream, *Pagellus acarne* (Figure 5.16a,b), but one of the best examples of predator–prey cycles comes not from these shallow waters, but from 4100 m in the northeast Pacific. Bailey et al (2006) towed camera sleds along the seabed at about 0.8 m s⁻¹, taking photographs every 5 or 6 seconds, for more than a kilometer. Marine fauna captured by the cameras were identified and counted. This photo sampling was carried out between October 1989 and February 2004, and Figure 5.17 illustrates the results found for the only abundant predatory fish, the grenadier, *Coryphaenoides armatus*, and its prey which consisted of a variety of echinoderms (sea cucumbers, sea urchins, and brittle stars), termed mobile epibenthic megafauna (MEM). Not surprisingly, the data in the graph are variable, but there is an undoubted cyclical pattern to the predator–prey interaction, with fish numbers peaking at or more frequently, somewhat later, than those of the echinoderms. It is impossible to detect which of the two combatants, predator or prey, is driving this pattern from these data alone, but Bailey et al feel that the grenadier populations are likely to be under "bottom-up" control (see later) by the grazing echinoderms.

Trophic cascades

A trophic cascade describes the ways in which perturbations in high trophic levels can significantly influence the population in the next trophic level down, and so on down the food chain to the primary producers. These cascades are certainly significant ecologically and also have important implications for marine reserve management (Mumby et al 2007) (see Chapter 10). So, for example, the removal of large predators such as hogfish, pufferfish, and spiny lobsters increases attacks by the gastropod snail *Cyphoma gibbosum* on gorgonian corals such *as Pseudopterogorgia americana* in the Florida Keys (Burkepile & Hay 2007). These authors found that if large fish predators were removed, the resulting increase in attacks by the snail produced an eightfold increase in damage to gorgonians. A broader example involves the links between killer whales, sea otters, sea urchins, and kelp (Estes et al 1998). Urchins keep kelp in check by grazing off the young germlings, sea otters eat urchins, enabling more kelp to recruit and grow, but whales eat otters so that more urchins survive and hence less kelp is produced. The end point of this cascade, if urchins are allowed to proliferate, is known as a "barren," where numerous sea urchins keep recruiting kelp (and many other sessile species) in check (Guidetti & Dulcic 2007). Barrens, as the name suggests, consist mainly of bare substrate covered with encrusting coralline algae and not much else. Crucially though, they do usually have the potential to be recolonized by more luxuriant species if grazing pressure is reduced by increased predation.

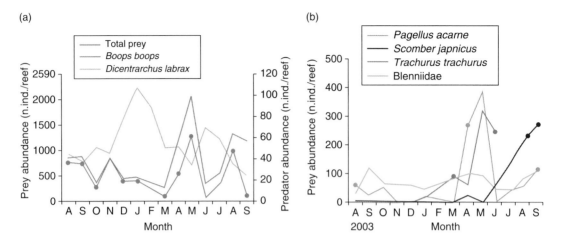

Figure 5.16

(a) Monthly density variation of European bass, *Dicentrarchus labrax* (predator), and the bogue fish, *Boops boops* (principal prey), plus overall reef fish prey. (b) Mean abundance of other prey species: *Pagellus acarne* = seabream; *Scomber japonicus* = mackerel; *Trachurus trachurus* = horse mackerel. Filled circle = presence of a given prey in sea bass stomachs in a given month. (From Leitao et al 2008; reproduced with permission of Springer.)

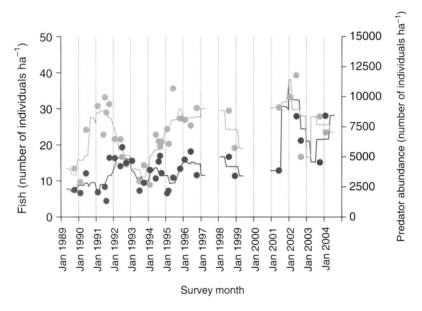

Figure 5.17

Monthly abundances of mobile epibenthic megafauna (MEM) (light blue circles) and grenadier fish (dark blue circles) recorded by a towed camera in the northeast Pacific at 4100 m depth. All data are monthly averages, with 13-month centered running means for MEM (light blue lines) and fish (dark blue lines). (From Bailey et al. 2006; reproduced with permission of Wylie Interscience.)

An example of a trophic cascade, and the effects of human intervention on stability, is provided by Dulvy et al (2004). They studied the links between predatory reef fish (111 species in all), the crown of thorns starfish, *Acanthaster planci* (Figure 5.18), that eats live coral, and the benthic cover of the reefs including corals, algae, and other sessile taxa. This study was undertaken over a series of 13 oceanic islands in the Fiji group in the Pacific Ocean. The islands were chosen as showing different levels of fishing pressure from local villages. The "natural" cascade involves a high species richness of predatory reef fish eating the starfish,

which in turn allows live hard corals to proliferate and algae to be relatively rare. Figures 5.19 and 5.20 show that as fishing pressure increases (more fish of all sorts, including the big predators, are harvested), the density of crown of thorns starfish increases. This is directly related to a clear decline in the percentage cover of living hard coral, and an almost mirror-image increase in encrusting filamentous algae, on heavily fished reefs.

The large-scale exploitation of sharks around the world is causing great concern, not just because of their own vulnerability to over-exploitation and population collapse,

Figure 5.18

Crown of thorns starfish, *Acanthaster planci*. (Photograph Martin Speight.)

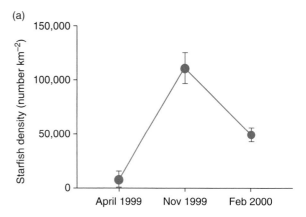

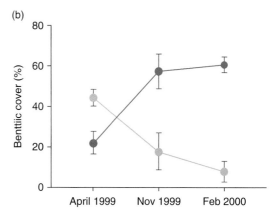

Figure 5.20

The change in (a) starfish density and (b) percentage cover of hard coral (light blue circles) and microfilamentous algae (dark blue circles) at three time intervals at Kabara Island, which has the second highest fishing intensity in Figure 5.19, 43.3 people per kilometer of reef front. (From Dulvy et al 2004; reproduced with permission of Wiley-Blackwell.)

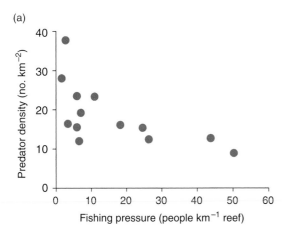

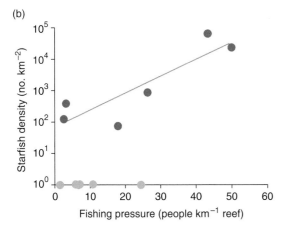

Figure 5.19

(a) The relationship between fishing intensity and average density of predatory fish on 13 oceanic islands around Fiji. (b) The relationship between fishing intensity and average density of crown-of-thorns starfish. Dark blue circles = islands where starfish were present; light blue circles = islands without starfish. (From Dulvy et al 2004; reproduced with permission of Wiley-Blackwell.)

but also because of their role as "apex" predators. As the name suggests, an apex predator is at the top of food chain – very few things eat adult tiger shark. Figure 5.21 shows the abundances of species in three tropic levels over time, great sharks prey on a variety of smaller elasmobranch predators such as sharks and rays, which in turn eat scallops. There has been a very clear and indeed alarming exponential decline in all the major great sharks including tigers, hammerheads, and bulls, with a clear increase in the mesopredators (Myers et al 2007). One of these in particular, the cow-nosed ray, has had such an impact on the third trophic level in the cascade that the scallop-fishing industry in North Carolina Bay has ceased. Other consequences of this type of interference with trophic cascades in the sea can easily be imagined.

Cascades do not always function in the way they seem. Results of work carried out by Guidetti & Dulcic (2007) in the Adriatic Sea are shown in Figure 5.22. Both graphs show positive relationships between predation pressure on urchins by fish, and grazing pressure on algae by urchins, from which it is tempting to assume cause and effect.

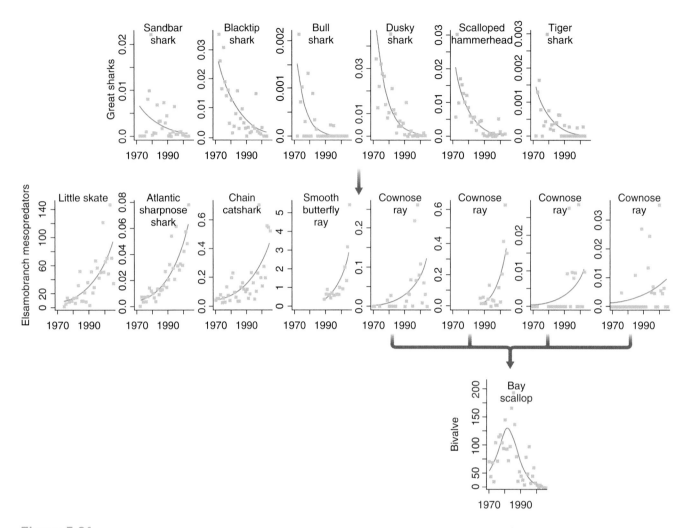

Figure 5.21

Change in time in species within a trophic cascade from the northwest Atlantic ocean (North Carolina coast), for three trophic levels, i.e. top predators, intermediate (meso) predators, and scallops. (From Myers et al 2007; reproduced with permission of *Science*, AAAS.)

However, it seems that predation pressure at the rate measured in this study was unlikely to be strong enough over wide spatial scales to have an overriding control of urchin populations. A cascade effect might be more important if fishing pressure could be reduced within a no-fishing zone. This brings us to another fundamental question in population ecology, alluded to earlier, that of "who is doing the driving."

Population regulation

Population regulation, a fundamentally density-dependent process, can be either top-down, from a trophic level above the current one, or bottom-up, from within the current trophic level or in the ones below. In the trophic cascade examples above, it seems clear that predators top-down regulate their prey, and in turn, the prey top-down regu-

lates its own food supply. Thus, in a simple top-down concept, reef fish regulate crown of thorns starfish, which regulate coral/algal interactions. Alternatively, killer whales regulate sea-otters who regulate urchins and so on. Removing an entire functional group such as the great sharks illustrates nicely how top-down effects can have an impact.

In contrast to the rather obvious examples above, much recent work suggests that many marine food webs are in fact regulated by a mixture of top-down and bottom-up processes. Fleeger et al (2008) worked in salt marshes on the northeast coast of the USA where they studied the benthic infauna and their predators. They looked at a variety of associated sites, from mudflats to creek walls and several different types of *Spartina* grass communities. The effects of removing predatory fish and/or adding fertilizers (nutrients) to the systems were variable. Evidence for top-down regulation was hard to find. The authors suggest that reducing the numbers of predatory fish might in fact have stimulated an enhanced predation pressure on annelids by grass shrimps, whereas increased nutrients, a proxy for bottom-up effects, may have been beneficial to taxa such as ostracods (see Figure 2.7) and copepods that fed on the increased primary production of epiphytes in the salt marsh. There is certainly no clear direction of control.

A convincing example of where predators at two different levels are in turn regulated by bottom-up processes comes from the North Sea off the coast of southeast Scotland (Frederiksen et al 2006). Seabirds such as puffins and guillemots feed on sandeels, *Ammodytes* spp. (Figure 5.23), which in turn prey on copepods and crustacean nauplii, which graze on diatoms and other phytoplankton, so when persistent declines in the intermediate predators occur as shown in Figure 5.24, it is tempting to blame these declines for the decreasing numbers of puffins and other seabirds around the coasts of Britain. Bottom-up regulation would appear to dominate in this system. Phytoplanktonic primary production (see Chapter 3) may be linked to nutrients and climate-driven factors such as temperature, and the survival of larval sandeels, which feed on phytoplankton, may be the key to this rather precariously balanced system. Add to this the fact that sandeels themselves are the basis of a commercial fishing industry (see Chapter 8), it isn't surprising that top predators such as puffins are responding to what happens in trophic levels below them.

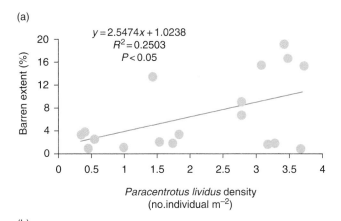

(a)

$y = 2.5474x + 1.0238$
$R^2 = 0.2503$
$P < 0.05$

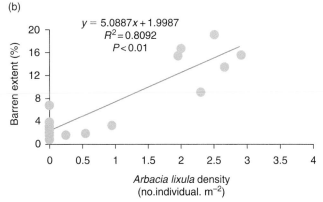

(b)

$y = 5.0887x + 1.9987$
$R^2 = 0.8092$
$P < 0.01$

Figure 5.22

Relationships between (a) density of two sea urchin species and the extent of "barrens" (areas denuded of macroalgal growth by urchin grazing pressure) and (b) density of predatory fish feeding on juvenile urchins. (From Guidetti & Dulcic 2007; reproduced with permission of Elsevier.)

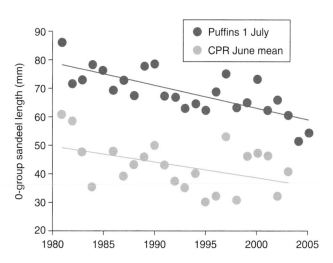

Figure 5.24

Annual mean length of first year sandeels either removed from the returning food of puffins standardized to July 1, or collected by the continuous plankton recording system (CPR) in June. (From Frederiksen et al 2006; reproduced with permission of InterResearch.)

Figure 5.23

The common sandeel *Ammodytes tobianus*. (Photograph Peter Henderson.)

Predation in the deep-sea

Most of our examples of predator–prey interactions have so far been from shallow-water marine ecosystems, mainly because a great deal less is known about organisms in the deep. Undoubtedly, large, if not giant, predators do occur in the deep-sea. Giant squid for example are no longer creatures of myth and legend, and the deep-sea mesopelagic squid, *Taninga danae*, has been observed hunting and killing prey in waters up to 940 m deep off Japan (Kubodera et al 2007). This species reaches an impressive 2.3 m in length and approximately 60 kg in weight, but its cousin, the true giant squid, *Architeuthis dux*, is huge and may achieve 8 or even 10 m in length. This species was first caught live on film by Kubodera & Mori (2005), but dead specimens (or parts thereof) have been noted in whale stomachs or on beaches for many years. Giant squid are undoubtedly active predators and the gut contents have been found to contain squid, including members of its own species (Bolstad & O'Shea 2004). However, even these predators are not really what could be called deep-sea. Landman et al (2004), for example, estimated that the species was unlikely to venture below 500 m.

In fact, large predators such as sharks do not appear to penetrate all that far down into the sea. Granted, sampling gets much more problematic as we go deeper, so that obtaining accurate records is difficult. However, assuming it is equally difficult to see any type of fish at depth, the data shown in Figure 5.25 portray a real difference in the occurrences of the two major groups of fish with depth (Priede et al 2006). Chondrichtyes (elasmobranchs) contain the sharks and rays, whilst Actinopterygii are, in the main, the bony fish (teleosts). As can be seen, the sharks and rays have an average body length of around 100 cm (but note the log scale), and very few have been recorded below 3000 m. The bony fish on the other hand tend to be considerably smaller on average, but have been seen, albeit rarely, in the very deepest parts of the oceans. It may be that all top predators such as sharks are extremely rare simply because of the lack of nutrients at the base of extremely oligotrophic deep-sea food chains to support something as large as a shark. Priede et al also suggest that this oligotrophy is unable to sustain an oil-rich liver which sharks depend on for buoyancy. So the large mobile predators that do occur in the deep-sea tend to be bony fish.

Large single units of meat, alive or dead, are clearly very rare in the deep-sea. One source is dead bodies sinking from the surface layers producing whale (or other large marine creature) fall. The majority of animals that colonize these sources of bounty do so as a succession of scavengers (see Chapter 7), and a true predator, requiring live prey, consumes these scavengers (Kemp et al 2006).

Hydrothermal vents are the one relatively productive region of the deep-sea (see Chapter 3), and it is here that predation occurs as it does at the surface. Predatory crabs,

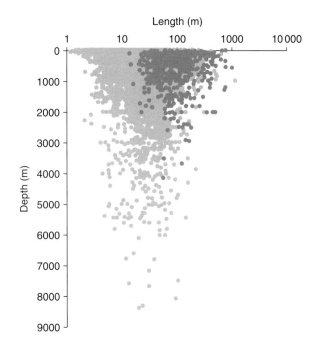

Figure 5.25

Maximum depth occurrences of Chondrichthyes (cartilaginous fish = dark blue circles) and Actinopterygii (bony fish = light blue circles), in relation to body length. (From Priede et al 2006; reproduced with permission of The Royal Society.)

snails, and fish abound on vents, and prey species such as limpets have been found that bear the scars of attacks by deep-sea whelks (Voight and Sigwart 2007). Most of these predators do appear to be rather specialized. Life on a vent is not as easy as a comparable habitat in shallow seas, and general predators tend not to wander into a vent community, probably because of the unusual temperature and chemical conditions. Thus, predatory fish for example are specific to vent communities, such as the eelpout, *Thermarces cerberus*. Sancho et al (2005) suggest that this species is a specialist predator of gastropod molluscs such as *Lepetodrilus elevatus*. Even here, we can see the evidence of top-down control via a small trophic cascade, since *Lepetodrilus* is an important grazer on sessile vent species, including the all-crucial vestimentiferan vent worms such as *Riftia pachyptila*. If we remove large numbers of eelpout, perhaps by deep-sea fishing, snails will proliferate and remove the very primary producers that sustain the whole vent ecosystem.

Perhaps because of the fundamental lack of copious quantities of live prey in the deep-sea, animals have evolved some very unusual tactics to locate and catch prey. *Acesta bullisi* is a large bivalve mollusc which lives permanently attached around the tube opening of the sessile cold-seep tubeworm, *Lamellibrachia luymesi* (Jarnegren et al 2005). The plume of the tubeworm actually sits inside the bivalve's mantle cavity. All evidence points to the fact that *Acesta* is a predator on the eggs of the tubeworm, which are produced throughout the year in an otherwise very sparse habitat for a sessile predator. Most compelling is the fact that the bivalve is almost never found attached to male worms. Perhaps the weirdest adaptation to predation in the deep-sea involves sponges in the group Cladorhizidae. One member of this group is the deepest known sponge in the world, *Asbestopluma occidentalis*, which has been found at over 8800 m, and it appears to have evolved a sit-and-wait predatory life style (Vacelet & Duport 2004). These sponges have lost the ability to filter feed, presumably because the amount of organic seston in these regions is extremely low. Instead, the sponge produces filaments maybe several millimeters or more long, at the tip of which hook-like spicules entrap swimming prey such as tiny crustaceans. The authors point out that this example of evolving predation from a filter-feeding way of life may have had important consequences for the basic evolution and radiation of metazoan animals.

Figure 5.26

A large isopod parasite or fish louse attached to the underside of a large mullet, *Liza ramada*. (Photograph Peter Henderson.)

Parasitism

According to Palm & Klimpel (2006), parasitism is one of the most successful modes of life on earth. Marine parasitology is, like so many other elements of marine ecology, tricky to unravel because of the difficulties of access for research in many parts of the oceans. Nonetheless, as these authors point out, the stability of marine habitats on an evolutionary scale means that there has been great potential for radiation and speciation within parasite taxa. For example, there are just over 29,000 species of marine fish so far described, but Rohde (2002) estimate around 100,000 fish parasite species, getting on for four species of parasite for every one of fish. This average is frequently exceeded. All taxa of animal parasites can be found in the sea, ectoparasites such as sealice (Figure 5.26) and leeches, and endoparasites such as flukes, nematodes, and tapeworms. As might be expected, the latter group holds the world record for the size of a parasite; tapeworms recovered from the guts of adult sperm whales have been measured as over 30 m in length. There are even a few really weird and secondarily specialized parasites, such as the sea anemone *Edwardsii lineate*. Anemones are of course a large group of anthozoan cnidarians, typical sit-and-wait predators all over the world. This species however has evolved to become an endoparasite of ctenophores such as *Beroe ovata* and its own prey species, *Mnemiopsis leidyi* (Reitzel et al 2007).

We can illustrate this remarkable diversity of parasites in the sea with some more detailed examples. Zander (2007) sampled three- and nine-spined sticklebacks (*Gasterosteus aculeatus* and *Pungitius pungitius* respectively) in the Baltic Sea. In total, they found 18 ecto- and endo-parasitic species

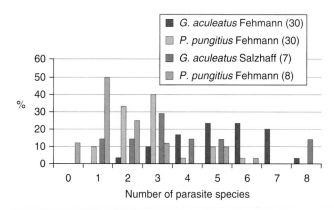

Figure 5.27

Ratios of numbers of parasite species from two species of stickleback from two localities in the Baltic Sea. (From Zander 2007; reproduced with permission of Springer.)

infecting these fish; Figure 5.27 shows the distribution of parasite species between the two species of fish at two different localities. Note that the maximum percentage infection by a parasite was as high as 50%, but varied considerably between sites. Some of these parasites were somewhat host generalist, being known to occur in a fairly wide range of fish species, whilst others were much more host-species specific. For example, one species of sea bream, *Dilpodus sargus*, was found to harbor an average of 10 monogenean parasites per individual, all 11 different species observed belonged to one genus (*Lamellodiscus*) (Sasal et al 2004). Host-specificity in fish parasites tends to be the norm. Species of tapeworm in the common marine cestode genus, *Pedibothrium*, have been found in the guts of nurse sharks, *Ginglymostoma cirratum*, and its very close relatives only. Caira & Euzet (2001) suggest that this very

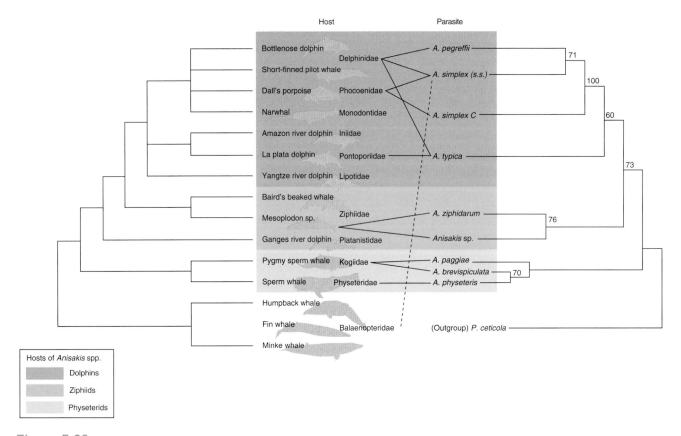

Figure 5.28

Phylogenies of *Anisakis* spp. and their cetacean hosts. Relationships inferred from mtDNA *cox*2 sequence analysis. Lines depict the observed host–parasite co-speciation events; dotted line indicates a possible host-switching event. (From Mattiucci & Nascetti 2008; reproduced with permission of Elsevier.)

close association has been in existence for at least 2 million years, and indeed, it is possible that there has been speciation of the parasites within the single small group of orectolobiform sharks (Caira & Jensen 2001).

It is not just fish that host parasites. Beron-Vera et al (2007) studied the endoparasites of common dolphins, *Delphinius delphis*, and discovered four species, a nematode, an acanthocephalan, and two digeneans, in various sections of the stomach. As we shall see later in this section, many invertebrates including crustaceans and gastropod molluscs such as periwinkles, not only have their own parasites, but also act as intermediate hosts for the complex lifecycles of vertebrate parasites (Thieltges & Buschbaum 2007). Animals from the deep-sea are unable to escape parasitism. Vent crabs (family Bythograeidae) from hydrothermal vents in the Pacific Ocean have been found to be infested by nemertean worms situated on various parts of their bodies (Shields & Segonzac 2007). The details of this association, in particular the impact on the host, are so far unknown.

The most important aspects of marine parasite–host relationships, as for the predator–prey associations discussed earlier in this chapter, concern the impact on the host and potential for density-dependent regulation. Crucially, the necessarily complex lifecycles of marine parasites, in particular their host-finding abilities, relate to both these topics. We shall briefly consider these subjects from the points of view of both endo- and ecto-parasites in turn.

One of the best studied endoparasitic associations in the sea involves the nematode *Anisakis simplex*. Anisakid nematodes in general are parasites of the guts of mainly marine vertebrates and, as can be seen in Figure 5.28, the host range for parasites within the genus *Anisakis* is confined to cetaceans, though the distribution of species is not necessarily host specific. These parasites have complex (so-called indirect) lifecycles (Mattiucci & Nascetti 2008). Cetaceans (whales and dolphins) and pinnipeds (seals and walruses), as well as fish-eating birds, are the main (definitive) hosts. Fish, squid and other macroinvertebrates are the intermediate (paratenic) hosts and crustacea such as shrimps the first intermediate host (Figure 5.29; Butt et al 2004). Clearly, *Anisakis simplex* is a host generalist, since it not only manifests itself in marine food webs, but can also jump onto land

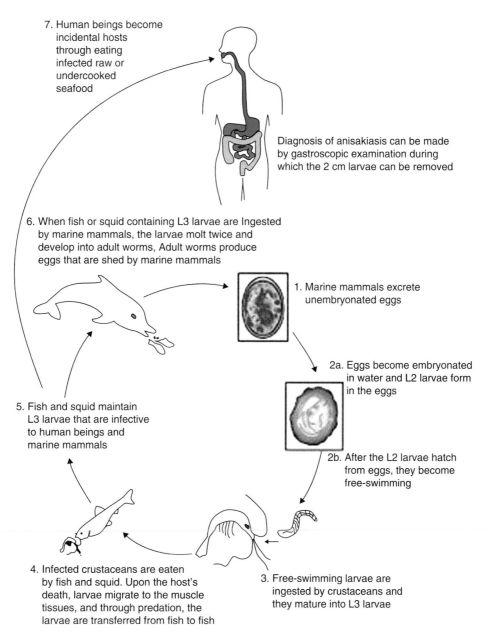

7. Human beings become incidental hosts through eating infected raw or undercooked seafood

Diagnosis of anisakiasis can be made by gastroscopic examination during which the 2 cm larvae can be removed

6. When fish or squid containing L3 larvae are Ingested by marine mammals, the larvae molt twice and develop into adult worms, Adult worms produce eggs that are shed by marine mammals

1. Marine mammals excrete unembryonated eggs

2a. Eggs become embryonated in water and L2 larvae form in the eggs

5. Fish and squid maintain L3 larvae that are infective to human beings and marine mammals

2b. After the L2 larvae hatch from eggs, they become free-swimming

4. Infected crustaceans are eaten by fish and squid. Upon the host's death, larvae migrate to the muscle tissues, and through predation, the larvae are transferred from fish to fish

3. Free-swimming larvae are ingested by crustaceans and they mature into L3 larvae

Figure 5.29

Lifecycle of *Anisakis simplex*. (From Butt et al 2004; reproduced with permission of Elsevier.)

via the consumption of raw fish by humans, resulting in a medical condition known as anisakiasis or anisakidosis. Victims present with stomach pains, diarrhea, nausea and vomiting, and allergic responses such as asthma (Butt et al 2004). Perhaps unsurprisingly, less is known about the effects of this nematode on whales and dolphins.

Undoubtedly endoparasites of all taxa can cause serious damage to their hosts. Take the example of bull sharks, *Carcharhinus leucas*. Borucinska & Caira (2006) found serious damage to the digestive tracts of these sharks caught off Borneo. Large numbers of cestode tapeworms were

found in shark in the pyloric gut and the stomach, with most direct damage to the host done by the worm scolexes burrowing deeply into shark tissue. This resulted in "acute necrotizing" and "chronic granulomatous gastroenteritis" amongst other pathologies. And this of course is over and above the predictable competition for food between the tapeworm and the shark. Clearly, host fitness is likely to be significantly reduced.

A halfway-house between endo- and ecto-parasitism of marine animals involves the highly specialized barnacle group, the Rhizocephala. Though closely related to (and descended from) sessile, filter-feeding barnacles, this group is entirely parasitic on other crustaceans such as anomuran (hermit) and brachyuran ("true") crabs (Glenner & Hebsgaard 2006). The commonest genus found parasitizing shore crabs for example is *Sacculina* (Figure 5.30). The larvae of rhizocephalans such as *Sacculina* are perfectly normal nauplii, which metamorphose into cyprids, just like all barnacles and other crustaceans. However, when a female cyprid alights on a new crab host, it metamorphoses into the so-called kentrogen larva whose tissues invade those of the host. Eventually, *Sacculina* spreads roots throughout the crab's abdominal tissues, forming the interna. This then breaks out of the still living crab body, forming a bulbous externa which can be fertilized by male *Sacculina* cyprids from which new eggs are formed. Young crabs tend to be parasitized rather than older ones. In Figure 5.31, it can be seen that only small (young) individuals are infected, and in this case at least, only about 10% of them. Both males and females appear equally infected (Lopez-Abellan 2002). The impact on host crab population dynamics is not easy to predict. Both males and females are sterilized (parasitic castration) by *Sacculina*, and since it is young individuals that are mainly infected, then the parasite essentially removes a proportion of new recruits into the reproductive population of crabs. There is however no evidence of a density-dependent process here, and because shore crabs and their relatives tend to be highly fecund, parasitism by *Sacculina*, though visually impressive, is unlikely to have a significant regulatory role on crab populations (Innocenti & Galil 2007). There is, perhaps, one indirect benefit of *Sacculina* infections in shore crabs. Because the hosts tend not to molt, they retain their original carapace for much longer than normal for a crab, which offers hard substrate for sessile weeds and invertebrates to settle, thus providing extra habitat richness in mixed sandy shallow waters (Mouritsen & Jensen 2006).

Marine ectoparasites are by their very nature much more obvious than the ones we have discussed so far. All are invertebrates apart from a very few such as the sea lamprey, *Petromyzon marinus*, a jawless primitive fish which rasps holes in the skins of larger fish and some cetaceans (Nichols & Hamilton 2004) (see also Chapter 8). Some of the most important fish ectoparasites are platyhelminth monogeneans, and crustaceans such as isopods (family Gnathiidae) and copepods (families including Pennellidae and Caligidae). Large parasites can be found attached to the scales, skin or fins of their hosts, whereas smaller ones are frequently found on the gills. Monogeneans fall into the latter category, being fairly protected from strong currents and abrasion. As with endoparasites, several species can be found on one host species, as in the case of the South Pacific grouper, *Epinephelus maculates*. Justine (2007) found 16 species of ectoparasites on the gills of this species, 12 of which were monogeneans and the remainder crustaceans. Note that all 16 species could be found on one individual fish, quite a feat of interspecifc coexistence.

Some copepods also infest the gills of fish often with 100% prevalence, especially in tropical waters (Scott-Holland

Figure 5.30

Crab infected with *Sacculina*. (Photograph courtesy of Paul Naylor.)

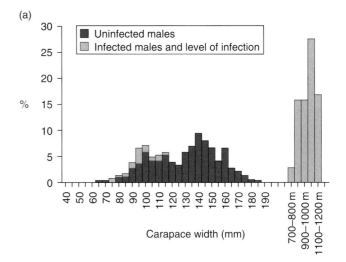

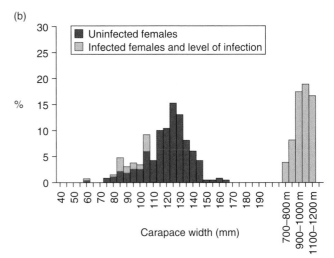

Figure 5.31

Length–frequency distributions and *Sacculina* incidence in the red crab *Chaceon affinis* at various depths off the Canary Islands. Light blue bars represent percentage of crabs infected in relation to carapace width, and all individuals caught within each depth stratum. (From Lopez Abellan et al 2002; reproduced with permission of Elsevier.)

(a)

(b)

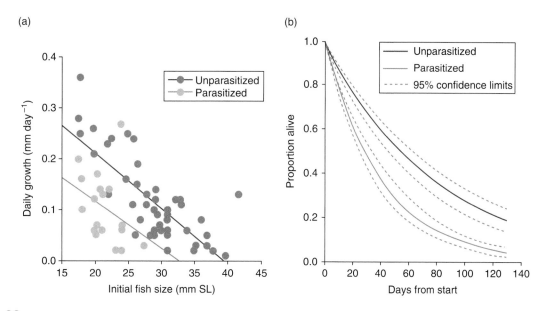

Figure 5.32

(a) Growth rates of unparasitized (dark blue circles) and parasitized (light blue circles) bridled gobies related to initial size of fish at date of tagging. SL = standard length (length of fish excluding tail fin). (b) Survival curves for bridled gobies parastized some or all of the time, or never. Dark blue = unparasitized; light blue = parasitized; dashed line = 95% confidence intervals. ((a,b) From Finley & Forrester 2003; reproduced with permission of InterResearch.)

et al 2006), and, although small, can in high densities have very deleterious effects on their hosts. Finley & Forrester (2003) looked at the effects of a copepod parasite, *Pharodes tortugensis*, on the gills of the small bridled goby, *Coryphopterus glaucofraenum*, in the British Virgin Islands (Figure 5.32a,b). Undoubtedly, the daily growth rate of the host was significantly retarded by the presence of parasites, as was survival rate (measured as fish still present, since emigration was thought not to be important). For example, 50 days after the start of the field experiment, about 50% of the unparasitized fish were still alive, compared with less than 30% of the parasitized individuals. Most importantly, the authors detected a highly significant, positive, relationship between the percentage of gobies infected with the parasite and their density; crowding increased parasitism.

It is the relatively massive, truly ectoparasitic, copepods/branchiura that catch the attention of ecologists and aquaculturalists. Probably the most infamous of these crustacean ectoparasites are sea lice (Figure 5.26), mainly in the genera *Caligus* and *Lepeophtheirus*. Sea lice are hugely important in global aquaculture industries, especially in the farming of salmonid fish such as salmon and trout. In 2004, Johnson et al estimated that economic losses in world aquaculture due to fish lice infection came to more than US$100 million annually, including treatment costs, reduced fish growth rates, management costs, and so on. Unlike some of the very complex lifecycles seen in endoparasites, ectoparasitic sea lice have rather simple lifecycles (Boxaspen 2006). Juvenile and adult stages occur on the host fish, and there are three or four free-living planktonic larval stages, the last one of which settles back on the host. As with many parasites and pathogens, it is the dispersal and migration movements of

the host that often dictate new inoculation rates, and the mechanisms involved are of crucial importance in aquaculture (Krkosek et al 2007). These authors studied the outbreak dynamics of two species of fish louse, *Lepeophtheirus salmonis* and *Caligus clemensii*, in stocks of young salmon off the coast of British Columbia. As Figure 5.33 shows, the parasitism rate varied between year and sea louse species, as well as during one season, with infection rates peaking in July. Around 50% of all fish caught in July were infected, with some samples showing as high as 80% infection rate. Based on this field data, Krkosek et al modeled effects of sea louse infection rate on abundance, reproductive rate and survival of juvenile salmon according to the amount of exposure to parasites (Figure 5.34). As can be seen, even relatively small increases in parasite load (relative to that found naturally in salmon populations) and exposure time can have a critical effect on salmon population dynamics and stock survival, which has particular consequences for maintaining fish at unnaturally high densities in aquaculture systems. In general, it seems clear that many parasites of marine animals can act in a top-down regulatory manner on their hosts.

Pathogens

Put simply, pathogens are biological agents that cause disease (Munn 2006). In essence, they are a particular type of parasite, not really very different to a sea louse or tapeworm, except in size and the aforementioned link to diseases in the host animal or plant. Marine animals are attacked by as wide a variety of pathogens as are their counterparts on land; for example, fish suffer from DNA and RNA viruses, fungi,

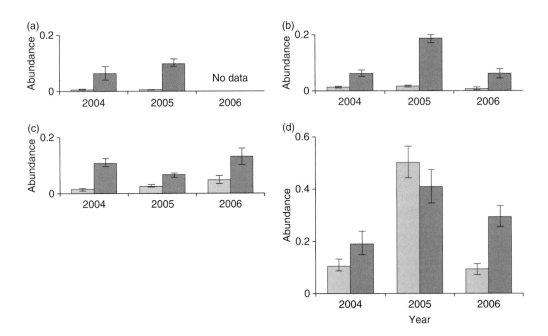

Figure 5.33

Sea lice abundance for all parasite development stages combined (±95% CI) in (a) April, (b) May, (c) June, and (d) July for *Lepeophtheirus salmonis* (light blue bars) and *Caligus clemensi* (dark blue bars) over 3 years. (From Krkosek et al 2007; reproduced with permission of The Royal Society.)

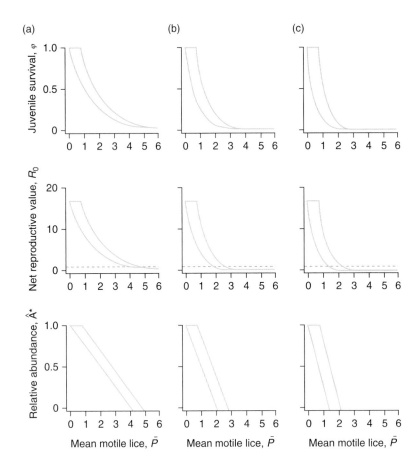

Figure 5.34

Model results showing the effects of increasing sea lice infection of juvenile salmon over and above the abundance at natural sea lice levels. (a) 1 month, (b) 2 months and (c) 3 months exposure to the parasites (predictions of two models provided). Horizontal dotted line = point below which salmon populations collapse. (From Krkosek et al 2007; reproduced with permission of The Royal Society.)

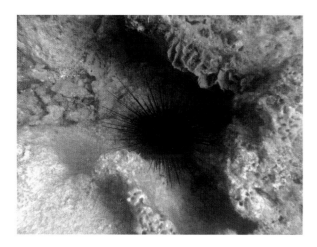

Figure 5.35

Diadema antillarum. (Photograph courtesy of Jonathan Shrives.)

protozoa, bacteria, and so on, whilst amongst the invertebrates, seasquirts (tunicates) have bacterial pathogens, bivalves such as clams and oysters can be infected with viruses and bacteria, as do horseshoe crabs, shrimps, and starfish (Mydlarz et al 2006).

One of the problems with marine pathogens is establishing the links between host animal health and environmental stresses. Many if not most examples of pathogen–host interactions in the sea may be explained by host debilitation as a result of suboptimal water quality (temperature, salinity, chemistry, etc.) leading to secondary infections of normally fairly benign microbial organisms in the sea (Lesser et al 2007). One notorious example involves the spiny sea urchin, *Diadema antillarum.*(Figure 5.35). Lessios et al (1984) described the rapid and devastating spread of mass mortality of this species throughout the Caribbean (Figure 5.36). Dead urchins were first noticed in Panama in January 1983,

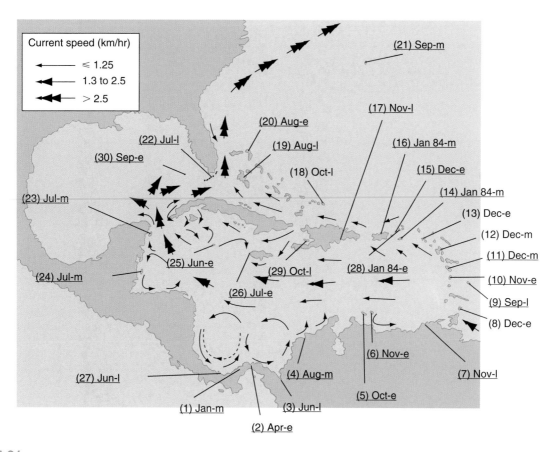

Figure 5.36

Spread of *Diadema antillarum* mass mortality through the Caribbean and the western Atlantic in 1983/84. Underlined dates indicate the first time mortality was noted at each locality. Dates without underlining indicate the last time lack of mortality was verified in unaffected areas. e, m, and l indicate the early, middle, and late part of the month. Unless otherwise noted, dates refer to 1983. Numbers denote the following localities: 1 = Galeta Point, Panama; 2 = San Blas Archipelago; 3 = Puerto Obaldia, Panama; 4 = Santa Marta, Colombia; 5 = Curaçao; 6 = Bonaire; 7 = Venezuela; 8 = Tobago; 9 = Barbados; 10 = St Lucia; 11 = Martinique; 12 = Guadeloupe; 13 = St Kitts; 14 = St Croix; 15 = St Thomas and St John; 16 = Puerto Rico; 17 = Santo Domingo; 18 = Grand Turk; 19 = Andros and New Providence; 20 = Grand Bahama; 21 = Bermuda; 22 = Florida Keys; 23 = Cancun, Mexico; 24 = Belize; 25 = Grand Cayman; 26 = Jamaica; 27 = Cahuita, Costa Rica; 28 = Tortola, Virgin Gorda, and Salt Island; 29 = Gulf of Gonave; 30 = Dry Tortugas. (From Lessios et al 1984; reproduced with permission of *Science*, AAAS.)

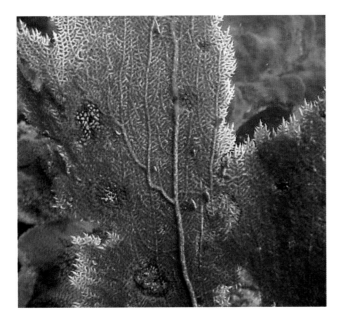

Figure 5.37

Seafan *Gorgonia ventalina* infected with aspergillosis. (Photograph courtesy of NOAA.)

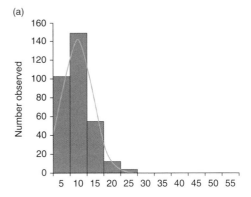

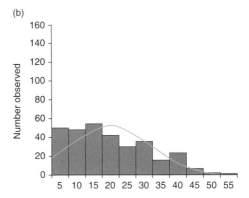

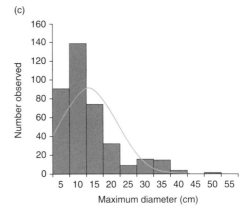

Figure 5.38

Size–frequency distributions of the star coral *Dichocoenia stokesi* in the Florida Keys in 1998 (a) and 2002 (b), in association with white plague (WP), and on Lee Stocking Island in the Bahamas in 2002 (c) where WP was less prevalent. Curves are the predicted normal distribution. (From Richardson & Voss 2005; reproduced with permission of InterResearch.)

but within a year, all of the Caribbean and the western Atlantic were affected, with *Diadema* populations reduced to merely 1–6% of previous densities over a geographic area of around 3.5 million square kilometers. The patterns of spread of the mortality as shown in Figure 5.36 suggest a water-borne pathogen which is carried over areas of open water on the currents until it reaches coastal regions, whereupon it spreads more locally with eddies and tides. To this day, the actual pathogen responsible has not been identified, though it is likely to have been a bacterium of some sort (Mydlarz et al 2006). This situation has been very slow to recover in many parts of the Caribbean, probably due to the very low recruitment rate of juvenile urchins linked to low adult densities (Miller et al 2007). Numbers are increasing again in various places such as the US Virgin Islands (Miller et al 2003), Mexico (Jordan-Garza et al 2008), and Costa Rica (Myhre & Acevedo-Gutierrez 2007). As urchin densities grow, algal cover has declined and coral cover increased. It may be that the catastrophic decline and slow recovery of *Diadema* populations is not all that unusual. Uthicke et al (2009) suggest that certain echinoderms, including *Diadema*, routinely exhibit these so-called "die-off" events; pathogen interactions would then be part of long-term natural cycles.

Gorgonian seafans are known to be attacked by fungal pathogens in the genus *Aspergillus*, which provides the name of the disease aspergillosis (Figure 5.37). *Gorgonia ventalina* and *G. flabellum* are mainly infected (Cróquer & Weil 2009). The specific pathogen responsible for the disease was thought to be *Aspergillus sydowii* (Weir-Brush et al 2004), a terrestrial fungus whose spores, it was

hypothesized, could be blown on the wind to the Caribbean from Africa. More recently, however, *A. sydowii* could not be detected in any samples of windborne African dust examined (Rypien 2008), and indeed it has proved difficult to isolate this particular pathogen from diseased gorgonians, leading Toledo-Hernandez et al (2008) to conclude that seafan disease is a product of opportunist pathogens and general host stress linked to water temperature and chemistry.

Most worrying in many parts of the tropical world are the diseases of scleractinian corals (Borger 2005). As mentioned above, there is controversy about whether these diseases are indeed caused by primary pathogens such as bacteria, or are really only manifestations of reactions to physiological stress in the host which becomes infected as a secondary problem (Somerfield et al 2008). Either way, a whole vari-

Figure 5.39

Black band disease (BBD). (Photograph courtesy of Jonathan Shrives.)

ety of "diseases," "plagues," and "syndromes" are now recognized in corals from the Caribbean to the Great Barrier Reef, but active and specific pathogens have only been identified for a few of these (Lesser et al 2007). A case in point is white syndrome (WS) which affects corals in the genus *Acropora* in the Indo-Pacific region, including the Great Barrier Reef. Coral tissues rapidly disappear at rates of up to more than 100 cm² per day, leaving exposed coral skeletons, in a manner reminiscent of programmed cell death (Ainsworth et al 2007). These authors were unable to find any pathogenic microorganisms associated with WS that might be the primary cause of the syndrome, but Sussman et al (2008) did isolate a unique group of bacteria in the *Vibrio* family from corals with WS.

In the Caribbean, two coral diseases, white plague (WP) and black band disease (BBD), are widespread. Various coral genera are infected, including *Montastraea*, but the elliptical star coral, *Dichocoenia stokesi*, is one of the most susceptible to WP, and as can be seen in Figure 5.38, many large colonies disappeared on the Florida Keys between 1998 and 2002 where WP was prevalent (Richardson & Voss 2005). In this case, a primary pathogen does appear to be associated with the disease, a member of the alphaproteobacteria, *Aurantimonas coralicida* (Cróquer et al 2005). BBD (Figure 5.39) affects a variety of coral genera and species, including *Diplora strigosa, Siderastrea siderea, Colpophylla natans,* and various *Montastraea*, and is the

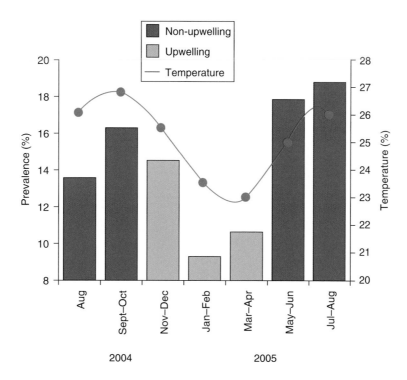

Figure 5.40

Prevalence of black band disease (BBD) in Venezuela over 2 years and the sea surface temperature (circles). (From Rodríguez & Cróquer 2008; reproduced with permission of Springer.)

most widespread of all coral diseases across the world (Voss et al 2007). As might be expected, the pathogens involved with BBD have proven difficult to identify, but it seems that a consortium of microorganisms are involved, again including members of the alphaproteobacteria (Sekar et al 2008) as well as filamentous cyanobacteria and sulfur-oxidizing bacteria (Voss et al 2007). All these pathogens congregate in the dark band at the leading edge of tissue death, giving the disease its name. It is not yet known how these pathogens are spread or vectored around the oceans, but it is once again clear that environmental factors or stressors play an important role in predisposing corals to pathogen attack (Aeby & Santavy 2006). Figure 5.40 shows the situation on the northeast coast of Venezuela, where it can be seen that BBD is most prevalent when the mean monthly water temperatures are highest (Rodríguez & Cróquer 2008). Note that there appears not to be an association with upwelling events that might be predicted to influence disease incidence via nutrient loads.

There is a final ray of hope. Vollmer & Kline (2008) studied White Band Disease (WBD) in Panama, and discovered that 6% of staghorn coral, *Acropora cervicornis*, genotypes were resistant to WBD infection. If this resistance remains stable, and the mechanisms become better understood, we may look forward to the recovery and restoration of some of our coral reefs in the future (see also Chapter 10).

Chapter 6

Competition and succession

Unlike predation, parasitism and grazing, competitive interactions frequently take place within a trophic level. Individuals either compete with members of their own species (intraspecific) or with members of a different species (interspecific). Either way, competition occurs for a resource that is in short supply such as space, food, refuges, mates, territories, or other mutually required item. So for example, barnacles and seaweed may compete for space on a hard substrate with members of their own species, and also with each other. The barnacles however may not only compete for space, but also for food with another filter feeder such as a sponge or a fanworm. Both inter- and intraspecific competition usually occur simultaneously, but for the sake of simplicity we shall consider them one at a time here.

Both intra- and in particular interspecific competition can be especially prevalent as organisms enter a new habitat and sequentially replace each other as ecological succession proceeds to the climax state. In this chapter, we first discuss competitive relationships within and between species, and then consider how such factors interact during community succession in the sea.

Intraspecific competition

Individuals of a single species frequently come into contact with one another, and as their density increases so does their competition for a limited resource. A simple example

is shown in Figures 6.1 and 6.2 which is from the work of Steen & Scrosati (2004), who investigated the growth of the settled germlings of serrated wrack, *Fucus serratus*, in experimental tanks in Norway. Macroalgal settlers often colonize rocks in huge densities and, as can be seen, grow much more slowly in high densities (Figure 6.1) and also suffer greater mortality (Figure 6.2). This mortality rate is significantly higher at high nutrient levels, because the young algae grow faster and compete more for space and, as they get bigger, probably for light too. The results of this intraspecific competition amount to a self-thinning process, where no matter what the initial settlement density, the carrying capacity of the substrate in terms of biomass or numbers of seaweed fronds is eventually achieved. A similarly simple example for an animal mimics the effects of intraspecific competition seen in algae. Amphipods such as *Monoporeira affinis* (Figure 6.3) can be enormously common on benthic substrates from the shallows to the deep-sea. Elmgren et al (2001) experimentally reared amphipods from the "young-of-the-year" individuals using a constant concentration of food material collected from 30 m deep in the Baltic Sea (Figure 6.4). Intraspecific competition for food resulted in exponential declines in growth rate as population density increased. Both the seaweed and the amphipod examples illustrate the density-dependent effects of intraspecific competition. The consequence of intraspecific competition influencing the growth rates and hence final sizes of adult marine organisms is most often reflected in the fecundity of the adults. Figure 6.5 shows the relationship between body weight and egg mass weight (a measure of fecundity) in the barnacle *Scalpellum stearnsii* (Ozaki et al 2008). Though rather variable, there is a positive relationship; big females have more offspring which themselves may also be more reproductively successful, especially in a lecithotrophic (nonfeeding as larvae) species where large adults are able to put more resources into each offspring, as well as produce a larger number (see Chapter 7). Thus, if intraspecific competition

Marine Ecology: Concepts and Applications, 1st edition. By M. Speight and P. Henderson. Published 2010 by Blackwell Publishing Ltd.

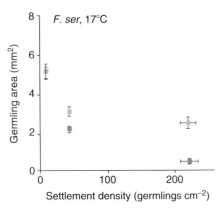

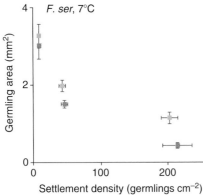

Figure 6.1

Fucus serratus (*F. ser*): mean area of germlings as a function of settlement density after 90 days of cultivation (±95% CI). Circles = normal seawater; squares = N-P enriched seawater. (From Steen & Scrosati 2004; reproduced with permission of Springer.)

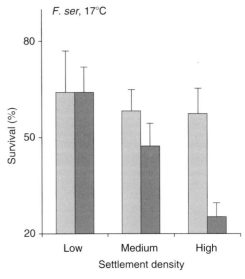

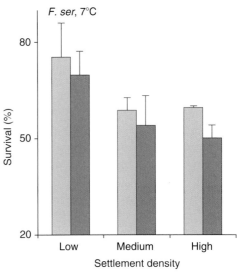

Figure 6.2

Fucus serratus (*F. ser*): mean survival of germlings as a function of settlement density after 90 days of cultivation in two levels of nutrients, normal seawater (light blue bars) and nitrogen-enriched seawater (dark blue bars). Low density settlement = 10 germlings per cm²; medium density settlement = 50 germlings per cm²; high settlement density = 250 germlings per cm²) (±95% CI). (From Steen & Scrosati 2004; reproduced with permission of Springer.)

for food, space and other resources reduces female size, then individual fecundity may also be reduced with significant consequences for the population.

Given that intraspecific competition frequently results in limitations to population growth, are these examples of bottom-up resource regulation (see also Chapter 5)? Wenngren & Olafsson (2002) also studied *Monoporeira affinis*, and fed the amphipods a mixture of naturally occurring algae at three resource levels, zero, medium (to represent a normal spring bloom of diatoms (see Chapter 3), and high (to represent double the normal amount of diatoms encountered in a spring bloom). As seen in Figure 6.6, food was not surprisingly important to juvenile amphipods – clearly they would grow rapidly and effectively during a spring bloom of diatoms, but note that doubling the food supply again did not result in an additional enhancement of the growth rate. Something else limits growth at very high densities of food. It can also be seen from the figure that adding adults to the experimental jars resulted in significant reductions in juvenile growth. This indicates that we must consider intraspecific competition acting not just within one cohort or age class of a

population of one species, but also between different age classes such as adults and juveniles.

Observation suggests that for many species adults and juveniles are unlikely to compete. Adult sea anemones for example may have tentacles many times larger than juveniles of the same species. Given their sit-and-wait predation method of foraging they will capture distinctly different sizes of live prey such as fish and prawns. It is frequently observed that fish use different parts of the habitat as juveniles and adults and may even undertake different seasonal migrations. For sessile organisms in particular, competition for space may become more intense with age, in other words, size increases competition.

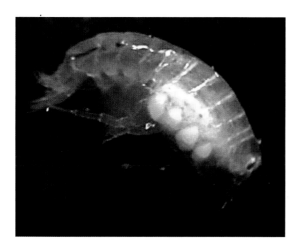

Figure 6.3

Monoporeia affinis. (Photograph courtesy of Rasmus Neiderham.)

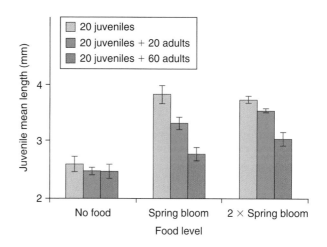

Figure 6.6

Mean (±s.e.) of juvenile *Monoporeira affinis* after 64 days in experimental jars when provided with three different food levels, and in the presence or absence of conspecifc adults. (From Wenngren & Olafsson 2002; reproduced with permission of InterResearch.)

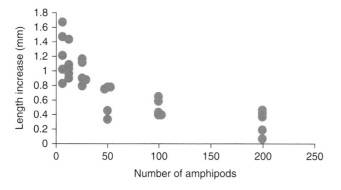

Figure 6.4

Increase in length of subadult amphipods, *Monoporeira affinis*, after 57 days, as a function of initial amphipod density (density = number of amphipods per 2l experimental jars). (From Elmgren et al 2001; reproduced with permission of InterResearch.)

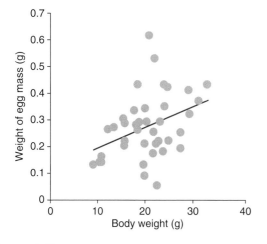

Figure 6.5

Relationship between the body weight of barnacles, *Scalpellum stearnsii*, and the egg mass within them. $P < 0.005$, $r^2 = 0.14$. (From Ozaki et al 2008; reproduced with permission of Cambridge University Press.)

Competition for limited resources in one habitat patch may not only result in reduced weight gain or survival. It is probably also a natural reaction to leave an area where life is not easy, in search of better opportunities. Moksnes (2004) showed how juvenile shore crabs, *Carcinus maenas*, leave patches of mussel beds as the density of the crabs increases. There is a suggestion here that emigration rate does not have a linear relationship with population density. Instead emigration only occurs above a threshold density. Leaving a patch of food and/or otherwise suitable habitat can be a risky business. Mussel beds provide food and shelter for young crabs, and leaving such a safe habitat would not be undertaken lightly. Only when the consequences of intense intraspecific interactions outweigh the risks of moving does emigration occur.

Ameliorating the effects of intraspecific competition may involve a partial shift in food type, or perhaps a switch to a new method of feeding, which by its very nature will bring in a different type of food for which intraspecific competition is less intense. The common (or edible) periwinkle, *Littorina littorea*, is a classic algal grazer (see Chapter 4), but Petraitis (2002) discovered that this snail can do very well as a scavenger feeding on dead animal material if necessary. Having alternative food is clearly an important survival trait when times are hard and the normal resources are in short supply. A final reaction to increasing intraspecific competition can be to reduce the consumption of the normal, heavily competed for, food supply and to switch to another which is less limited. To do this, the behavior of the animal may have to change. Take, for example, the sand dollar, *Dendraster excentricus* (Figure 6.7). Sand dollars are

Figure 6.7

Sand dollar, *Dendraster excentricus*. (Photograph courtesy of Carissa Thomas.)

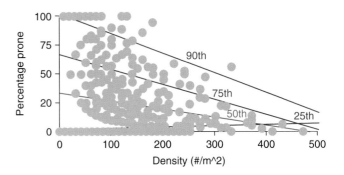

Figure 6.8

Relationship between the positioning (body posture) of the sand dollar *Dendraster excentricus* and intraspecific density. Data are from 401 0.1-m³ quadrats pooled from field surveys in February, May, August, and November on the coast of Baja California. Solid lines show various regression quantiles, all of which are statistically significant. (From Fodrie et al 2007; reproduced with permission of Elsevier.)

dorsoventrally flattened sea urchins (Echinoidea) that feed on surface deposits, or, more rarely, on organic material suspended in the water column. In order to switch between deposit and suspension feeding, the animal has to alter its body posture. Deposit feeding involves the normal posture, flat on the substrate with the entire ventral surface in contact with the substrate. This is the "prone" position. Alternatively, the front end can be buried in sand, whilst the rest of the body is lifted into the current so that suspended material can be removed from the water. Surface deposits tend to be used up rather easily, and take some time to be replaced by sedimentation, whereas suspended material is easily and rapidly renewed by the currents (assuming that current flow and content is maintained). Therefore sand dollars should be less food limited when filter feeding (Fodrie et al 2007). These authors studied the incidence of prone (deposit) feeding versus inclined (suspension) feeding in relation to sand dollar density. In Figure 6.8 the pooled data shows that as the density of animals increases, the percentage of time spent prone decreases, indicating that intraspecific competition for food increases with sand dollar density exhausting surface deposits and necessitating their switch to suspension feeding. This can be a risky tactic, if currents are too strong the inclined animal is lifted out of the substrate and carried away. Deposit feeding is a safer strategy at low population densities.

Quite how marine organisms detect the increasing density of con-specifics varies with species. Simple visual or tactile stimuli will inform some species of the density within a patch. While the lack of food for an individual may not obviously be linked to the number of neighbors the result (reduced growth for example) is clear. Some competitors may release chemicals which signal increasing density. Larvae of the marine lamprey *Petyromyzon marinus*, for example, are thought to release a chemical compound into the water that has a negative effect on growth rate. This system has a density-dependent regulatory effect on juvenile and then hence adult size. In this case, it is not necessarily related to food supply which may not be limiting. Density-dependent regulation via intraspecific competition may only occur in the juvenile stage, but can have subsequent impacts on the adult population. This phenomenon was demonstrated by Jenkins et al (2008) who worked on populations of the barnacle, *Semibalanus balanoides*, in North Wales. Barnacle recruitment and subsequent survival and growth was studied using digital photo recording over 2 years, tracking the changes in barnacle populations in small (5 × 5 cm) quadrats. Some of the results are shown in Figure 6.9. When plotted against recruit density, adult density 2 years later increases for a while, and then declines thereafter. This is evidence of strong overcompensatory density-dependent mortality caused by intraspecific crowding. Instantaneous barnacle mortality increases significantly with barnacle density, with a curvilinear relationship being a better fit to the observations than a simple linear one. It seems that the level of recruitment (number of barnacle settlers) plays an important role in determining adult abundance. This relationship is positive at low recruit numbers, and negative at high ones.

To summarize, intraspecific competition regularly occurs and can maintain, in the long-term, populations around their carrying capacity. Such processes can be entirely intraspecific and take place in the absence of other species.

Interspecific competition

Interspecific competition is competition between different species and is likely to result in reduced survival or fitness for the competing individuals. Most examples of interspecific

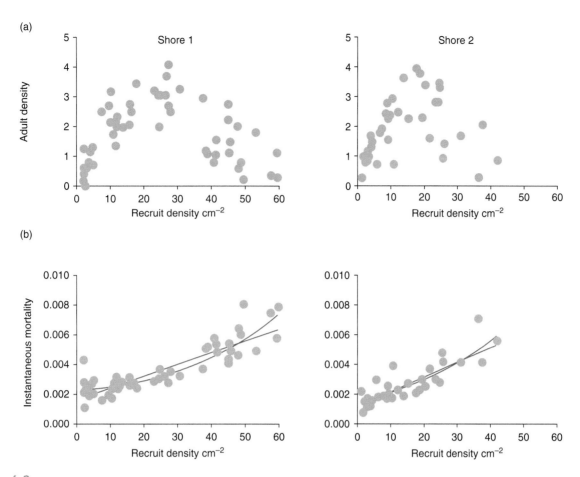

Figure 6.9

The relationship between the density of manipulated recruits of the barnacle *Semibalanaus balanoides* in June 2002 and (a) the density of surviving adults in May 2004, and (b) the instantaneous mortality rate over the same period. Significant linear and curvilinear regression lines are shown. (From Jenkins et al 2008; reproduced with permission of Wylie Interscience.)

competition in the sea feature two or more species trying to avoid confrontation by utilizing different resources whenever possible. It is of little importance for two different species to use the same food source when the resource is abundant, but as this food becomes limiting, resource partitioning can limit adverse impacts. Interacting species should differ more in resource use where they live together compared to when they live apart (Limbourn et al 2007). For a resource such as space or food the range of utilization when a species lives alone defines the fundamental niche. The reduced breadth of this resource in the presence of another potentially competing species is the realized niche. Below we describe various consequences of interspecific competition between marine species, and examine the methods employed to coexist.

First, we will consider two examples of resource partitioning as a result of interspecific competition, one where the whole habitat is partitioned up between closely related species, and the other where the food resource alone is partitioned. We start with two elasmobranch fish from

Australia, the common stingaree, *Trygonoptera testacea*, and the kapala stingaree, *Urolophus kapalensis*. (Figure 6.10). Stingarees are bottom-dwelling rays found in mainly subtropical waters feeding on benthic invertebrates and small fish. These two species are similar in size and shape, and live on sandy bottom habitats in the same sites and at the same depths. There is thus great potential for interspecific competition. Marshall et al (2008) compared the diets of these two species in southeast Queensland by trawling for them in depths of up to 112 m, and then examining the stomach contents. Their results are shown in Figure 6.11. By comparing the bar charts the details of the food composition can be seen. There is some overlap in food types, but as the graphs show, *T. testacea* eats mainly polychaetes, whereas *U. kapalensis* eats crustacea such as shrimps and amphipods, a classic example of food resource partitioning. As long as there is a plentiful supply of both polychaetes and crustacea, these two species of fish should be able to successfully coexist in the same locality at the same time.

Figure 6.10

Kapala stingaree, *Urolophus kapalensis*. (Photograph courtesy of Andy Murch.)

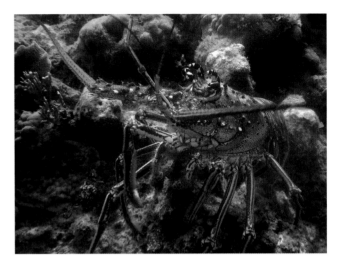

Figure 6.12

Caribbean spiny lobster, *Palinurus argus*. (Photograph courtesy of Jonathan Shrives.)

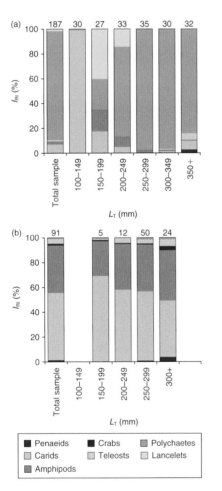

Figure 6.11

Percentage index of relative importance of seven main broad dietary categories for the total sample, and six different length classes, of (a) *Trygonoptera testacea* and (b) *Urolophus kapalensis*. L_T = total length of animal (to nearest 1 mm); I_{RI} = index of relative importance. (From Marshall et al 2008; reproduced with permission of Wiley-Blackwell.)

Broadening the scale of the resource from food type to a whole habitat level, Lozano-Alvarez et al (2007) studied the abilities of two species of spiny lobster, *Palinurus guttatus* and *P. argus* (Figure 6.12), to coexist on the coral reefs of Mexico. Spiny lobsters (actually crayfish) live in crevices or dens and venture out at night to feed on a variety of living invertebrates including gastropod molluscs, bivalves, and crabs and any dead organisms they may encounter. They are food generalists and if these resources are limited, interspecific competition could ensue. However, as can be seen in Figure 6.13, the two species tend to inhabit different parts of the reef complex. *P. guttatus* was more abundant on fore-reefs all through the year, although it could also be found on back-reefs. In contrast, *P. argus* appears more of a habitat specialist, staying mostly on back-reef sites. It is on these latter sites that both lobster species appeared to coexist, which Lozano-Alvarez et al suggested was possibly achieved by predator pressure and/or the availability of dens where each species could shelter.

Even small sections of a habitat may be partitioned by two potential competitors whose food requirements overlap. Gobies *Gobius fallax* and *G. auratus* are closely related fish and are thought to have only relatively recently separated from a common ancestor (Herler & Patzner 2005). Both species inhabit rocky substrates in the northern Mediterranean, where they feed on benthic crustaceans such as small shrimps and amphipods. Though their food and general habitat niches overlap, they can be found in different parts of the same rocky ecosystems (Figure 6.14). *G. fallax* most commonly occurs at shallow depths (2–10 m) and prefers gradual slope angles of between 0 and 30 degrees. *G. auratus*, on the other hand, was found mainly below 6 m deep, extending all the way to 30 m and beyond, resting on slopes with steeper gradients between 30 and 90 degrees. While obviously there is some overlap between the two species, they do reduce their competition.

(a)
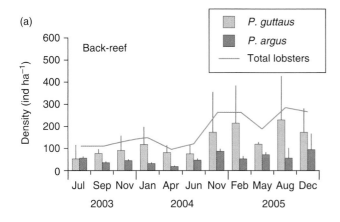

(b)
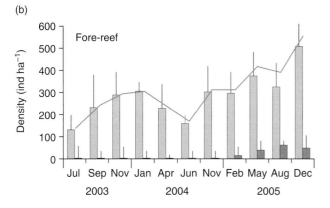

Figure 6.13

Mean density (±s.d.) (individuals per hectare) of spiny lobsters, *Palinurus guttatus* and *P. argus*, and of both species combined in (a) back reef zone and (b) fore reef zone, in Mexico. (From Lozano-Alvarez et al 2007; reproduced with permission of Springer.)

(a)
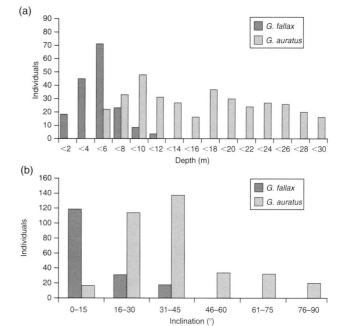

(b)

Figure 6.14

Frequency distributions of gobies, *Gobius fallax* and *G.auratus*, within (a) 2-m depth intervals and (b) six classes of substrate inclination, counted visually in 20 belt transects in the Adriatic Sea. (From Herler & Patzner 2005; reproduced with permission of Wylie Interscience.)

Figure 6.15

Blackspot tuskfish, *Choerodon schoenleinii*, on a southern Queensland wreck. (Photograph courtesy of Ian Banks, divingthegoldcoast.com.au)

In many cases, interspecific competition occurs between multiple species, especially if they are closely related, such as congeneric coral reef fish. Fairclough et al (2008) used underwater visual surveys (UVC) to study the ways in which five different species of tuskfish in the genus *Choerodon* coexisted within various habitats in Shark Bay on the west coast of Australia. Two of the five species, *C. rubescens* and *C. cephalotes*, were found to be habitat specialists (living mainly on entrance channel reefs and seagrass beds respectively). The three other species, *C. schoenleinii* (Figure 6.15), *C. cyanodus*, and *C. cauteroma*, could all be found in inner gulf reefs and around rocky shorelines, as well as seagrass in the case of *C. cauteroma*. (Figure 6.16). In this example, the partitioning of the habitat resources is not completely distinct, unlike the previous example, but undoubtedly this division of the various habitats in one locality has the potential to reduce interspecific competition for spatial resources. Other methods such as food type partitioning will have evolved to fine tune the system within one habitat type.

Looking at reef fish from different feeding guilds (see Chapter 4), we can see more evidence of resource partition-

ing involving many species. Pratchett (2005a) studied the dietary preference of 20 species of butterfly fish (Figure 6.17) on the Great Barrier Reef. In general, butterfly fish feed on hard (scleractinian) corals, and can frequently be observed "grazing" on coral heads, digesting the polyps and other organic tissue and excreting the white coral dust as "sand." Some species, however, prefer to feed on soft-corals and gorgonians, whilst others do not eat coral at all, instead

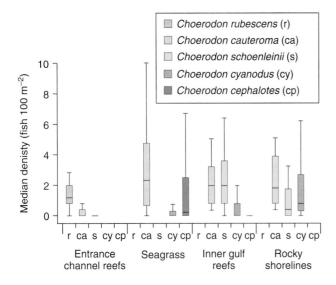

Figure 6.16

Box (25th, 50th, and 75th percentiles) and whisker (10th and 90th percentiles) plot of the densities of five species of *Choerodon* in four habitats, recorded using underwater visual surveys (UVC) in Shark Bay, Western Australia. (From Fairclough et al 2008; reproduced with permission of Elsevier.)

Figure 6.17

Four-eye butterfly fish (*Chaetodon capistratus*). (Photograph courtesy of Jessica Harm.)

consuming parts of zooanthids, polychaete fan worms, and even sea cucumbers. Divers followed individual butterfly fish from the 20 species noting what they ate over 3-minute periods. The results are presented as a canonical discriminate analysis (CDA) plot in Figure 6.18. CDA is a multivariate statistical method used to identify similarities between objects (species in this case) and identify the explanatory variables causing the pattern revealed. Four separate species clusters were identified, each containing between three and seven species differing in the dominant dietary components. There is considerable distinction in the major types of diet amongst co-occurring butterfly fish

species on a single reef system. The food available has been partitioned into separate subsets, within which reduced interspecific competition would be expected. Other mechanisms to reduce competition within these dietary subsets would be anticipated, such as differing seasonality, choice of habitat structure, and different size ranges for each species.

The more mobile the species, the bigger the spatial scale over which competition with other species may occur. Two cetaceans examples illustrate how large spatial scales can be dealt with in the context of reducing interspecific competition. According to Praca & Gannier (2008) sperm whales, *Physeter macrocephalus*, pilot whales, *Globicephala melas*, and Risso's dolphins, *Grampus griseus*, all prey either exclusively or preferentially on cephalopods such as bathypelagic squid. Praca & Gannier employed a technique known as Ecological Niche Factor Analysis (ENFA) using data from the northwestern Mediterranean where all three species co-occur to investigate their niche separation. Data were collected from 10 years summer cruising, searching visually and acoustically for cetaceans. For each sighting, various oceanographic measurements were taken or estimated, including sea state, water depth, sea surface temperature, salinity, marine primary production in the area, slope topography, and distance to the 200 m contour underwater. All these eco-geographical variables (EGVs) were put into ENFA models to produce habitat suitability predictions for each cetacean species. In addition, the authors employed discriminant analysis to compare the ecological niches of each of the three species, by relating the habitat space defined by the EGVs to the observed distributions of the animals. Figure 6.19 shows habitat suitability maps for the three species, which can be seen to be reasonably discrete even on a two-dimensional plot. Extending these habitats to include depth data as well, it was concluded that sperm whales had a core habitat (hunting ground) on the continental slope over much of the whole area including the Balearic and Corsica Islands, with a mean depth of around 1650 m, mean slope angle of 2.1 degrees, and a mean summer sea surface temperature (SST) of 21.5°C. Pilot whales showed a principal habitat in oceanic waters in the central Ligurian Sea and Provencal basin, with a mean depth of 2500 m, a mean slope of 0.55 degrees, and a mean SST of 21.7°C. Finally, Risso's dolphin seemed to prefer coastal fringe habitats on the upper part of the continental slope with a mean depth of 640 m, 3.6 degrees of slope, and a mean SST of 22.2°C. This highly sophisticated analysis clearly demonstrates that three large predator species all of whom require the same food sources, are able to partition the resource (in this case a huge volume of the Mediterranean Sea) using subtle but highly functional niche attributes. Assuming squid are abundant over all these habitats, these cetaceans should coexist with ease. If, however, the prey species become limited in numbers and/or distribution, by overfishing for example (see

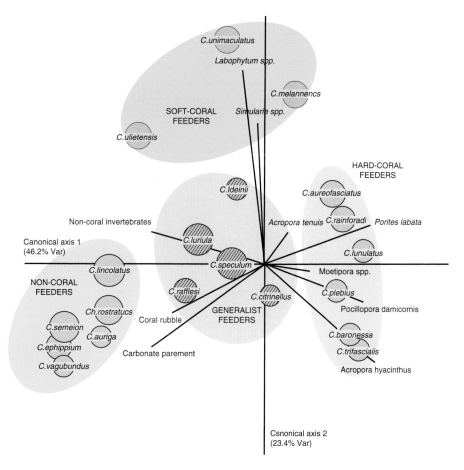

Figure 6.18

Canonical discrimination analysis (CDA) showing interspecific variation in the dietary composition of 20 species of butterfly fish. Circles show 95% confidence limits around each species or genus; shading and ellipses indicate guilds. (From Pratchett 2005; reproduced with permission of Springer.)

Chapter 8), then this coexistence could break down. One additional point is that in an equally elegant analysis, Azzellino et al (2008) found that sperm whale and Risso's dolphin add an extra sophistication to their division of the Mediterranean continental slope by segregating their activities seasonally.

So far in this discussion of interspecific competition, we have mainly considered mobile species. Sessile species however might have to endure interspecific competition without recourse to moving apart, so mechanisms to reduce or remove interspecific competition are likely to have evolved on many occasions. In New South Wales, Australia, populations of native rock oysters, *Saccostrea glomerata*, have begun to interact with the introduced exotic Pacific oyster, *Crassostrea gigas* (Krassoi et al 2008). Unlike the previous examples, where coexistence between potentially competing species has developed as a co-evolutionary process, the appearance of a new competitor is likely to result in competitive exclusion of the less competitively fit species. Unfortunately, the less fit species tends on these occasions to be the native one. Krassoi et al manipulated the densities of both oyster species on artificial perspex plates located at dif-

ferent parts of the intertidal region and measured proportional mortalities over 18 months in the field. The exotic Pacific oyster was able to overgrow the native rock oyster at low and mid-intertidal levels because of its 60% faster growth rate. However, the advantages of the exotic over the native species were negated at the high intertidal level, where the Pacific oyster suffered up to 80% mortality. The overall result is that the native species is likely to be replaced by the exotic in the lower intertidal, but the native *Saccostrea* can hold its own in the high intertidal where it inhabits competition-free space. This scenario is reminiscent of the classic story elucidated by Connell (1961) where the two temperate barnacle species *Semibalanus balanoides* and *Chthalamus stellatus* become zoned relative to the low tide mark on temperate rocky shores according to their ability to compete with each other while surviving different environmental stresses.

Returning to coral reefs, the literature is full of studies of how different species of coral compete with each other and with completely unrelated species such as algae. Reefs, as we have said in Chapter 2, are enormously species rich and how so many different sessile species coexist. Essentially, coral reefs are naturally low productivity systems (oligotrophic),

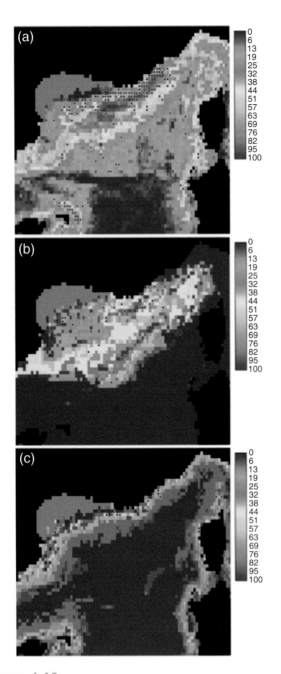

Figure 6.19

Habitat suitability maps for (a) sperm whale, (b) pilot whale, and (c) Rhisso's dolphin, based on ecological niche factor analysis (ENFA). Grey areas not used in the analysis; black dots are actual presence records for each species. (From Praca & Gannier 2008; reproduced with permission of Ocean Science.)

Fiugre 6.20

Porites cylindrica – yellow and purple colour morphs. (Photograph Martin Speight.)

encounter interspecific competition for a suitable hard substrate, in other words, space limitation.

Porites is a large genus of scleractinian coral found throughout the world (Figure 6.20). Its growth habitats can be very variable, from a squat or lobed shape, to a fairly branched form. Idjadi & Karlson (2009) studied competitive interactions between *Porites lobata*, a rather slow-growing lobed species, and *P. rus*, a faster-growing branching species, in French Polynesia. Various combinations of corals were attached to plates in an area of reef and the mean growth rate of *P. lobata* was measured over a year (Figure 6.21). Overall, these authors found that competition reduced the growth rate of a single coral colony by over 60%. Intraspecific competition is certainly powerful, even in the absence of other species. Whether this is because of food or space limitation, or some physical or chemical inhibition, is undetermined.

Interspecific competition between corals and other sessile animals is commonplace on reefs, especially between those phyla that have similar niches (sessile, colonial, active filter feeding, etc.). Bryozoa for example are often found competing with corals, and the results of such an interaction depend on the health of both competitors. Bryozoa tend to be able to overgrow corals when bleaching events stress the corals (Fine & Loya 2003). Corals and sponges in the main do not compete for food since at least in shallow, clear water one is a symbiotic photoautotroph while the other is a filter feeder. However, they can sometimes be mortal enemies when it comes to space on a reef. Sponges in particular can be quite aggressive on occasion, producing secondary metabolites which can kill the zooxanthellae in coral polyps and cause bleaching (Pawlik et al 2007). Once this dead space is created, some sponge species rapidly overgrow the coral surface (Vermeij 2006). It is not

and in many cases, food limitation may force strong intraspecific competition and result in weak interspecific competition. If we enhance the productivity of a reef using fertilizers from the land, we see a rapid reduction in coral diversity (see Chapter 9), suggesting that species coexistence is at least in part a function of oligotrophy. We must not forget of course that even if all corals are equally food limited, they may still

surprising then to find that many corals and sponges are frequently found as antagonists on reefs (Maliao et al 2008). Twenty reef sites around the Florida Keys were assessed for coral and sponge cover, as well as algal and chlorophyll (a measure of primary productivity) incidence.

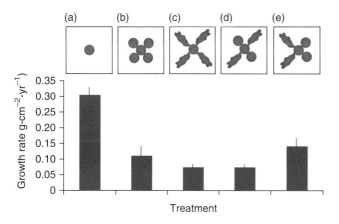

Figure 6.21

Mean (±s.e.) skeletal growth of small colonies of the coral *Porites lobata* in each of five experimental treatments after one year. From left to right: (a) growth of single colony (control), (b) central colony surrounded by four conspecifics, (c) central colony surrounded by four congenerics, (d) nonaggregated group consisting of two conspecifics and two congenerics, and (e) aggregated group consisting of two conspecifics and two congenerics. Circle = *P. lobata*; elongate shape = competitively superior *P. rus*. (From Idjadi & Karlson 2009; reproduced with permission of *Ecology*, ESA.)

Some coral species such as *Mycetophyllia alicia* and *Agaricia fragilis* tended to co-occur on the same reefs. Other species, however, such as *Montastraea annularis* and *Porites astreoides* are mutually exclusive with sponges. Coral–sponge interactions can be thought of as standoffs, where neither competitor wins or loses over an extended timeframe (Aerts 2000). It is only when something else causes reduced vigor or stress to one of the competitors, usually the coral, that the sponges then dominate.

So, competition between algae and some corals is widespread, and may consist of a simple overgrowth system where algae are able to grow over the top of corals, reducing the light reaching the photosynthetic zooxanthelae, or a more complex chemical interaction where algae poison the corals or the symbionts (Titlyanov et al (2007). McCook et al (2001) list various different ways in which corals and algae may compete with each other (Table 6.1); Figure 6.22 provides just one example. Jompa & McCook (2002) removed any unwanted effects of herbivory on a reef in Australia using exclusion cages and manipulated the interactions between coral, *Porites cylindrica*, and algae, *Lobophora variegata*. The presence and overgrowth of algae caused significant mortality of coral tissue, but in return, coral was able to inhibit the growth of algae to some extent. On the whole though, algae were markedly better competitors than coral. Note that reduced herbivory resulted in faster algal growth and consequent increased overgrowth and coral mortality via interspecific competition, Clearly, removing fish and urchins can have significant effects (see

Table 6.1 Mechanisms for competition between corals and algae (from McCook et al 2001) (credit Springer)

ALGAL INHIBITION OF CORALS	CORAL INHIBITION OF ALGAE	INCLUDES:	CATEGORIES
Overgrowth	Overgrowth	Smothering;	Direct, interference, overgrowth
Shading*	Shading	Overtopping;	Indirect, exploitative, consumptive
Abrasion	Abrasion	Whiplash[†]; including sweeper tentacles and polyps, mesenterial filaments	Direct, interference, encounter
	Stinging, etc.		Direct, interference, encounter
Chemical[‡]	Chemical	Allelopathy	Direct, interference, chemical
Pre-emption / Recruitment barrier[§]	Space pre-emption		Direct, exploitative, consumptive
Epithelial sloughing[¶]	Mucus secretion		Defensive mechanism

* Shading or overtopping may include establishment of dense canopy, with numerous effects on the chemical and physical conditions, hydrodynamics, etc
† Whiplash, often cited as damaging corals, will generally also be very detrimental to the softer algal tissue
‡ Allelopathic chemical effects have been demonstrated on soft corals
§ Canopy-forming macrophytes will actually occupy little of the substrate, but may still form an effective barrier to coral settlement
¶ Epithelial sloughing and mucus secretion are defence mechanisms against epibiotic colonisation, rather than mechanisms for expansion

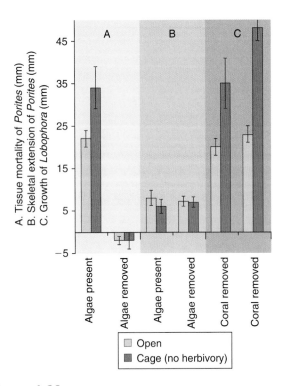

Figure 6.22

Effects of competition and herbivory on the growth and mortality of the coral *Porites cylindrica* and the brown alga *Lobophora variegata* on the Great Barrier Reef. (From Jompa & McCook 2002; reproduced with permission of Elsevier.)

Chapter 4). It is crucial to note that in most coral/algal interactions, the decline of coral tends to promote the ascendancy of the algae.

Finally, we need to look at complex marine communities where competitors at different trophic levels interact with each other to share out a variety of different resources. Such systems are notoriously tricky to unravel. One study was carried out by Purcell & Sturdevant (2001) who investigated the composition of zooplankton prey in two jellyfish species, *Aurelia labiata* and *Cyanea capillata*, one ctenophore species, *Pleurobrachia bachei*, one hydromedusan species, *Aequorea aequorea*, and juveniles of four fish species, walleye pollock *Theragra chalcogramma*, Pacific sandlance *Ammodytes hexapterus*, Pacific herring *Clupea pallasi*, and pink salmon *Oncorhynchus gorbuscha*, in Prince William Sound in Alaska. Gut content analysis revealed that interspecific competition between predator species may be reduced by there being two trophic groups. Group 1 (*A. labiata*, *P. bachel* and pollack, herring and sandlance) ate primarily crustacea, whilst Group 2 (*A. aequorea, C. capillata* and salmon) ate mostly larvaceans. Within a predator group, the diets of similar sized invertebrate and vertebrate pelagic predators overlapped (50% ± 21%), and since these species co-occur spatially and temporally, the potential for competition for prey still existed. The authors

concluded that competition could be avoided by (a) fish such as salmon migrating away, perhaps in response to declining food supplies; (b) fish such as sandlance (a sand eel, see Figure 5.23) switching from prey in the zooplankton to feed on epibenthic species instead; and (c) fish such as herring partitioning food by depth, or by feeding at different times of day.

Succession

According to Hambler (2004), succession is the process whereby areas devoid of life become colonized by a sequence of species. These areas may be devoid of life because they have recently come into being, as in the case of a natural volcanic eruption, or an artificial new substrate such as dredging waste or a sunken warship. Other areas may have once been full of life, but have been stripped of their biota by storms or pollution, for example. Whatever the starting point, ecological succession describes the sequence in which life arrives in the new area. Originally, succession was seen as an "orderly and directional process resulting from the alteration of the environment by the developing community" (Bram et al 2005), which leads eventually to a stable finale, the climax community. However, it has sometimes proved difficult to illustrate this simple system in nature, and Smith & Baco (2003) describe the three types or models of ecological succession first proposed by Connell & Slatyer in 1977, namely facilitation, tolerance, and inhibition. Note that all three mechanisms are based on competitive interactions.

Facilitation describes the ways that early-arriving species modify the existing habitat making it suitable for later species – paving the way for new-comers. In doing this, the early-arrivals are killed or displaced via competition from the new arrivals. Tolerance describes the situation where early-stage species are less able to tolerate limited resources (food, space, light, etc.) than later-comers, and are outcompeted by the latter species. Inhibition describes individuals of species from all stages, early and late-comers, who once they have managed to establish, inhibit the use of the habitat by others. Perhaps the best way to illustrate succession in the sea is to use artificial habitats which can be studied to provide simple but realistic models of natural ecosystems undergoing succession. It is important to point out that though succession in artificial sites provides very useful insights into what may be expected from natural assemblages, there can be distinct and long-lasting differences between natural and artificial equivalents. For example, Bolam et al (2006) studied the re-establishment of macro- and meiofaunal communities in soft sediments produced by dredging (see Chapter 9) off the southeast coast of Essex, England, and found that community structures were persistently different to those in neighboring but undisturbed (i.e. natural) substrates.

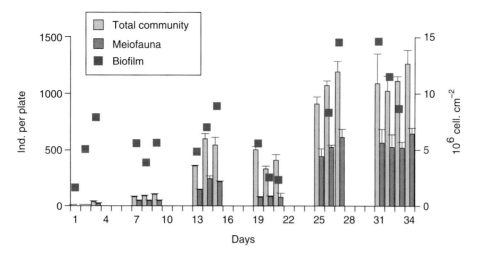

Figure 6.23

Total community (meio- and macrofauna) (light blue bars), meiofauna (dark blue bars), and biofilm (purple squares – right-hand *y*-axis) density variations over 34 days on aluminum plates off Brazil. (From da Fonseca-Genevois et al 2006; reproduced with permission of Springer.)

Nonetheless, we start this section by considering artificial situations. We will discuss in detail the importance of artificial reefs (ARs) in Chapter 10 when we describe marine conservation and restoration. Here we look briefly at two types of AR, those put in place accidentally or purposely to add habitat diversity and complexity to marine environments such as fish aggregation devices and shipwrecks, and those which are located for other purposes, such as pier piles and oil rig legs.

In a simple and short set of trials, da Fonsêca-Genevois et al (2006) suspended 10 cm² aluminum plates in around 3 m of water off the coast of Brazil near Rio de Janeiro, and studied the organisms which appeared on the plates over 34 days. As Figure 6.23 shows, biofilms consisting mainly of bacteria, phytoflagellates, and diatoms appeared on the plates very rapidly, less than 24 hours after the plates were put in position. Meio- and macrofauna composed of many species of barnacle, nematode, copepod, bivalvia, nudibranchia, hydrozoa, and so on also appeared rapidly and numbers increased dramatically over 3 or 4 weeks. It proved possible to partition these species into three separate groups. These groups referred to time (days) from the start of the experiment, and a clear progression was detected for the meiofaunal taxa from the initial phase of 2 days (stage 1), a second phase from days 3 to 10 (stage 2), and a third phase from day 13 onwards (stage 3). Nematodes also showed three phases, though over somewhat different time periods. In essence, what is happening is that very early and rapidly colonizing microbes such as coccoid bacteria appeared first, with slower-colonizing *Pseudomonas*, for example, coming in later. With respect to the fauna, harpocticoid copepods are early arrivers, whilst nematodes increased their density and diversity all the way to the end of the experiment. Relatively large species such as barnacles, even after

Figure 6.24

Encrusting organisms colonized concrete pier legs, southern Sulawesi. (Photograph Martin Speight.)

a short period of time, began to modify the physical environment on the artificial plates by altering water flow over the surfaces and thus changing the habitats for new colonizers.

Anything with a hard and clean substrate if left in the sea for a little while will have organisms settling on it, a topic we shall return to in Chapter 10. Pier piles (Figure 6.24), harbor walls, breakwaters all provide surfaces on which ecological succession can progress. Bram et al (2005) investigated the succession of taxa on the legs of an offshore oil platform in the Santa Barbara Channel, California. Oil platforms or rigs are frequently located in areas of continental shelf that are naturally devoid of

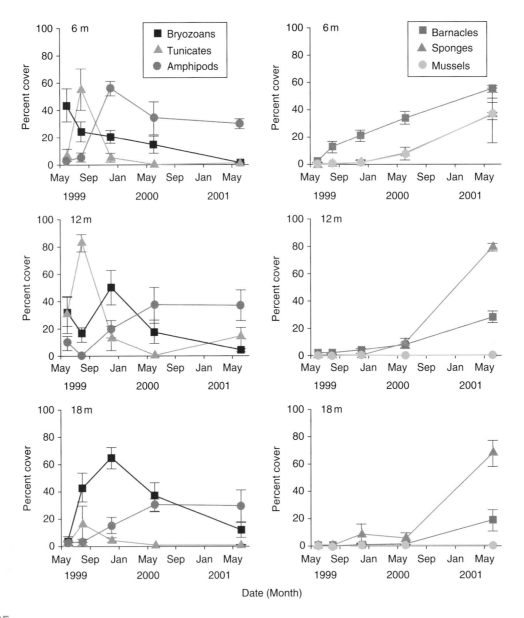

Figure 6.25

Mean percent cover (±s.e.) of various taxa found on tiles attached to an oil platform off the California coast. The tiles were submerged in April 1999 at depths of 6,12, and 18 m, and sampled 2, 4, 6, 12, and 24 months after submersion. (From Bram et al 2005; reproduced with permission of Elsevier.)

complex vertical habitats. Adding the structural and support legs of a rig or platform increases complexity with a concomitant boost in floral and faunal communities (see Chapter 2). In addition, platforms are usually left *in situ* for years (and in fact may be difficult to remove at the end of their useful lives), thus providing plenty of opportunity for succession to eventually reach a stable end point, if there is one. Bram et al attached tiles to their platform as experimental sample units and looked at the succession of faunal assemblages through time and with depth. Some of their results are shown in Figure 6.25. Early colonists included colonial tunicates and encrusting bryozoans, both of which declined after 6 months or so

to be replaced by tube-dwelling amphipods. Between 12 and 24 months after submersion, barnacles, sponges, and mussels were predominant, having increased their cover fairly steadily over the successional process. Notice also that though depth influenced the density and distribution of certain taxa, it did not alter the sequence of succession. The development of the late assemblage on the tiles was predictable and once the barnacles/sponges/mussels stage had been reached (after about 2 years). This stage was also fairly stable, a successional climax had been achieved.

Rather than artificial structures erected in the sea for commercial purposes such as piers and breakwaters, some

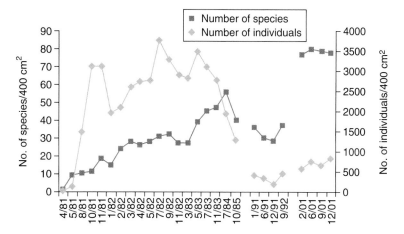

Figure 6.26

Temporal trend in species richness and abundance on concrete artificial reefs in the Tyrrhenian Sea. (From Nicoletti et al 2007; reproduced with permission of Springer.)

are placed deliberately to provide extra habitat for biodiversity conservation and recreation (see Chapter 10). Here we present the results of a 20-year study on the benthic assemblages found on a purpose-built artificial reef in the Tyrrhenian Sea off the shore of Italy. This AR consisted of 2-m concrete cubes arranged in pyramids (four basal and one on top) to create hollows and cavities in an otherwise relatively simple marine habitat (Nicoletti et al 2007). Epibenthic samples were scraped off the pyramids at regular intervals, identified and counted. Some of the results are shown in Figure 6.26, where it can be seen that there was a gradual increase in species richness on the ARs over the sampling period, with a suggestion of an interim peak in the mid-years, whilst the number of individuals (abundance) peaked fairly early in the succession and declined thereafter to relatively low levels. The authors recognized five different phases in this benthic succession: (a) pioneer species recruitment (May 1981 to June 1981) where barnacles, polychaetes, and hydroids appeared; (b) mussel (*Mytilus*) dominance (August 1981 to November 1983) with large numbers of bivalves and barnacles; (c) mussel decline (July 1984 to October 1985) where the physical structure of the habitats encouraged the proliferation of localized soft sediments and their associated fauna; (d) mussel absence (1991 to 1992) with many soft-bottom species and some bryozoa; and (e) return of the bryozoans by 2001 (byrozoans such as *Schizoporella* and *Turbecellepora* which build rigid skeletons (Figure 6.27) and are important "bioconstructors", providing physical hard substrates once more). Perhaps the most important observation here is that there appears not to be a stable assemblage that could be labeled as climax in this successional process.

Figure 6.27

Bryozoan skeletons providing new settlements sites for succession. (Photograph Martin Speight.)

We shall now review some examples of succession in a variety of marine habitats to further illustrate the processes and their resultant communities.

Plankton

The succession of planktonic species in temperate and mediterranean regions shows annual cycles (Figure 6.28) (Aubry et al 2004). As the spring and summer progress, species of phytoplankton replace one another; diatoms, for example, seem to be abundant throughout the year, whereas dinoflagellates were only abundant in June and July. This system is influenced by river run-off, and three periods of growth could be identified, in February, April and June. As might be

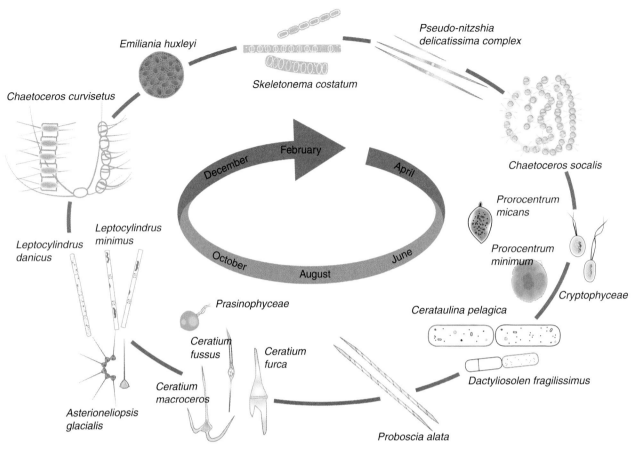

Figure 6.28

Diagram illustrating annual species succession in plankton communities in the Adriatic Sea. (From Aubry et al 2004; reproduced with permission of Elsevier.)

expected, growth was dependent on nutrient levels (see Chapter 3), and though differences in plankton abundance was found according to site, the community structure of phytoplankton was the same, indicating a robust and predictable successional system.

Soft sediments

Succession in soft substrates has been shown to follow another fairly predictable sequence (Pearson & Rosenberg 1978), as illustrated by Nilsson & Rosenburg (2000). These authors studied succession in soft-bottom benthic communities in Sweden, and found that in general, disturbance promotes early successional stage processes, and whilst species richness increases with successional stages, the abundances of single species are most pronounced earlier in succession (Figure 6.29). Notice also that total species richness and biomass is not maximal at the end of a succession (i.e. at climax), but instead peaks at an intermediate level.

Succession in sands and muds can be inhibited by certain groups of organisms (just as sheep or rabbit grazing can

inhibit succession on grasslands) via a process called bioturbation. Lugworms, *Arenicola marina*, are sediment-feeding polychaete annelids that live in U-shaped tunnels in intertidal beaches (see Chapter 4). They can be extremely abundant, as evidenced by their twinned inhalant holes and excreted "casts" at low tide on sandy shores, and in order to produce their "homes" in the sand they tunnel extensively, thereby turning over large amounts of mud and sand sediments which would otherwise zone out into profiles, relatively low in taxa (Volkenborn et al 2007). In doing so, *Arenicola* decreases the accumulation of fine particles and thus maintains the suitability of soft sediments for colonization by taxa from intermediate successional stages. In this context, therefore, lugworms are very important bioengineers, inhibiting succession and promoting species and ecosystem diversity.

Hard substrates

In some situations, resetting the successional stage back to time zero by storms or oil pollution for example will result

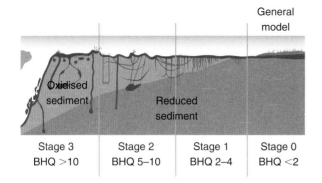

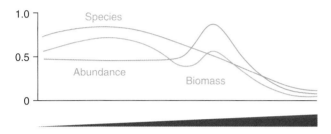

Figure 6.29

Distribution of benthic infaunal successional stages along a gradient of increasing environmental disturbance (succession proceeds from right to left; disturbance from left to right). (From Nilsson & Rosenberg 2000; reproduced with permission of Marine Ecology Progress Series – InterResearch.)

in the system being rebuilt over time to replace the disturbed assemblages just as before. So, we can confidently expect that a rocky intertidal community in England or Wales damaged by oil pollution will recover via rapid and very transitory succession stages to a situation where it is impossible to see that anything ever occurred (see also Chapter 9). However, ecological succession does not necessarily follow exactly the same linear progression to an end point, no matter what the specific starting point is in a particular marine situation. Instead there appear to be multiple stable states available as end points to succession depending on relatively subtle changes in initial conditions (Petraitis & Methratta 2006). Petraitis & Didgeon (2005) explored the concept of divergent succession and alternative states on rocky shores in the Gulf of Maine, by setting up various sizes of experimental plot from which the dominant brown seaweed *Ascophyllum nodosum* was removed. The abundance and cover of the most common species appearing in the cleared plots was measured over 6 years (Figure 6.30). To quote the authors of this work, succession was "very idiosyncratic." Remember that if succession is predictable in that it will always return to the same end state, then an *Ascophyllum*-dominated community should be regained. However, there was no sign of this in these experiments. Instead, large clearings in particular became dominated by the brown weed *Fucus vesicu-*

losus and the barnacle *Semibalanus balanoides*. Crucially, this dominance appeared permanent.

Coral reefs

Succession on coral reefs is a vital competent of restoration and rehabilitation (see Chapter 10), and in some circumstances we are able to predict the end-point of succession with some confidence. Species of reef fish for example come and go during succession on reefs on the Kenyan coast (Indian Ocean) (McClanahan et al 2007a). Labrids (wrasses) and scarids (parrotfish) show a rapid rise initially, followed by a more gradual increase in balistids (loaches) and acanthurids (surgeonfish and tangs). Species richness of fish on recovering reefs reached an asymptote at about 10 years, at which point the community could be called a stable climax. The appearance (or re-appearance) of coral species on a degraded reef depends on the complex associations between herbivore pressure and algal competition as well as the influence of fishing pressure (McClanahan 2008). Figure 6.31 depicts the progressions through time (so-called trajectories) from 1992 to 2004 for three types of Kenyan coral reef, unfished or protected reefs (P), reefs where fishing has ceased at the start of this works in 1991 (MSA) and fished or unprotected reefs (UP). Detrended correspondence analysis (DCA) is a type of multivariate analysis which produces an ordination of samples and species which is particularly suited to data forming a gradient or sequence. It tends to place close together sites with similar species and species with similar habitat requirements. It can be seen that the fished reefs (UP) remain fairly constant over the time period, whereas both the newly protected and long term protected reefs show considerable assemblage change with time from the initial change in management. With fishing pressure removed, large herbivores start to return and communities dominated by fleshy or turf algae give way to hard and soft corals and coralline algae.

Note how important the local environmental conditions can be in determining the direction and ultimate fate of succession on a reef. Increasing nutrients from anthropogenic sources such as river runoff, for example, can fundamentally change the end-point of succession and the species which are ultimately dominant. In Figure 6.32, McClanahan et al (2007b) show the results of a study on the secondary succession of algae on coral skeletons on the Belize Barrier reef, using four fertilizer treatments (pure nitrate – N, pure phosphate – P, nitrate/phosphate mixture – NP, and an untreated control – C). Dead, clean plates of acroporan coral were allowed to be colonized over 56 days during the Caribbean summer, and the final assemblages of algae were compared, again using DCA. Assuming that the control treatment represents the natural successional end point, it can be seen that frondose brown algae (*Dictyota* and *Padina*) should dominate. Nitrogen treatment had the most

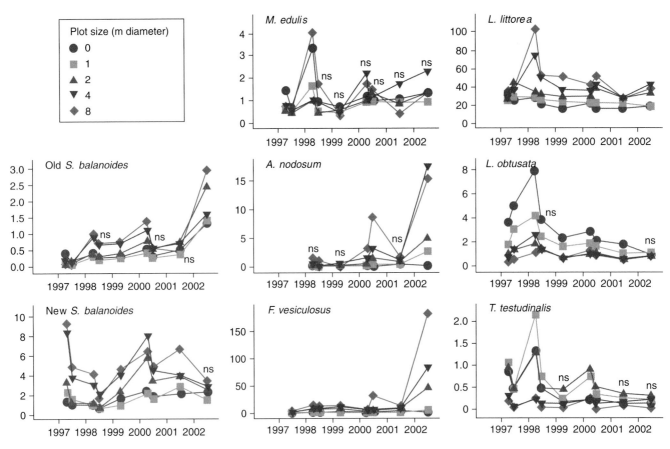

Figure 6.30

Mean densities of barnacles (*Semibalanus balanoides*), mussels (*Mytilus edulis*), seaweeds (*Fucus vesiculosus* and *Ascophyllum nodosum*), periwinkles (*Littorina littorea* and *L. obtusata*), and limpets (*Tectura testudinalis*) over 6 years in various sized (0- to 8-m diameter) plots initially cleared of *A. nodosum*, in the Gulf of Maine. All densities except barnacles are numbers per 0.25 m²; barnacles are numbers per 4 cm². ns = no significant effect of plot size. (From Petraitis & Dudgeon 2005; reproduced with permission of Elsevier.)

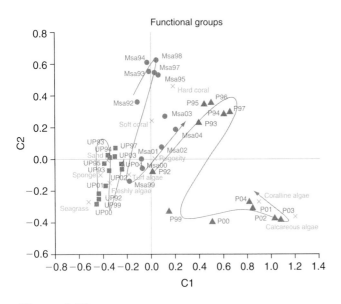

Figure 6.31

Detrended correspondence analysis of major functional taxon groups found in unprotected fishery closures (P), new enclosures (MSA), and fished or unprotected reefs (UP). Lines and arrows indicate the direction over time of the three groups. Numbers indicate years. (From McClanahan 2008; reproduced with permission of Springer.)

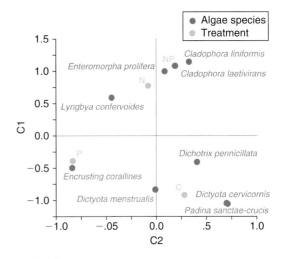

Figure 6.32

Detrended correspondence analysis of the dominant algal taxa in four experimental treatments. C = control; P = phosphorus addition; N = nitrogen addition; NP = nitrogen and phosphorus addition. (From McClanahan et al 2007a; reproduced with permission of Elsevier.)

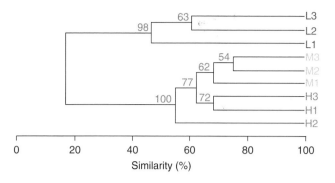

Figure 6.34

Cluster analysis of pair-wise Bray-Curtis similarity coefficients between *Riftia*-mimicing artificial aggregations of PVC tubes along a productivity gradient. H1, H2, H3 = high productivity zone; M1, M2, M3 = intermediate productivity zone; L1, L2, L3 = low productivity zone. (From Govenar & Fisher 2007; reproduced with permission of Wylie-Blackwell.)

distinctive effect on algal succession, with the assemblages under N (and NP) treatments becoming dominated by green turfs consisting of *Cladophera* and *Enteromorpha* species. Organic pollution of reef water can clearly be expected to alter significantly the likely success of coral reef restoration programs (see Chapter 10).

Deep-sea

Succession in the deep-sea has had a fascinating if short history of research. Hydrothermal vents and cold seeps are both regions of immense novelty, being hard to access, highly endemic, isolated geographically and genetically, and at least in the case of vents, very short lived. Hydrothemal vents are typified by the large vestimentiferan worms such as *Riftia* (Figure 6.33) and *Ridgia* species, or vent mussel beds, *Bathymodiolus* species, which have been labeled as foundation species because they provide so much of the community structure, habitat diversity, and primary production (Dreyer et al 2005; Govenar & Fisher 2007).

The establishment of vent worm communities on new or cleared hydrothermal vent substrates is thought to be a clear example of facilitation succession (see above) (Govenar et al 2004). In 1998, these authors cleared all *Riftia pachyptila* individuals from sites on the East Pacific Rise. After only one year post clearance, the sites were colonized again by *Riftia*, even though other early successional species such as the bivalve *Bathymodiolus* had an equal opportunity to establish. Another 24 invertebrate species had become associated with the tube worms, which included various gastropods, many polychaetes, plus a few crustacea and bivalves. *Riftia* had modified the vent habitat, facilitating the colonization and subsequent succession of other species. The aggregations of species associated with *Riftia* communities vary with productivity, so that for a fixed physical complexity (determined by the vent worm provisions of ecospace),

the density and diversity of species assemblages increases with high vent productivity (Govenar & Fisher 2007). Artificial PVC tube clusters which mimicked young *Riftia* worm communities were placed in areas of varying productivity determined by the distance from active diffuse hydrothermal flow, and hence the levels of chemosynthetic primary production (see Chapter 3). Results after 1 year are presented in Figure 6.34 and show that communities varied both in species composition and abundance, and that low productivity communities were particularly distinct from the others. Clearly, the end point of succession after a year is based not just on physical complexity, but also nutrient load, as we also showed earlier in the case of coral reefs.

We now return to the concepts of facilitation and inhibition in ecological succession, as exemplified by hydrothermal vent communities. Mullineaux et al (2003) studied vent sites at approximately 2500 m on the East Pacific Rise using taxa accumulating on cubic basalt blocks placed on sites of varying productivity (as in the previous example) using the DSV *Alvin*. This work suggests that facilitation of sessile species may be due to settlement cues and/or refuges from disturbance, whilst inhibition of mobile species may be due to disruption and competition from other mobile individuals, impediments to grazing imposed by other sessile individuals, and/or direct effects of fish predation. Put simply, primary succession on the basalt blocks showed that biological interactions during the early development of vent communities significantly modified initial patterns of settlement, a process leading to predictable and uniform trends in habitat characteristics and community structures. Finally, Cordes et al (2006) studied succession on hydrocarbonseep sites on the upper Louisiana slope of the Gulf of Mexico, and concluded that succession is typified by a reduction in biomass and a shift in trophic structure from endemic primary consumers to nonendemic higher-level predators. In essence, though early stage settlement and establishment may consist of multiple community types,

Figure 6.35
Composite picture of a whale-fall community. (Photograph courtesy of Monterey Bay Aquarium MBARI.)

these authors suggest that succession progresses consistently towards similar late stage systems no matter what the composition of the initial colonizers.

Our final discussion of succession in the sea involves the hugely enigmatic phenomena of shelf, bathyal or abyssal food falls (megacarrion, or whale-falls) (Kemp et al 2006) (Figure 6.35). Whale-falls are clearly intense point sources of organic enrichment on the deep-ocean floor (Dahlgren et al 2004), since abundant sources of food are likely to be few and far between in most of these environments. Smith & Baco (2003), for instance, suggest that the body of a 40-ton whale which sinks to the bottom of the sea is equivalent to the general rain of detritus sinking from the surface onto one hectare of sea floor over 100–200 years. In other words, an enormous super-abundance of food for any organism in the desert of the ocean floor that is able to (a) locate and (b) utilize these riches, whilst out-competing all others. Once a large whale arrives on the sea floor, species accumulate on the carcass over a period of months and indeed years (Figure 6.36) (Braby et al 2007). Although the accumulation of species on whale-falls seems from the graphs to be fairly smooth, scientists have recognized three (or may be four) fairly predictable successional stages, as described by Smith & Baco (2003). First, a mobile-scavenger stage lasts for a few months to a year, where sleeper sharks, hagfish, rat-tail fish and various invertebrate scavengers remove whale soft tissue at rates of between 40 and 60 kg per day. Second, an enrichment opportunist stage which again may last from months to years. Sediments organically enriched by the actions of stage one animals (facilitated) are colonized by dense (up to 40,000 per m²) errant polychaetes and crustaceans. Third, a sulfophilic stage is reached which can last for decades. Here, a large and species-rich assemblage of taxa lives on the whale skeleton as it emits sulfides derived

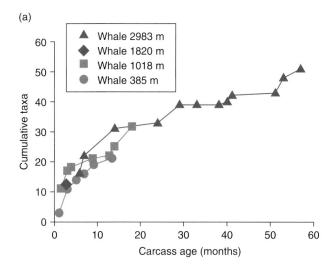

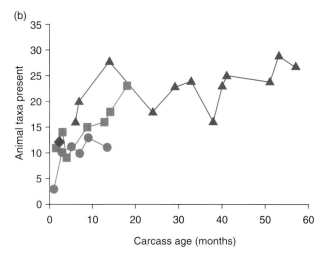

Figure 6.36
Taxonomic diversity of whale-fall communities. Numbers refer to depth of carcass. (From Braby et al 2007; reproduced with permission of Elsevier.)

from the breakdown of bone lipids. A fourth and final successional stage has been proposed for whale-falls, a reef stage, where once all the organic material has been utilized, the remaining physical structure of bones provides a hard substrate for the settlement of suspension feeders which benefit from altered current flows around and through the skeletal structure.

Rather surprisingly, the fourth, reef, stage has yet to be found in the deep-sea, but it has been located in shallower whale-fall. If cetaceans die in relatively shallow water, their bodies will sink into regions of the sea that are not as deep as conventional whale-falls, but nonetheless appear to demonstrate similar successional progressions to those in deeper water. Fujiwara et al (2007) studied the succession on sperm whale-falls in around 250 m of water in the northwest Pacific, and found that the sulfophilic stage was in evidence a mere year and a half after the whale carcass arrived, as evidenced by the huge biomass of bivalves of the genus *Adipicola*. These molluscs harbor chemosymbiotic microbes which utilize the sulfides produced by the whale carcasses. Within 2 or 3 years though, suspension feeders such as crinoids, basket stars and cnidarians have colonized the whale skeleton, representing a clear reef stage.

We have perhaps jumped a little ahead in this chapter by discussing succession in relation to competition. Before organisms can colonize and undergo ecological succession they have to get to their settlement sites, and the next chapter therefore discusses dispersal.

Chapter 7

Dispersal and settlement

Ecologists tend to focus on adult organisms, in part because they are able to reproduce and thus more likely to contribute to future generations than juveniles, but, perhaps also because they are bigger, more easily caught and counted, and often more attractive than larvae and juveniles. However, both on land and in the seas, propagules in their vast variety of forms very much dictate the success of a species or individual population. Figure 7.1 illustrates some of the major types of developmental life history. As can be seen, there is a range of tactics and strategies employed by marine organisms for egg production, juvenile dissemination, settlement and subsequent recruitment into the next generation of adults. The larval stage may be long, short, or nonexist. In this chapter, we discuss the different systems that have evolved for fertilization of gametes, release of eggs and larvae, larval ecology in terms of time scales and locations, metamorphosis into a juvenile form and habit, and finally how the new adult habitat is located and selected.

Dispersal

The movement of members of a species or population from one place to another occurs for a number of important reasons. The survival of a marine species in fluctuating, often unpredictable, environments has genetic, ecological and evolutionary consequences. There are many reasons to disperse, which include genetic mixing and outbreeding, the seeking out of pastures new as all habitats are eventually lost, the departure from ephemeral or dwindling resources, the reduction in competition, colonization, adaptive radiation and the avoidance of predators or the buildup of parasite numbers.

Many marine species have pelagic larvae which spend variable times in the plankton (Strathmann 2007), whilst others have no free-living larval stage. Direct developer adults produce offspring that are essentially small versions of themselves. Female octopuses, for example, lay eggs that hatch into small octopuses. Dog whelks lay large eggs, relative to their body size, which again hatch into little whelks; there is no larval stage of any sort (Colson & Hughes 2004). Many sharks such as tiger sharks and hammerheads give birth to live, fully formed pups, whilst others such as cat shark and dogfish lay eggs within which the young sharks develop and grow externally, eventually hatching once again into small versions of the adult. Some species of marine organisms however produce eggs which are fertilized either inside the maternal body or in the water column, which hatch into very different creatures from the adult. These larvae may settle back down again almost immediately, as with many sea squirts, for example, or alternatively, may spend protracted periods in the plankton as do the larvae of many crabs and barnacles. As mentioned above, some larvae may be endowed with energy sources from the parent, whilst others have to forage in the plankton to survive and grow. The time spent in the plankton will to some extent dictate the distances over which the larvae can disperse.

Dispersal may take individuals great distances from where they started, never to return, or it may remove them from their points of departure only temporarily, to experience new and profitable habits and habitats before returning to where they were born. All these tactics have their advantages and disadvantages. Having no larval stage at all

Marine Ecology: Concepts and Applications, 1st edition. By M. Speight and P. Henderson. Published 2010 by Blackwell Publishing Ltd.

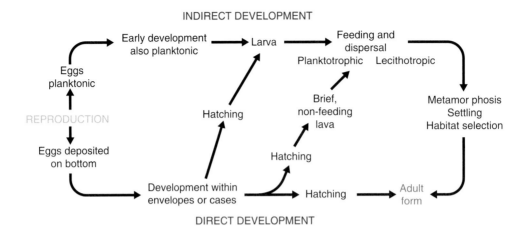

Figure 7.1

Patterns of development in marine animals.

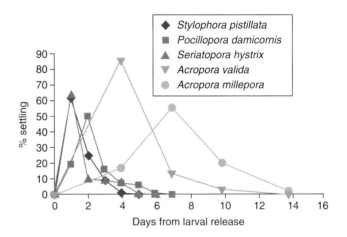

Figure 7.2

Brooders versus broadcasters – time from release of larvae to settlement in five coral species. Diamonds, squares, and triangles = brooders; Inverted triangles and circles = broadcasters. (From Ayre & Hughes 2000; reproduced with permission of Wylie Interscience.)

means that the risk of dying from predation or getting lost before settling is removed, but the likelihood of competing and indeed breeding with close relatives is much enhanced. Even closely related species can show very different strategies. For example, Ebert (1996) compares the modes of reproduction and development for various species of starfish. Of the seven species studied, two were asexual, two retained the eggs in their bodies, releasing live young (brooders), one species laid eggs which remained benthic until the young starfish hatched, one species produced eggs which hatched into larvae which entered the plankton but carried with them yolk from the parent (lecithotrophic), whilst the final species also had planktonic larvae but without a parental energy supply, having instead to feed on other members of the plankton (planktotrophic). All these different strategies may have evolved more than once even in closely related taxa (Byrne 2006).

Ronce (2007) recognizes three stages in movement by dispersal: (a) departure stage (emigration); (b) vagrant stage (traveling); and (c) settling stage (immigration). We shall consider each of these in turn.

Departure

Brooders and broadcasters

Marine species that employ some form of dispersal mechanism in their offspring, even over very short distance and time scales, can be divided into two categories depending on how they care for their young before departure from the parent. These are brooders and broadcast spawners. In Figure 7.2, Ayre & Hughes (2000) illustrate the time to settlement from the plankton onto benthic habitats of five species of coral. Three species, *Stylophora pistillata*, *Pocillopora damicornis*, and *Seriatopora hystrix*, show peak settlement

one or at the most two days after release. The eggs and the larvae hatching from them have in fact been brooded inside the parent coral until almost ready to settle as sessile young corals. In contrast, *Acropora valida* and *Acropora millepora* do not show larval settlement for several days or even more than a week after release by the parent. Broadcast spawners in general may either release just-hatched larvae, or alternatively fertilized eggs which hatch in the water, or they may even shed sperm and unfertilized eggs separately into the water where fertilization takes place away from the adult. All combinations seem possible. Whatever happens, the resulting larvae may spend a considerable time in the water gradually maturing, and those that do are much more likely to travel long distances before settling. We shall return to the topic of planktonic larval duration later in the chapter.

Eggs

Eggs which are not brooded inside the parent are either released into the water (pelagic) or laid close to or attached to substrates (demersal). Pelagic eggs tend to be produced by species in the open ocean, where long-range dispersal of propagules has been selected for. Demersal eggs on the other hand tend to be produced when dispersal is limited and indeed, disadvantageous. For example, Lazzari (2001) reports on a complex of fish species including herrings (*Clupea* spp.), gunnels (*Pholis* spp.), sculpins (*Myoxocephalus* spp.), and shannys (*Ulvaria* spp.) that live around the Gulf of Maine. All these fish lay demersal eggs, and use protected areas of coastline as nurseries. In this system, pelagic eggs would be removed from the local area by currents and the integrity of the nurseries would be lost. Larvae that hatch from both types of egg tend to be equally good swimmers (Leis et al 2007), but those from pelagic eggs tend of course to be much more at the mercy of currents.

Fecundity would be predicted to differ between species that produce demersal rather than pelagic eggs. Huge losses might be expected during a pelagic life, both from predation and also simply because they enter unsuitable habitats, so large numbers of eggs must be produced to ensure that a few at least survive the larval stage to become juvenile recruits. Barnacles, such as *Semibalanus balanoides* for example, produce between 20,000 and 30,000 eggs per adult (Hills & Thomason 2003). In comparison, other crustacea such as Norway lobster (or Dublin Bay prawn or scampi or langoustine), *Nephrops norvegicus* (Figure 7.3), has a fecundity 10 times less than barnacles, between 2000 and 3000 per adult depending on adult female body weight (Figure 7.4) (Briggs et al 2002). However, both species have a planktonic larval stage (nauplii in barnacles, zoeae in decapod crustacea) which in both cases can last 4–8 weeks. Unlike *Semibalanus*, however, *Nephrops* has large larvae capable of relatively high swimming speeds (dos Santos & Peliz 2005). As a deep-sea example, Kosobokova et al (2007) studied the reproductive biology of calanoid copepods from the Arctic Ocean, and found little difference between the fecundity of egg brooding and broadcast spawning species. Clutch sizes for both were found to be in the range of 1 to 90, much less than in the shallow water examples. In general, it was found that deepwater broadcasters produced much larger eggs than similar species occurring near the surface, which the authors attribute to the reduction or even elimination of the need to feed during the larval stage spent in deep water where planktonic food can be very scarce. In another example, Tyler et al (2008) measured the fecundity of various species of gastropod molluscs on hydrothermal vents in both the East Pacific and the Mid-Atlantic Ridge. Fecundities varied from around 200 to a maximum of 1800 depending on species, but with no discernable pattern that could be related to ecology. So, it appears that fecundities of marine animals

Figure 7.3

Dublin Bay prawn, scampi, *Nephrops norvegicus*. (Photograph Paul Naylor.)

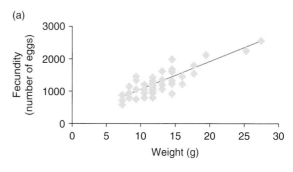

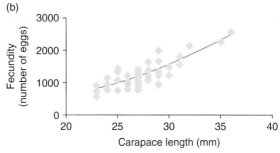

Figure 7.4

(a,b) Fecundity of female *Nephrops norvegicus* in captivity in relation to size of females. (From Briggs et al 2002.)

vary tremendously, and the success of each strategy probably depends more on the biology and ecology of the hatching larvae than on the eggs themselves.

Larvae

It has been argued that larval stages are the key to understanding the success of marine species, especially when it comes to conservation and exploitation (Levin 2006). As

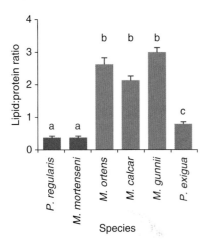

Figure 7.5

Mean egg lipid:protein ration (±s.e.) of six species of star fish with different types of larvae. Blue bars = planktotrophic; green bars = planktonic lecithotrophic, purple bars = benthic lecithotrophic. a,b, and c denote significant statistical differences (Tukey–Kramer test). (From Prowse et al 2008; reproduced with permission of Springer.)

we mentioned earlier, marine larvae can be split into two groups, the planktotrophs who have to feed for themselves, and the lecithotrophs who do little or no feeding, relying on energy stores supplied by the adult. For the former group, life in the plankton concerns not just movement from release to settlement, but also an often protracted period of feeding and growth prior to metamorphosis into a new young recruit for the adult population. For the latter group, larval development and the success mainly depends on the size of the egg from which the larva hatches, since this is directly linked to the available energy supply (Marshall & Bolton 2007). Prowse et al (2008) investigated maternal provisioning by various species of asterinind star fish. *Meridiastra mortenseni* eggs contain very little lipid relative to protein, whereas the planktonic lecithotrophic eggs of *M. oriens, M calcar,* and *M. gunnii* contain much more lipid (Figure 7.5). The benthic lecithotrophic species, *Parvulastra exigua*, is intermediate between the first two extremes. Energetic lipids, in particular triglyceride, are required in abundance by nonfeeding lecithotrophic larvae in the plankton, whilst similar, but benthic, species need less since their larval period is likely to be shorter. Planktotrophic larvae on the other hand have to be supplied with much more protein with which to construct their more complex, actively feeding morphologies.

Planktotrophy is likely to be the ancestral state, from which lecithotrophy has evolved, probably many times. As long as a suitable settlement site can be reached before the fuel runs out, lecithotrophs don't have to run the same risks of prolonged food web engagements in the plankton as do planktotrophs, and since they do not feed, the former group should be able to survive in nutrient-poor or very cold environments such as polar regions. This is still, however, a matter of conjecture (Pearse & Lockhart 2004).

Spawning and hatching

Whether or not the offspring are brooded or broadcast, supplied with energy stores or expected to forage for themselves, the timing of their production and subsequent release from the parents is another important evolutionary choice for marine species. There are many examples of mobile species such as fish, squid, sea stars and spider crabs that form large aggregations of adults who simultaneously release eggs and sperm for external fertilization into the water. Claydon (2005), for instance, reports that spawning aggregations are known from 164 species in 26 families of coral reef fish, and cites various hypotheses for the evolution of such aggregations, from the reduction of predation on individual fish and their newly released gametes, to maximizing the success of fertilization and the ensuing outbreeding vigor. The timing of these aggregations tends to be determined by environmental cues such as the phase of the moon, tidal height, and water temperature. *Lutjanus cyanopterus*, the cubera snapper, is a species of fish in the Caribbean that forms impressive shoals of between 4000 and 10,000 individual adults at maximum (Heyman et al 2005). These schools form tight rising spirals, releasing dense clouds of eggs and sperm into the water. These aggregations occur at fairly regular intervals (at or around sunset) from March to September as shown in Figure 7.6. Spawning is linked to lunar phase and cycle.

If eggs are laid demersally rather than planktonically, the timing of their hatching is also defined environmentally. For example, Figure 7.7 shows the hatching of egg masses of the damselfish *Chromis hypsilepis* in New South Wales (Gladstone 2007b). This species of fish also spawns in large aggregations of adults (up to 33,000 individuals in this study), with the fertilized demersal eggs brooded in nests by the territorial males. After 4 or 5 days, the eggs hatch 3 to 7 hours after sunset, seemingly whatever the state of the tide. So in this case, although spawning patterns appear linked to new and full moons, the release of larvae into the plankton is less determined by lunar cycle or tides, but more by the time of day or night. One advantage of releasing larvae at night is their protection from visually hunting predators, whilst a high tide would take the larvae away from their release site (Park et al 2005).

For sessile species, the release of eggs, sperm and larvae is also often climatically determined. Many coral species for example employ external fertilization, and so have to broadcast their unfertilized eggs and sperm in unison in mass spawning events (Rosser & Gilmour 2008). These spectacular releases tend to occur at one particular time of year. Carroll et al (2006) reviewed the literature on this type of coral spawning, and concluded that gamete release from scleractinain corals usually took place in a relatively brief annual season. On the Great Barrier Reef in Australia, for example, over 140 broadcast spawning species are reported to take part in mass spawning events which occur

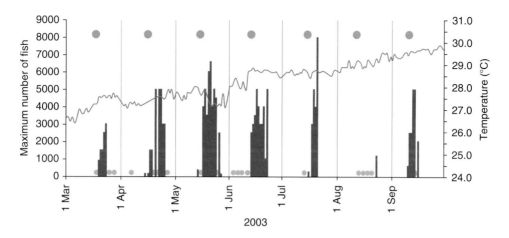

Figure 7.6

Daily maximum abundance of cubera snappers on the Belize Barrier reef in relation to mean daily seawater temperature (line) and full moon (green circle). Blue circles indicate observations with no aggregations. (From Heyman et al 2005; reproduced with permission of Wylie Blackwell.)

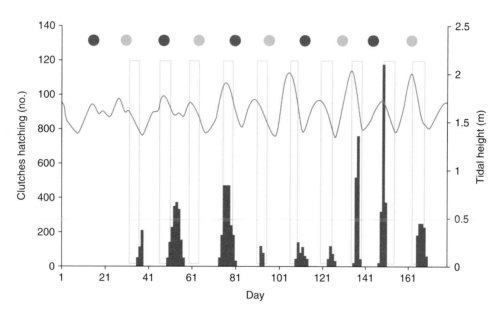

Figure 7.7

The number of egg masses of the damselfish *Chromis hypsilepis* in Australia hatching in relation to maximum daily tidal height and moon phase. Boxes enclose days on which high tide occurred 3 to 7 hours after sunset (times of hatching). Day 1 = September 1, 2004; peaks in tidal height represent spring tides and troughs represent ebb tides. (From Gladstone 2007b; reproduced with permission of Springer.)

in late spring and early summer. Similar events occur in the Caribbean between July and September.

Larval release for brooders may occur annually or, more rarely, monthly. The bryozoan species *Schizobrachiella sanguinea* from the Mediterranean only releases larvae once per year in May and June (Figure 7.8) as water temperatures begin to rise (Mariani et al 2005a). The brooding coral *Madracis senaria* (Figure 7.9.) in Curaçao in contrast releases larvae every month, with maximum release around the last quarter moon (Figure 7.10) (Vermeij et al 2003). This behavior is thought to be unusual amongst corals, which tend in the main to release their larvae on seasonal

rather than lunar cycles (see above). It is possible *M. senaria* is exhibiting a relatively recently evolved system.

Traveling

Once larvae have been released from the parent, or hatched from the egg, their role is to live long enough in order to disperse for a sufficient distance in the right direction to arrive at a suitable place for settlement and recruitment as a juvenile. Several factors combine to achieve this goal, including larval duration, horizontal and vertical movements

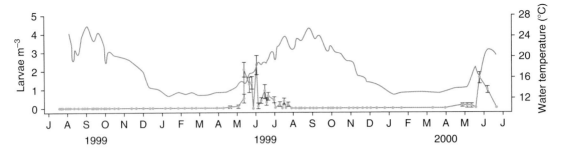

Figure 7.8

Fluctuations in water temperature and abundance (±s.e) of larvae of the bryozoan *Schizobrachiella sanguinea* in the Mediterranean. (From Mariani et al 2005a; reproduced with permission of Springer.)

Figure 7.9

Brooding coral *Madracis senaria*. (Photograph courtesy of E. Weil.)

in the water, connectivity where one habitat or ecosystem leads to another, and retention where larvae are kept in a locale close to where they were released.

Planktonic larval duration

Levin (2006) discusses the crucial links between how far larvae disperse and the time they spend in the plankton (planktonic larval duration, or PLD). Clearly whatever system has evolved will be a compromise between dispersal and survival. As Lesoway & Page (2008) put it, the larvae of marine benthic invertebrates are "small organisms in a vast ocean." PLD is extremely variable. Some species as we have said earlier do not have a larval stage at all, so their PLD is zero. Sponges almost universally have larvae which exist for less than 2 or 3 days (Mariani et al 2006), brooding corals settle perhaps within 2 days of release (Hwang & Song 2007), the flatfish plaice *Pleuronectes platessa* has a PLD of 40 days (van der Molen et al 2007), whilst the

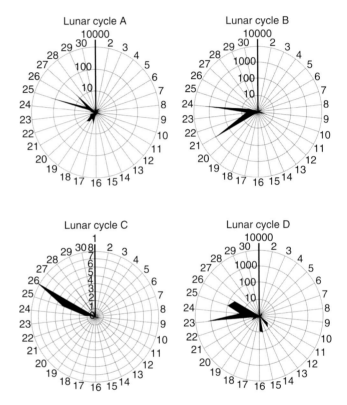

Figure 7.10

Timing of larval release in the brooding coral *Madracis senaria* in Curaçao for four exemplar lunar cycles. New moon = Lunar Day 1; first quarter = Lunar Day 8; full moon = Lunar Day 17; last quarter = Lunar day 23. Logarithmic scale indicates number of larvae spawned by all coral colonies per day. (From Vermeij et al 2003; reproduced with permission of InterReseach.)

larvae of a gastropod mollusc from the northeast Pacific was reared in culture for an astonishing 4.5 years from hatching to metamorphosis (Strathmann & Strathmann 2007). Even within fairly closely related species, PLDs can vary considerably. Bay et al (2006) studied eight species of pomacentrids (damselfish) on the Great Barrier Reef, where their PLDs varied from 0 to an average of 28 and a maximum of 32 days. Over a month at sea gives ample potential for long range dispersal.

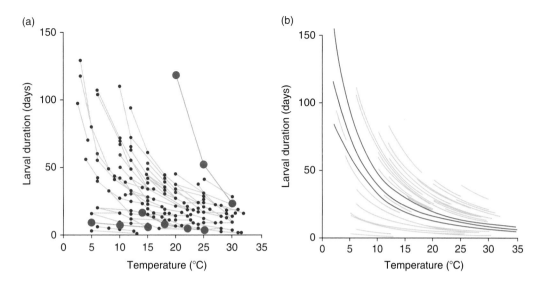

Figure 7.11

Relationship between water temperature and planktonic larval duration (PLD) based on published studies of 72 species (6 fish and 66 invertebrates). (a) Mean recorded larval duration at each temperature for each species; (b) population-averaged (purple) and species-specific (light blue) model fits. (From O'Connor et al 2007; copyright 2007 National Academy of Sciences, USA.)

PLD within one species depends on various environmental factors. Food for planktotrophs will dictate how long they take to reach a stage sufficiently advanced for metamorphosis, and water temperature will greatly influence development rate and larval duration. Figure 7.11 shows the relationship between water temperature and PLD for 72 species of fish and invertebrates derived from the literature (O'Connor et al 2007). It follows that because warm water accelerates larval development, everything else being equal, dispersal distance decreases rapidly as temperature increases. Sanford et al (2006) looked at the relationship between PLD and sea surface temperature of just one species, the fiddler crab, *Uca pugnax*, around Cape Cod in Massachusetts. In decapod crustacea, the main feeding and dispersive larval stage is the zoea, which metamorphoses into the megalopa larval when ready to settle in a suitable habitat (see below). The authors found that the duration of the zoea stage was around 15 days at 26°C, 20 days at 22°C, 26 days at 20°C, and 35–40 days at 18°C. Below 18°C, very few larvae survived at all. The conclusion of this study was that the most northerly limit to the range of the crab was determined not by the ecology or biology of the adult, but by the influence of cold water on the larval stage. Such a relationship with cold water may be an advantage in special circumstances however. Hydrothermal vents are characterized by small and isolated patches of warm water, separated by vast tracts of very cold water (see below). In order for sessile organisms such as vent barnacles to disperse between separate vents, a very protracted larval stage is required, and the cold water is ideally capable of extending PLD as much as possible (Watanabe et al 2006).

The influence of PLD on population genetics is fairly easy to predict. A long PLD provides the potential for long-distance

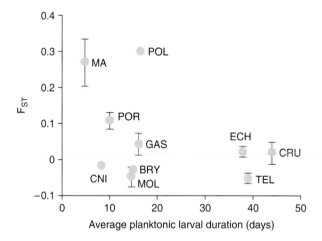

Figure 7.12

Relationship between planktonic larval duration (PLD) and genetic distance for marine organisms. MA = macroalgae and seagrass; POR = porifera; ECH = echinoderms; POL = polychaetes; CNI = cnidarians; MOL = bivalve molluscs; BRY = bryozoans; GAS = gastropod molluscs; CRU = crustacea; TEL = teleosts. (From Bradbury et al 2008; reproduced with permission of The Royal Society.)

dispersal, and so genotypes are able to mix freely. An extended PLD may reduce the likelihood of genetic distinctiveness between populations of a species. As we have seen in Chapter 2, and will again below, scientists have frequently used genetic analyses to comment on dispersal abilities for species. In one example, Bradbury et al (2008) looked at 10 major taxa and related their PLDs to estimates of genetic relatedness using F_{ST} measurements (Figure 7.12). F_{ST} relates the genetic variation in a subgroup or population to the overall genetic variation in that species, and it can vary

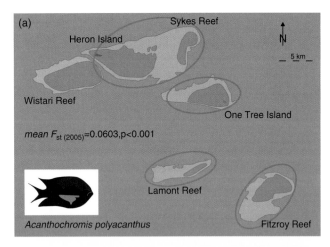

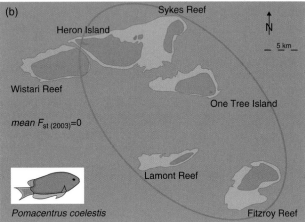

Figure 7.13

Genetic substructure of two damselfish populations across five test reefs on the Great Barrier Reef. (a) *Acanthochromis polyacanthus* which lacks a planktonic larval stage, and (b) *Pomacentrus coelestes* with a three-week PLD. Red circles enclose genetically different populations. (From Gerlach et al 2007; copyright 2007 National Academy of Sciences, USA.)

between zero and one. A higher value indicates more genetic variability. The figure shows clearly that as PLD increases, genetic variability declines almost exponentially, with macroalgae and seagrasses showing least genetic variation, and crustacean and teleost fish the largest. This type of genetic variability can be detected at a much closer level of relatedness between species, as shown by the work of Gerlach et al (2007) (Figure 7.13). Using F_{ST} measures as in the previous example, they analyzed the genetic differences between populations of two species of damselfish. *Acanthochromis polyacanthus* does not have a dispersing larval stage, and showed genetic diversity on four different islands separated by a matter of a few kilometers only. In contrast, *Pomacentrus coelestis* has a planktonic larval stage that can last as much as 3 weeks; the populations of this species on the different islands showed no genetic diversity at all. This is not the whole story, since a third fish species was studied which also had a PLD of 3 weeks. *Ostorhinchus doederleini* however

showed the same level of genetic diversity as *A. polycanthus*, suggesting that its larvae, though planktonic for a considerable time, were drawn back to the reefs of their birth, possibly by chemical cues (see later in this chapter).

Currents

Eggs are of course completely passive in water currents, and even larvae, with some swimming ability, tend to get carried with the ocean currents. The fate of these propagules, at least in terms of getting to a suitable habitat for settlement, is thus determined by current speed and direction. The hope is that wherever spawning or egg/larval release occurs, settlement or nursery grounds will be situated down-current. Behavioral, seasonal and ontogenetic adaptations, such as those discussed above, ensure that this happens. Currents can act in an essentially horizontal fashion, keeping to the same depths. Alternatively they can have a powerful vertical component, moving over considerable depths but in much the same horizontal, geographic, location. They can also combine horizontal and vertical movement.

Take the example of the Cape hake, *Merluccius capensis* and *M. paradoxus*. These two species spawn from June to October on the western Agulhas Bank off the coast of South Africa, when offshore currents are relatively weak (Grote et al 2007). Both eggs and larvae were found commonly in horizontal currents flowing in a northwesterly direction, as indicated in Figure 7.14. At each sampling station on the transect (labeled 1–20 in the figure), larvae were collected and their lengths measured. Back-calculations were then carried out using previously determined growth rates and current velocities prior to sampling to estimate the date of spawning and the distance up-stream of the event. So, for example, larvae sampled at station 16 were estimated to have been spawned on October 4, 1996, approximately 75 km up-current. The nursery grounds for these fish are situated off Cape Columbine, just off the northern range of the map in Figure 7.14. Grote et al conclude from these results that Cape hake have developed an efficient spawning strategy to avoid offshore losses of their eggs and larvae; they employ currents to take the larvae to the nursery grounds, but only at times when the currents are relatively weak.

A similar example comes from various spring-spawning fish in the Irish Sea. Van der Molen et al (2007) simulated the dispersal of eggs and larvae of fish such as cod, *Gadus morhua*, and plaice, *Pleuronectes platessa*, using particle tracking methods and current flow models. Field sampling was also used to verify the modeled data. Firstly, it was clear that the models fitted the observations rather well, and so as the authors suggest might be used for predictions of fish population dynamics and recruitment by fisheries scientists. Secondly, it was found that some plaice larvae tended to move eastwards across the Irish Sea from spawning grounds

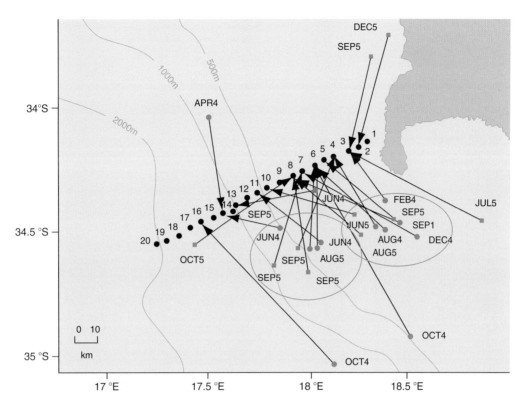

Figure 7.14

Modal spawning distance during the months of high larval abundances for Cape hake, *Merluccius* spp., off South Africa, in 1996/1997 (circles) and 1997/1998 (squares). Numbers indicate stations on the sampling transect; arrows indicate spawning distance (note 10-km scale) and current direction; ellipses indicate concentrations of larvae originating on shelf and upper slope. (From Grote et al 2007; reproduced with permission of Elsevier.)

off Ireland to nursery areas off the west coast of northern England. A second spawning area could be identified closer to the English coast. This study concluded that eggs and larvae of these fish species typically remained within 160 km of their spawning origins, though some traveled with the currents for up to 300 km.

Horizontal currents do not just occur in shallow waters. Deep-sea currents are likely to be extremely important in the transport of propagules between a variety of ephemeral and widely scattered phenomena such as whale falls, hydrothermal vents, and cold seeps. Such habitats can be short-lived so that the duration of a vent may be no more than the lifespan of an organism inhabiting it (Shank & Halanych 2007). Further, the environment of the deep-sea between the vents can be very inhospitable, so highly efficient dispersal systems have had to be evolved to move from one vent to the next, at regular and frequent intervals. Deep-water currents show predictable trajectories and velocities. Figure 7.15 shows current vectors calculated from circulation models at three depths on the northeast Pacific Ridge system (Young et al 2008). The relatively strong and unidirectional stream is fairly constant at depths of 2000–3000 m, whilst a counter current with a similar speed moves north up the coast at around 1500 m. Thus, the distance and direction of transport of propagules of marine organisms will

depend on their depth. Marsh et al (2001) estimated the lifespan of the larvae of the giant vent worm, *Riftia pachyptila*, from hydrothermal vent communities at 2500 m on the northeast Pacific rise. From knowledge of the rate of energy use during early larval development, a typical larva could potentially survive for 38 days (±4). Current velocities and directions were measured using current meters positioned at around 175 m above the bottom on the ridge axis, a height at which vertical plumes of convecting warm water would reach, bringing larvae up from their parental release sites. At this height, larvae are picked up by the horizontal currents whose residual flow direction is along the ridge axis (in this case SSE) (Figure 7.16). These horizontal currents undergo periodic flow reversals, which coupled with the estimated mean lifespan of larvae indicates that the maximum dispersal distance from one vent community would be around 100 km. Because of the current flow reversals, most larvae appear to travel fairly small distances, so that they tend to be retained to repopulate local vent sites. To assess the reality of these dispersal mechanisms, scientists have employed genetic analyses (see also elsewhere in this book). Shank & Halanych (2007) took the genetic fingerprints of 29 *Riftia pachyptila* from five vent fields over a distance of up to around 5000 km along the East Pacific rise. They found six separate assemblages (clades)

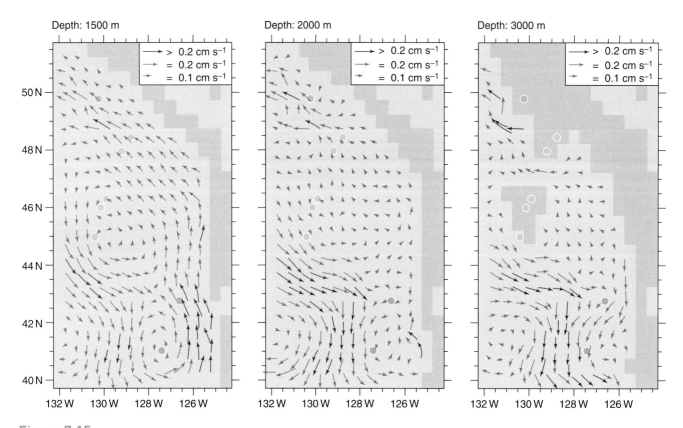

Figure 7.15
Current vectors calculated from circulation model at three depths in the region of the northeast Pacific ridge system. (From Young et al 2008; reproduced with permission of Wiley-Blackwell.)

which related to sampling localities and vent regions (Figure 7.17). The results suggest that genetically distinct vent worm communities can exist less than 10 km apart, implying that larval dispersal does not usually occur over large distances. It might be possible that closely related vent larvae from a single vent could be carried (entrained) together in a warm water plume, and eventually settle together on a new vent system still in the same genetic assemblage.

Diel vertical migration (DVM) is very common amongst marine planktonic organisms. Zooplankton of many kinds are able to swim up and down in a water column according to environmental conditions such as light, temperature, phytoplankton abundance, and the presence of predators. Krill, *Meganyctiphanes norvegica*, are an excellent example of DVM behavior, as can be seen in Figure 7.18. Onsrud et al (2004) used echosounders to study the distribution of krill and their fish predators with depth over 24-hour cycles in Norway. It is clear that krill spend the daylight hours in deeper water (in this case around 100 m), and rise to the surface at sunset, probably as a form of anti-predator behavior. This example doesn't involve a particular dispersal mechanism, but if we combine DVM behaviors with currents in the sea which move vertically themselves, it is possible to determine the distance and direction of larval dispersal, as shown for the larvae of the shore crab, *Carcinus maenas*, off the Atlantic coast of Spain and Portugal

(Queiroga et al 2007). The larvae make diurnal vertical migrations controlled by the times of sunrise and sunset, spending the day on the bottom and the night near the surface, in a similar fashion to the krill example above (Figure 7.19). Larvae on the surface tend not to be much affected by upwelling or downwelling events which take place at night, and so they tend to stay where they are relative to the shore line. However, an upwelling which takes place during the day will gather up crab larvae on the bottom and carry them upwards and inshore, thus retaining them in coastal waters. A downwelling during the day will do the reverse and take them out into deeper water. As shore-living animals, the latter process, especially towards the end of the planktonic larval stage, is to be avoided if possible.

Vertical currents also occur in the deep-sea, frequently driven by the warming and cooling effects of water circulation around hydrothermal vents. Larvae released from benthic organisms on the vents are carried some distance vertically (Mullineaux et al 2005), where they may be collected by horizontal deep-sea currents which we discussed earlier in this chapter. However, in most of the deep-sea there are no such helpful convection currents, and the larvae of benthic adults have to make their own way to the surface waters if they want to take advantage of the abundant food supplies there. In these situations, larvae have to begin life as lecithotrophs because they need sufficient

(a)

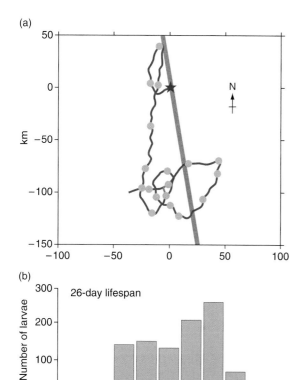

(b)

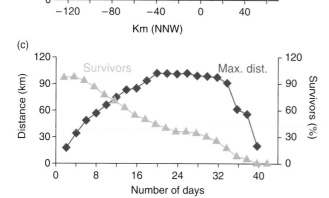

(c)

Figure 7.16

Modeled dispersal potential of the larvae of the vent worm, *Riftia pachyptila*, on the 9° northeast Pacific Rise. (a) Plan view of the trajectory of a water parcel with its origin (star) at the location of the current meter and its path calculated from hourly averaged velocities (filled circles indicate positions at weekly intervals; grey line denotes the ridge axis). (b) Along-axis distances (NNW is positive) traveled by larvae with a selected 26-day lifespan. (c) Maximum distance dispersed and percentage of survivors for each specific larval life span. (From Marsh et al 2001; reproduced with permission of *Nature* – MacMillan.)

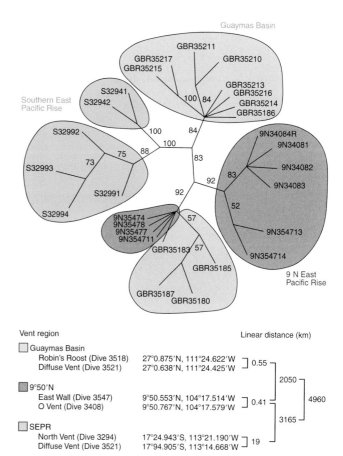

Figure 7.17

Parsimony amplified fragment length polymorphism (AFLP) fingerprint relationships amongst individuals of *Riftia pachyptila* sampled from two localities in each of three regions on the East Pacific Rise. (From Shank & Halaynch 2007; reproduced with permission of Wylie Interscience.)

energy to swim to the surface layers before they can switch to planktotrophy. As long as the water is cool (as in spring), and temperature boundary layers have not yet developed, these larvae are quite capable of making vertical migrations of 1000 m or more. Such is the case of the deep-water brittle star, *Ophiocten gracilis*, found at depths of 800 m or more in the North Atlantic (Sumida et al 2000). Synchronous spawning of males and females occurred on the bottom in

late February to early April, and fecundity ranged from 23,166 to 50,820 tiny eggs (average diameter 56.6 μm). Young larvae rose to surface waters where they reached densities of around 1000 per m³), to take advantage of phytodetritus from the spring bloom of phytoplankton. Metamorphosis took place in the water column, and rapid sinking rates of about 500 m per day occurred during mid-May. Post-larvae settled back on the bottom with a disk diameter of 0.6 mm. So the dispersal of propagules in the sea is a combination of active swimming and passive drifting, of vertical and horizontal movements, and small- to large-scale distances covered.

Retention

In the light of all these different types of dispersal mechanism, it is important to consider whether propagules set out on the ocean currents, never to return to the natal region, or they return to their origins, or in some cases, never leave. When juveniles are observed in a habitat (Figure 7.20), it is

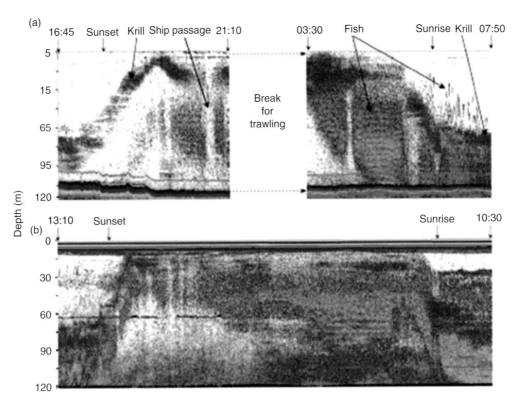

Figure 7.18

Acoustic recordings of the depth distributions of krill, *Meganyctiphanes norvegica* (blue traces), and their fish predators (green/yellow traces). (a) March 4–5, 1998; (B) November 13–14, 2000. (From Onsrud et al 2004; reproduced with permission of Oxford University Press.)

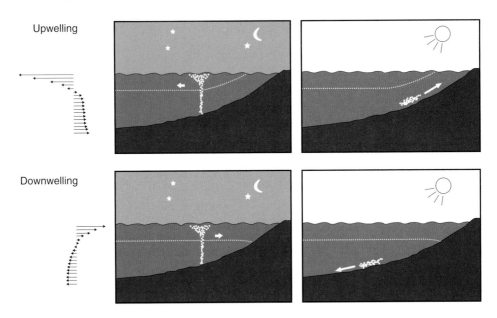

Figure 7.19

Model of across-shelf dispersal of shore crab, *Carcinus maenas*, larvae during upwelling and downwelling. White clouds indicate presence of larvae in day or at night; thick white arrows represent transport of larvae; arrows at left indicate vertical distribution of the across-shelf component of current velocity. (From Queiroga et al 2007; reproduced with permission of Elsevier.)

not clear whether they have arrived there as larvae from far away and recently settled or alternatively, have been there all the time. Indeed it seems that many marine fish species and corals for example remain in, or return to, the vicinity of their natal site (Mora & Sale 2002). This is a process known as retention, which can work via two distinct type of process. First, the offspring released by an adult population may have limited (or zero) dispersal abilities, so essentially they never leave the natal population. Alternatively, dispersal can be fairly prolonged, but the larvae through various behavioral and/or environmental mechanisms, return home after their planktonic wanderings. The planula larvae of the octocoral *Heliopora coerulea* are brooded inside the parent, and only spend a few hours after release before settling back on

Figure 7.20

Juvenile coral fish. (Photograph Martin Speight.)

the substrate. Harii & Kayanne (2003) showed that larval dispersal in this species was extremely limited beyond the limits of their home patch reef, despite the actions of wind, currents, and tides (Figure 7.21). Recruitment of juvenile corals was observed within a mere 350 m of the parent colonies at most, showing high retention. The majority of fish larvae spend at least some time in the plankton. Almany et al (2007) studied the vagabond butterfly fish, *Chaetodon vagabundus*, around Kimbe Island, Papua New Guinea, a tiny series of patch reefs around 0.3 km² in all. This species has an average planktonic larval duration (PLD) of around 38 days, providing considerable potential for large-scale dispersal. The authors injected adult fish with a barium isotope and released them around the island. These fish produced larvae with labeled (tagged) otoliths which could also be detected in newly settled juveniles. New recruits were collected on the reefs around the island, and their natal origins analyzed using the barium tagging. Figure 7.22 shows that a significant number of juvenile butterfly fish were found with tagged otoliths, indicating that they had returned to a very small target reef. Around 60% of the larvae were estimated to have been retained (self-seeded) in the natal reef system despite a long PLD. On a much larger scale, Bailey et al

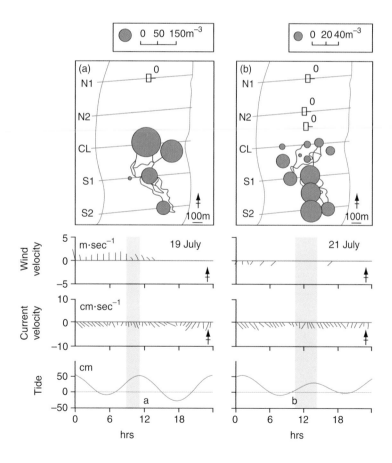

Figure 7.21

Larval dispersal and retention for the octocoral *Heliopora coerulea* in southwest Japan. Number of planulae per m³, currents, wind velocities and tides in July 1999. Bars for current and wind velocity indicate direction; circles indicate number of planulae. Shaded areas indicate periods of plankton netting. (From Harii & Kayanne 2003; reproduced with permission of Springer.)

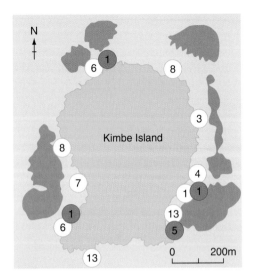

Figure 7.22

Diagram of Kimbe Island showing the locations of tagged (red circles) and untagged (white circles) juveniles of the butterfly fish *Chaetodon vagabundus*, resulting from tagging parent fish around the island 2 months earlier. The number in each circle indicates the number of juvenile fish collected from that location. (From Almany et al 2007; reproduced with permission of *Science* – AAAS.)

(2008) investigated the retention system for juvenile flatfish in the Gulf of Alaska. Fish such as the rex sole, *Glyptocephalus zachinus*, spawn off-shore in fairly deep water, at the edge of, or indeed beyond, the continental shelf. Their nursery grounds however are inshore in shallow water, where the adults also spend much of their lives. In Figure 7.23 the red squares in particular show that spawning occurs in a relatively narrow band offshore, coincident with the continental shelf break. The resulting eggs are clearly broadcast over wide areas of sea, but show concentrations in large patches. Larvae are even more diffuse, but the juveniles return to a few small locations closer inshore. Note also that if we assume equal catchability for all life stages, very considerable losses are incurred between eggs and juveniles. It is likely that oceanographic phenomena such as deep current flow onshore return the eggs and larvae from far and wide to their nursery grounds, retaining the fish populations in local coastal areas.

One example in the sea where retention, or almost guaranteed return to natal sites, involves deep-sea hydrothermal vents. As we have mentioned elsewhere, when these habitats are functioning actively, larvae that leave never to return, risk never finding another suitable site, so it is very important for

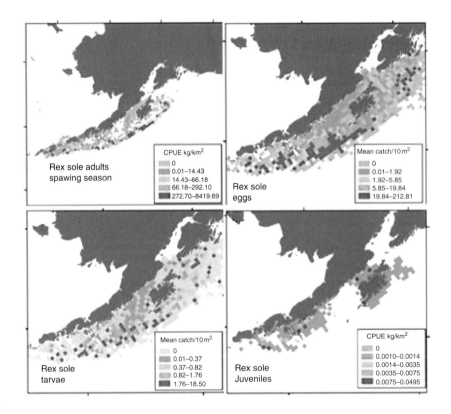

Figure 7.23

Distributions of life stages for the rex sole, *Glyptocephalus zachirus*, in the Gulf of Alaska. Different colors indicate catch per unit effort (CPUE) for each life stage in 20 km × 20 km squares. (From Bailey et al 2008; reproduced with permission of Wylie-Interscience.)

Figure 7.24

Bythograea thermydron. (Photograph courtesy of Susan Mills, Woods Hole.)

larvae of vent organisms to be able to find their way back to their natal origins, or, in the case of these sites disappearing, to find a very similar one (Dittel et al 2008). A nice example of larval retention around vent communities is shown by the behavior of megalopae larvae of the vent crab, *Bythograea thermydron* (Figure 7.24). As mentioned above, larvae from vent species tend to be lifted vertically away from the vent by warm convection currents, which only last for a few tens of meters before cooling to match the ambient abyssal water off-vent. It is crucial for vent crab larvae to find their way back to a vent to settle, and Figure 7.25 shows how they use water temperature variations to find their ways home (Epifanio et al 1999). In cold water, representative of off-vent abyssal conditions, crab larvae swim actively, as fast as the cold water allows, for minutes at a time, but as soon as they find warmer water, they stop swimming and settle on the bottom. The last thing they want having found warm water indicative of an active vent is to get swept away by another vertical current.

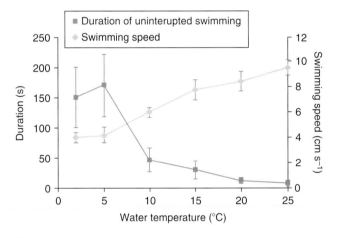

Figure 7.25

Effect of temperature at deep-sea vent communities on the swimming behavior of megalopae larvae of the hydrothermal vent crab *Bythograea thermydron*. (From Epifanio et al 1999; reproduced with permission of InterResearch.)

Connectivity

One final consequence of all the movement described in the previous sections is the potential for populations at isolated sites to be connected with each other. This connectivity has crucial significance for species and populations whose spawning, nursery and adult grounds are spatially separated, and especially for conservation, management and sustainability as lifecycles require each and every habitat stage to be functioning and interconnected (Cowen & Sponaugle 2009). In fact, it has been suggested that marine populations with dispersive stages (usually but not exclusively larvae) should be considered as metapopulations with interconnected subpopulations (Lipcius et al 2008). A graphical model of marine connectivity is provided by Hughes et al (2005) (Figure 7.26). In healthy ecosystems, many habitat patches are connected by dispersal of the larvae of both sessile and mobile adults of a species easily

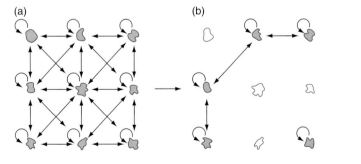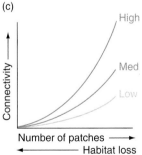

Figure 7.26

Graphic model of larval dispersal among patches of habitat. Arrows depict potential dispersal pathways between adjacent patches and self-seeding within a particular patch. (a) An intact system with high connectivity. (b) A damaged ecosystem showing reduced larval connectivity caused by habitat fragmentation. (c) Nonlinear relationship between habitat loss and the strength of larval connectivity for species with high, medium, and low dispersal abilities. (From Hughes et al 2005; reproduced with permission of Elsevier.)

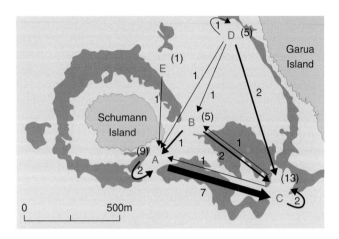

Figure 7.27

Local scale connectivity network of the clown fish, *Amphiprion polymnus*, in Papua New Guinea. The map shows distance and direction of small-scale dispersal of all juvenile fish settling within their natal population. The thickness of the arrows indicates the numbers of juveniles either moving between subareas A to E or returning to the subarea of their birth. Numbers in brackets = adult pairs in each subarea. (From Jones et al 2005; reproduced with permission of Elsevier.)

moving between patches, as well as "self-seeding" back onto natal patches. A fully functional coral reef system would be a good example of this. As individual patches disappear, for reasons such as pollution, disease and storm damage, connectivity declines, leaving one or two isolated patches outside the spatial range of connectivity. This is often the case with recent anthropogenic damage to coral reefs. The situation in a perturbed or degraded system is made much worse when species have poor powers of dispersal, or as Hughes et al put it, species with limited dispersal are more vulnerable to recruitment failure. Connectivity has been demonstrated in clown fish, *Amphiprion polymnus*. Jones et al (2005) labeled all the embryos produced by a population of clown fish on Schumann Island, Papua New Guinea, using tetracycline marking of their otoliths. Three months later, all the young fish settling in anemones were caught and inspected for otolith markers. The movements of young fish from their natal site to their settled site are shown in Figure 7.27, indicating a great deal of habitat (patch) connectivity in the distance range 500–1000 m.

Though the theoretical and conceptual bases for marine connectivity and metapopulation dynamics are beyond the scope of this book, we shall return to the practical and applied importance of connectivity in the restoration and rehabilitation of marine reserves in Chapter 10.

Settlement

So at last, propagules of young organisms are ready to settle in habitats where they can recruit into the adult population and begin the cycle again. Clearly it is crucial that this settlement takes place at the most appropriate time and in

the best environmental conditions to maximize survival, and in the most suitable specific habitats. In short, settlement is a risky business (Doherty et al 2004). For convenience, we shall discuss first settlement times and dates, and second, settlement cues and stimuli, though of course these two will act concurrently to promote recruitment success.

Settlement timings

Just as the release of eggs and larvae by an adult into the sea can be seasonally or diurnally controlled (see above), so is the timing of the settlement of the same marine larvae at the end of their journey, whether it is long or short. In the Dutch Wadden Zee, for example, there is a clear coincidence of settlement of three different genera of bivalve clam (*Macoma*, *Cerastoderma*, and *Mya*) onto soft substrates in the southern North Sea (van der Veer 1998). All three show peak settlement during July. Whatever cues and stimuli the larvae of these three mollusc species are using, the seasonal component at least is shared by all. A similar situation has been found in various species of brachyuran (true) crabs in Alaska (Figure 7.28) (Daly & Konar 2008). In all four crab species shown in the figure, there appear to be seasonal cycles of zoeae which are the planktonic, dispersive stage, and of megalopae that are responsible for finding the right time and place to settle. Clearly, there are again mechanisms in place to guide settlement success. These cues may be ecological (presence of predators or competitors), biological (specific habitat types), physical (wind, tides or currents) or chemical (water or substrate chemistry), or a combination of all three (Huijbers et al 2008).

Settlement cues and stimuli

As would be expected, settlement sites for marine larvae will vary tremendously in quality. Some will be completely unsuitable, others marginally acceptable, and just a very few may be ideal. In addition, quality involves not only the immediate, instantaneous degree of suitability for the specific requirements of settlers, but also the longer term, ongoing support of the growing juveniles once (and often irretrievably) settled. Shima et al (2008) compared a variety of environmental parameters on a coral reef on Moorea in French Polynesia to assess the site quality for settling six-bar wrasse, *Thalassoma hardwicke*. They included parameters such as coral and algal cover, depth, size, shape and isolation of reef patches, and the structure of fish species assemblages. Site quality itself was assessed as the survival time of juvenile wrasse on a particular reef section. In the end, two site parameters were found to be the most important, average density of arc-eye hawkfish, *Paracirrhites arcatus* (a predator of juvenile six-bar wrasse), and the amount (area) of the fine-branched coral *Pocillopora* sp. (a refuge from predation for juvenile wrasse). The fitted response surface in Figure 7.29 shows that the survival time (site quality) of six-bar wrasse is highest when predator density

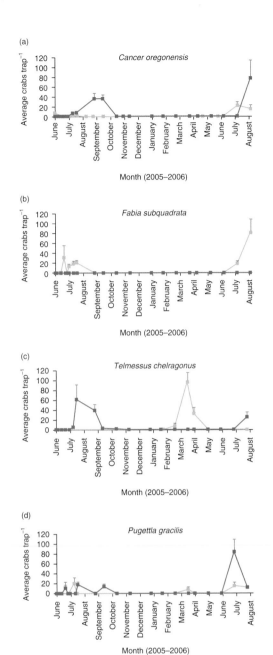

(a) *Cancer oregonensis*

(b) *Fabia subquadrata*

(c) *Telmessus chelragonus*

(d) *Pugettla gracilis*

Figure 7.28

Temporal variation in larval and post-larval abundances of four species of crab in Kachemak Bay, Alaska. Dark blue lines indicate megalopae; pale blue lines indicate zoeae (±s.e). (From Daly & Konar 2008; reproduced with permission of Springer.)

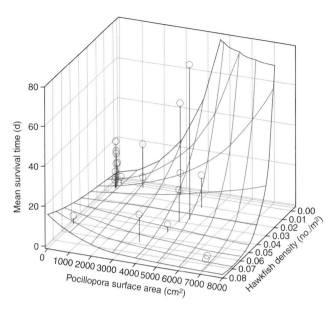

Figure 7.29

Mean survival times (a measure of site quality) of juvenile six-bar wrasse, *Thalassoma hardwicke*, in relation to surface area of branching coral *Pocillopra* sp,. and the local density of hawkfish, *Paracirrhites arcatus*, on reefs in Moorea. (From Shima et al 2008; reproduced with permission of *Ecology* – ESA.)

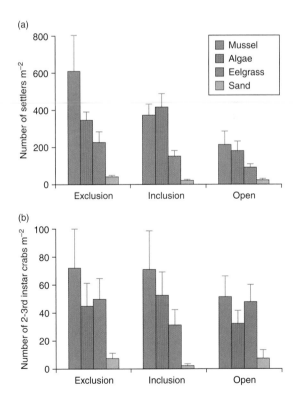

Figure 7.30

Mean (±s.e.). (a) Shore crab settlers (megalopae and first instars) and (b) second to third instar crabs, *Carcinus maenas*, sampled from different habitat types (blue mussel beds, filamentous algae, eelgrass, and sand) and cage treatments (predator exclusion cage, predator inclusion cage, and open uncaged plots). (From Moksnes 2002; reproduced with permission of Elsevier.)

is very low and coral cover very high. The "trick" which larval fish have to accomplish is to find and recognize sites of high quality, however defined, in the limited time available to them. Another example comes from shore crabs, *Carcinus maenas*, in Sweden (Moksnes 2002). In Figure 7.30 it can be seen that settling megalopae larvae favor complex habitats such as mussel beds or algae rather than simple ones such as sand. Recruiting to these types of habitat reduces the risk of predation on the young settlers by providing them with somewhere to hide. Notice that as the juvenile crabs begin to mature, there is some degree of redistribution of individuals between habitats, a luxury unavailable to sessile species such as corals and barnacles.

Once having reached a suitable habitat, substrate characteristics are crucial in the recognition of suitable sites by settling benthic organisms. Substrate types may be physical (sediment load, surface rugosity) or biological (biofilms, conspecifics) or chemical (scents and smells), which we shall discuss in turn. Some larvae have extremely sophisticated sensory and behavioral systems for locating and selecting specific settlement sites. For example, juvenile fish prior to recruitment have a complex array of visual, acoustic and olfactory senses at their disposal (Lecchini et al 2005), but even seemingly simple organisms are able to select the best substrates. Seaweeds (macroalgae) are essential ecospace providers in temperate shallow water ecosystems, but if they settle as planktonic spores on inappropriate substrates, they may not survive, or become unattached as they grow. For example, the success rate of attachment of two macroalgae, the brown seaweed *Hormosira banksii*, and the kelp *Durvillaea antarctica*, in southern New Zealand depends on the substrate (Schiel et al 2006). Even a light dusting of fine sediment on rocky surfaces reduces the attachment success, and heavy sediment, not surprisingly, permits no attachment at all. Sediment deposition on such substrates would be expected to increase in quieter habitats where current flow is reduced, but the sites where wave action clears all sediments from substrates may be too exposed for macroalgae to survive for long. Thus there has to be a compromise for germling settlers between too much sediment and too much wave action. There is also the added problem of grazers (see Chapter 4).

In their turn, algae frequently influence the settlement choices of marine animals, both mobile and sessile. The sponge *Haliclona caerulea* has a symbiotic relationship with the calcareous alga, *Jania adherens* in Mexico, the calcified fronds of the alga permeate the sponge tissue, serving as a skeletal system. It is clearly sensible that *Haliclona* larvae settle within the fronds of *Jania* (Avila & Carballo 2006) in preference to anywhere else. The abalone, *Haliotis rubra*, also seeks out and settles on various types of algae in preference to nonbiological substrates, but further selects certain types of algae above others, as shown in Figure 7.31 (Huggett et al 2005). Encrusting coralline algae and *Jaynia micrarthrodia* are particularly favored. It is possible that the algae themselves maintain and support biofilms of bacteria and other microorganism that are especially attractive to settling animals. Hugget et al (2006) also found that bacteria on the surface of coralline algae proved to be important settlement cues for larvae of the Australian sea urchin, *Heliocidaris erythrogramma*. A very similar story is provided by Dworjanyn & Pirozzi (2008) who studied the settling of the urchin *Tripneustes gratilla*, again in Australia (Figure 7.32). In this case, the seaweeds *Sargassum lineari-folium*, *Zonaria angustata*, and *Dictyota dichotoma* were the most preferred. These three species are not at all closely related to one another, and in fact it seemed once again that the settling urchins were more interested in the bacterial

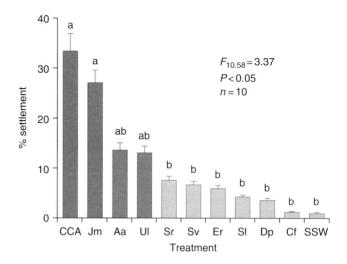

Figure 7.31

Haliotis rubra larval settlement in response to local algal species. Data are mean percent settlement (±s.e.). CCA = encrusting coralline algae; Jm = *Jaynia micrarthrodia*; Aa = *Amphiroa anceps*; Ul = *Ulva australis*; Sr = *Soleria robusta*; Sv = *Sargassum vestitum*; Er = *Ecklonia radiata*; Sl = *Sargassum linearifolium*; Dp = *Delisea pulchra*; Cf = *Codium fragile*; SSW = sterile seawater. Different letters and colors denote statistically significant differences. (From Huggett et al 2005; reproduced with permission of Springer.)

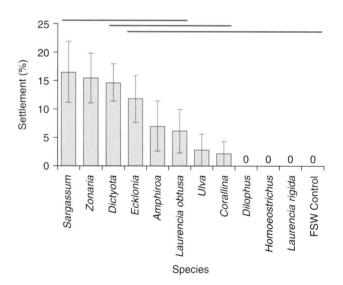

Figure 7.32

Mean (±s.e.) settlement and metamorphosis of larvae of the sea urchin *Tripneustes gratilla* in seawater conditioned (left overnight containing samples of seaweeds) by 12 seaweeds. FSW = fresh seawater control. Bars denote significant differences between treatments. (From Dworjanyn & Pirozzi 2008; reproduced with permission of Elsevier.)

biofilms carried on the surfaces of the seaweeds, rather than the weeds themselves. Recently metamorphosed urchins are likely to be using these biofilms as food, and hence a very useful adaptation is for the settling larvae to select sites (seaweeds) where their future food is likely to be abundant.

Figure 7.33
Barnacle cyprids settling. (Photograph courtesy of Jesus Pinada, Woods Hole.)

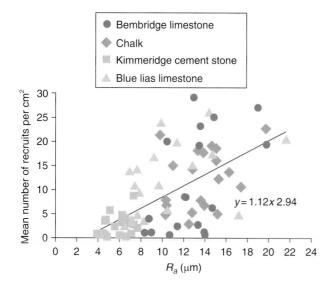

Figure 7.34
Relationship between roughness of rock surfaces (R_a – measured in µm) and the recruitment by cyprids of the barnacle *Chthalamus montagui*. Circles = Bembridge limestone; diamonds = chalk; squares = Kimmeridge cement stone; triangles = blue lias limestone. (From Herbert & Hawkins 2006; reproduced with permission of Elsevier.)

Perhaps the two animal groups whose settlement ecology and biology have been most studied are barnacles and corals. Barnacle cyprid larvae (Figure 7.33) employ both physical and chemical attributes of their rocky settlement sites to choose the most suitable substrates. Herbert & Hawkins (2006) conducted settlement experiments on the south coast of England with the intertidal barnacle, *Chthalamus montagui*, to see if the type of rock influenced settlement. Settlement tiles were constructed of four types of calcareous rock with different degrees of roughness (surface irregularity) and placed intertidally for up to 7 months post-settlement of larvae. Settlement rate and juvenile mortality was measured on the plates using photo analysis. As can be seen in Figure 7.34, the degree of roughness was linearly related to

the settlement rate (number of recruits) – the rougher or more irregular the rock surface, the more young barnacles settled. Roughness was directly related to the number of potential settlement sites. In order to find the best settlement site, barnacle cyprids (the stage between the planktonic dispersive zoeae and the settled juvenile) are able to alter their movement patterns in relation to the type of substrate they encounter whilst looking for somewhere to settle (Figure 7.35) (Berntsson et al 2004). These results conflict with those of Herbert & Hawkins because their species of barnacle, *Balanus improvisus*, preferred to settle on smooth surfaces rather than microtextured substrates. Cyprids were much more active on the less-preferred surface, indicating a higher degree of satisfaction and hence less "foraging" on the preferred surface. Berntsson et al concluded that *B. improvisus* uses settlement cues from surface topography above chemical cues. However, this is not to suggest that substrate chemistry has no role to play in barnacle settlement. All barnacles have internal cross-fertilization, which means that they have to be in close proximity – a female barnacle spatially isolated from a male doesn't get fertilized. Hence it is important that barnacles settle gregariously (notwithstanding the problems of intraspecifc competition for space later in life), and they have evolved mechanisms to locate places where members of their own species (conspecifics) have settled, recently or in the past. Dreanno et al (2007) describe a settlement-inducing protein complex (SIPC) which is a glycoprotein produced by barnacle cuticles, and which acts as a pheromone to induce settlement on newly arriving cyprids. There is a rapid increase in settlement rate of *Balanus amphitrite* as the concentration of SIPC begins to rise (Figure 7.36a), though the effect soon plateaus. The overall effect is that barnacles settle with members of their own species. As might be predicted, SIPCs from different, albeit closely related, barnacle species, do not elicit the same response (Figure 7.36b). Barnacles with uncommon substrate requirements certainly seem to use chemical cues for settlement. Larvae of the barnacle *Coronula diadema* have a particularly difficult substrate-locating task, in that they only settle on the skins of whales. Nogata & Matsumura (2006) suggest that this is a chemical process rather than a physical one.

Corals have similar site and substrate selection problems to barnacles. In Figure 7.37, Harrington et al (2004) show the effects of a variety of substrates and their orientations on the settlement of planula larvae of *Acropora* spp. on the Great Barrier Reef. They used five species of crustose coralline algae (*Neogoniolithon fosliei*, *Porolithon onkodes*, *Hydrolithon reinboldii*, *Titanoderma prototypum*, and *Lithoporella melobesioides*), both dead and alive, as well as the skeleton of the coral *Porites* sp., and some terracotta tiles. These choice experiments revealed very clearly that some species of crustose coralline algae, in particular *T. prototypum*, were much preferred by coral larvae as settlement substrates. In addition, living, horizontal substrates were chosen rather than dead, vertical ones. In a different study on Guam, Golbuu & Richmond (2007)

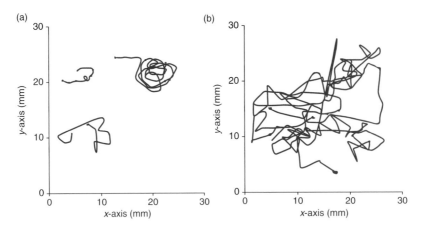

Figure 7.35

Movement paths of the barnacle *Balanus improvisus* on (a) smooth surfaces (preferred) and (b) microtextured surfaces. (From Berntsson et al 2004; reproduced with permission of InterResearch.)

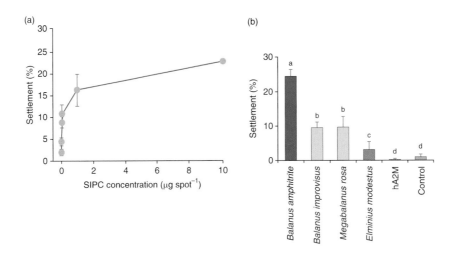

Figure 7.36

(a) Settlement response of cyprids of the barnacle *Balanus amphitrite* to different concentrations of *B. amphitrite* settlement-inducing protein complex (SIPC).
(b) Settlement response of cyprids of the barnacle *Balanus amphitrite* to SIPCs from various barnacle species and a general protein (hA2M). Different letters and colors denote significantly different means. (From Dreanno et al 2007; reproduced with permission of Elsevier.)

showed that larvae of two species of coral, *Goniastrea reti-formis* and *Stylaraea punctata*, selected substrates with encrusting algae or biofilms, but the two species showed distinctly different and almost nonoverlapping preferences (Figure 7.38). These complex observations need some explanation. Perhaps a living cover of encrusting algae sends a message to settling coral larvae about good light conditions, clear water, and so on. Larvae metamorphosing on to solid, clean rock may run the risk of being overgrown by aggressive fleshy or filamentous algae, which do not grow so well on their crustose relatives. Vertical substrates are likely to present self-shading problems as corals grow. Whatever the reasons, coral settlement would appear from these results to be anything but random, and furthermore, the "deterministic" system (Golbuu & Richmond) allows for a degree of resource partitioning and avoidance

of interspecific competition between settling larvae of different coral species.

We can use this nonrandom selection system to our advantage. As we shall discuss in Chapter 10, the restoration of coral reefs frequently entails the employment of artificial substrates, as does the "naturalization" of shipwrecks and other solid items which end up in the sea. Perkol-Finkel & Benayahu (2007) placed settlement plates on 17-year-old 15-m artificial reefs made of PVC and metal netting, and compared the settlement of sessile benthic species over 18 months with that on similar plates placed on natural reefs adjacent to the artificial ones. A large number of species were accumulated on the plates by the end of the experiment, consisting of both hard and soft coral species, as well as other benthic taxa such as sponges, bryozoa, and tunicates. MDS analyses show a clear distinction between communities

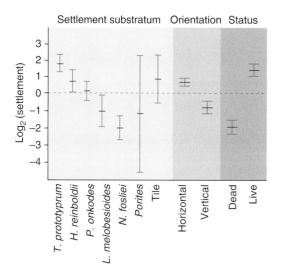

Figure 7.37

Partial effects plots of settlement densities of coral larvae, showing the effects of various substrata (live and dead crustose coralline algae species (CCA) and inorganic substrata) at horizontal and vertical orientation. The effects are plotted on log₂ scale, and thus an increase or decrease by 1 unit corresponds to a doubling or halving in settlement numbers, respectively (e.g. settlement on live CCA is 9.2 times as high as on dead CCA). Error bars represent $\pm$ s.e. (From Harrington et al 2004; reproduced with permission of *Ecology* – ESA.)

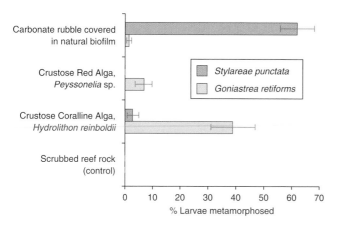

Figure 7.38

Substrate preferences for larvae of *Goniastrea retiformis* and *Stylaraea punctata*, as measured by percentage that undergo metamorphosis on four different substrates. (From Golbuu & Richmond 2007; reproduced with permission of Springer.)

settling on the artificial versus natural reefs. Settlement cues provided by the natural reef could be assumed to differ substantially from those of the artificial reef. We shall return to this important topic later in the book.

So far in this section on settlement cues, we have considered the biological and chemical processes that might influence larval choice, but we also need to think about the physical conditions in the sea, such as winds, tides, currents, and so on (Selleslagh & Amara 2007). All the organisms in the sea apart from a few marine mammals, large fish and birds

Figure 7.39

Paracentrotus lividus. (Photograph courtesy of www.louro.org)

are essentially poikilothermic; in other words, unable to regulate their body temperatures independently of their surrounding water, so that processes such as growth and swimming will be temperature dependent (Guan et al 2008). It makes adaptive sense therefore to settle out of the plankton to begin the adult phase as temperatures increase (in temperate regions at least) with the signs of spring and therefore food supply and associated growth rate increasing. The sea urchin *Paracentrotus lividus* (Figure 7.39) in the Mediterranean shows a good example of this (Figure 7.40) (Hereu et al 2004). Settlement of larvae on artificial substrates (yard brush heads in this case) showed a narrow (3-week) peak in May or June, and although the abundance of settlers varied from year to year, the annual settlement was clearly associated with increasing seawater temperatures. Note that seawater temperatures lag behind that of the air in temperate regions – the water takes several weeks to catch up with the air in spring and early summer.

Whatever the temperature, seawater is hardly ever still, and water flow (currents) on a large or small scale can have a profound effect on settlement and recruitment. Medium- to large-scale currents are wind or tide driven, and the settlement ecology of marine species has evolved to maximize the assistance given by such conditions to reach suitable adult habitats. Mangrove crabs are a case in point. It is clearly important to get crab larvae (megalopae) as far into the mangroves as possible before they settle, since young adults would have a much tougher time accessing their preferred habitat post-metamorphosis. The only way to do this is by using high tides with the wind behind them. Paula et al (2001) recognized two forcing factors in the settlement of larva of mangrove crabs in the genera *Sesarma* and *Uca*, on Inhaca Island in Mozambique (Figure 7.41). These factors

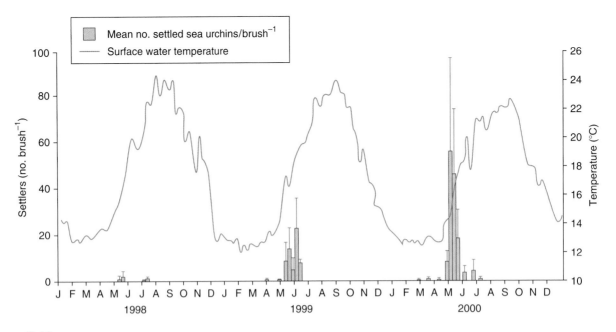

Figure 7.40

Weekly mean (±s.e.) numbers of sea urchin, *Paracentrotus lividus*, larvae settling onto artificial substrates in the Mediterranean, in relation to sea surface temperatures. (From Hereu et al 2004; reproduced with permission of Springer.)

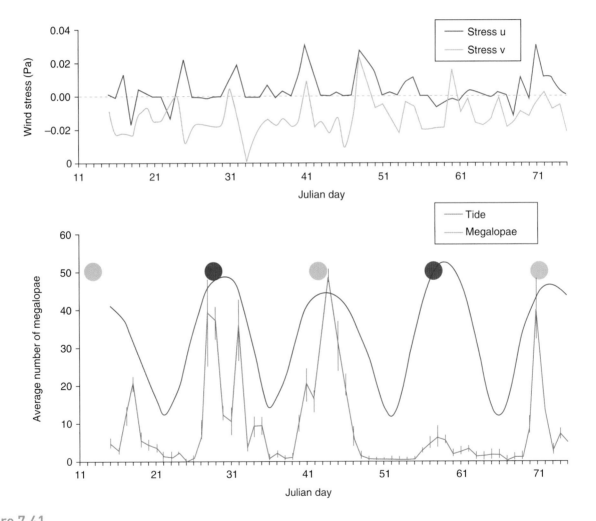

Figure 7.41

Abundances (±s.e.) of megalopae larvae of mangrove crabs, *Sesarma* spp. and *Uca* spp., in collectors in the mangroves of Inhaca Island, Mozambique, in relation to wind stress (power), tidal height, and moon phase. Stress u = westerly winds; Stress v = northerly winds; dark green circle = new moon; light green circle = full moon. (From Paula et al 2001; reproduced with permission of InterResearch.)

(a) Coral recruitment and survival

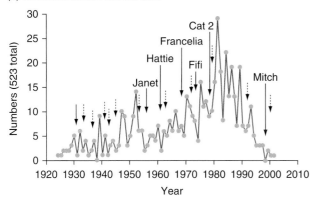

(b) Nonbranching coral recruitment

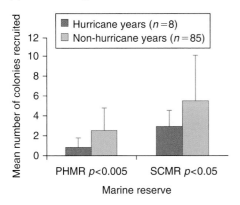

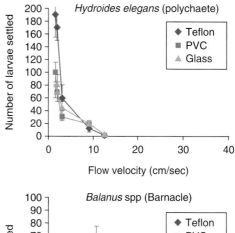

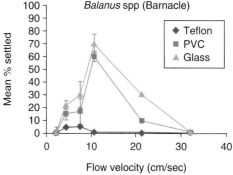

Figure 7.43

Settlement of invertebrate larvae on different substrates under varying flow rates. (Qian et al 2000; reproduced with permission of InterResearch.)

Figure 7.42

Settlement of nonbranching coral larvae in two marine nature reserves (Sapodilla Cayes Marine Reserve (SCMR) and Port Honduras Marine Reserve (PHMR)) in Belize. (a) Estimated recruitment dates of 523 individual corals in both reserves. Solid arrows = years of hurricanes; dotted arrows = years of tropical storms; (b) Significantly different coral recruitment in separate marine reserves between hurricane and no-hurricane years. (From Crabbe et al 2008; reproduced with permission of Elsevier.)

were wind stress (or power) and tidal height. As the figure shows, the peaks in mean number of megalopae caught coincided with the highest tidal ranges at full and new moons, except when the winds were quiet. Both high tide and an onshore wind are required in combination before crab larvae are able to reach far into the mangroves to settle as juvenile adults. At an extreme level, storms and even hurricanes are good examples of waves and winds which might influence larval settlement, as shown by Crabbe et al (2008) off the coast of Belize. Figure 7.42 shows how tropical storms, and their more violent relatives, hurricanes, are an almost routine phenomenon in the Caribbean. It might be expected therefore that coral communities have co-evolved with such physical dynamism, hence the large number of soft corals compared with some of the hard ones and the recruitment of nonbranching corals is significantly reduced in hurricane years. Presumably, the power of the winds and waves does not permit coral planulae to find their specific

settlement sites, or if they do settle it seems unlikely that they would be able to hang on for very long.

On a more gentle and local level, current flow can have a significant role to play in the settlement of marine larvae in specialized micro-habitats within a broader substrate tapestry. For a sessile filter feeder for example, your optimum settlement site (remembering that movement is impossible post-metamorphosis) will be a current flow not so strong as to wash them away, but not too weak that the water, and the food suspended in it is not refreshed sufficiently frequently. Certainly, larvae of the barnacle *Balanus improvisus* reject surfaces exposed to flows above 5 or 10 cm s[-1] (Larsson & Jonsson 2006), no matter what the chemical and biological suitability, and similarly, hydrodynamic conditions seem to outweigh those of other cues for the settlement of the velvet swimming crab, *Necora puber* (Lee et al 2004). Potentially competing species may use current flow selection as a means of avoiding micro-scale competition for space. In Figure 7.43, Qian et al (2000) show the settlement of three different taxa of sessile invertebrates on three different substrates, in relation to the velocity of the current flow to which they are exposed. Polychaetes settle only in very slow currents, bryozoans settle over much broader velocity ranges, whilst barnacles show peak settlement at medium velocities. Substrate preferences vary – teflon is most preferred by polychaetes and

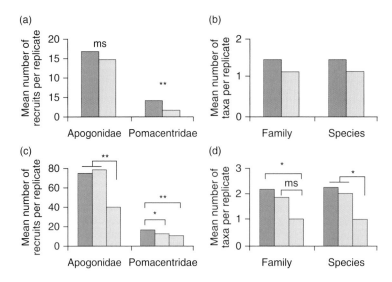

Figure 7.44

Catches of fish larvae from patch reefs on the Great Barrier Reef with different sound treatments. (a,b) Black = reef noise; white = silent reef. (c,d) Blue = high frequency sound from reefs; pink = low frequency sound from reefs; green = silent reefs. ms = $p < 0.1$. (From Simpson et al 2005; reproduced with permission of *Science* – AAAS.)

bryozoans, glass is least preferred. Barnacles favor glass followed by PVC. Obviously, none of these substrates is in any way natural, but the results of this work show that as we already know, substrates type is important, but niche separation can also be achieved by selecting subtle but optimum current velocities at the settlement site.

Finally in this section we consider a seemingly unlikely physical cue for settling reef taxa such as fish, sound. Marine ecologists have carried out experiments where they have recorded the background sounds which all divers and snorkelers have heard on a reef (the clicking of crustaceans, the chewing of coral feeding fish, and the general ebb and flow

of gentle currents on three-dimensional structures), and played these sounds back to fish larvae in search of a place to call home. Undoubtedly, the settlement of fish larvae can be enhanced by sound, not only used as a settlement cue but also as an orientation beacon for late stage larvae in search of somewhere suitable (Montgomery et al 2006; Tolimieri et al 2004). Simpson et al (2005) carried out experiments on the use of sound stimuli by larvae of several different families of reef fish (Figure 7.44). Though the results are somewhat complex, the message is clear – reef fish larvae can detect and recognize the sounds of their home reefs and swim toward them.

Chapter 8

The exploitation and maintenance of marine and estuarine fisheries

There is evidence from the South African coast that man was harvesting shellfish and other marine life 164,000 years ago. The discovery of harpoons, arrow heads, discarded shells and bones indicates that 10,000 years ago man was exploiting marine resources in many coastal areas. Given the ease with which man can collect shellfish and other invertebrates on an unexploited shore, it seems likely that early man would have readily searched beaches for food, and the abundance of oyster and other large animals in the intertidal zone of easily reached beaches must have soon declined. The push to build traps for fish, design better nets and hooks, and to move offshore and exploit more difficult to catch resources was inevitable and has continued and accelerated to the present. In the twentieth century the mechanization of fishing produced a massive increase in fishing pressure and the populations of species such as herring, which had once seemed too large to ever be reduced by man, started to decline and finally rapidly collapsed. Even now, with widespread concern about fishing impacts,

man is developing methods to catch fish from ever deeper waters. Our exploitation of marine fisheries has usually resulted in the degradation of the natural population both in terms of abundance and size or age structure. In many waters it is no longer possible to catch the largest sizes of fish, crabs, lobster, or shellfish which were once present. In many instances the activities of man have resulted in the collapse of fishing stocks to levels where fishing has become none economic. While in some cases it is clear that over-fishing has caused this collapse, the situation is often more complex and numerous additional factors may be acting contemporaneously to reduce fish populations. We will describe this complex interaction of factors using historical records from a particularly well studied estuary.

The over-exploitation of commercial fisheries

Clear examples of over-exploitation of commercial fishing stocks can be found in reports from all parts of the world. Because it is such a productive sea and has been under intense exploitation for more than 100 years, the North Sea (northwest Europe) offers particularly clear examples.

The clupeids and their relatives, the herrings, sardines and anchovies, are often open-water shoaling fish that can form extremely large populations. They are frequently easily caught in coastal waters and have therefore long been exploited by man. Unfortunately, there are numerous examples of the collapse of clupeid fisheries including North Sea herring (*Clupea harengus*), Pacific sardine (*Sardinops sagax*),

Marine Ecology: Concepts and Applications, 1st edition. By M. Speight and P. Henderson. Published 2010 by Blackwell Publishing Ltd.

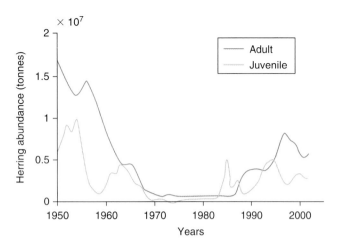

Figure 8.1

Abundance estimates of adult and juvenile Norwegian spring-spawning herring in the northeast Atlantic over the period 1950 to 2002. (From ICES 2003; reproduced with permission of Oxford University Press.)

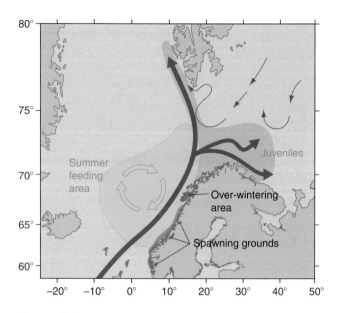

Figure 8.2

The seasonal movements of Norwegian spring-spawning herring. (From Tjelmeland & Lindstrøm 2005; reproduced with permission of Oxford University Press.)

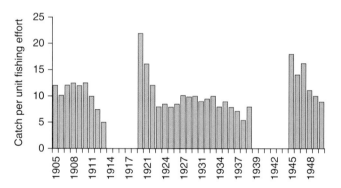

Figure 8.3

The catch per unit effort of haddock in the North Sea between 1905 and 1950 showing the effects of the First and Second World Wars.

Far Eastern sardine (*Sardinops melanosticta*), and menhaden (*Brevoortia tyrannus*) of North America. A typical example is the collapse of the Norwegian spring spawning herring which has been summarized by Tjelmeland & Lindstrøm (2005) and shown in Figure 8.1. Man has exploited this highly migratory herring stock for over a millennium. In the 1950s they were caught east of Iceland in the summer and off the coast of Norway in the winter and spring (Figure 8.2). The population underwent a dramatic collapse during the 1960s caused by over-fishing by purse-seiners. As the stock declined during the 1960s, improved fishing gear allowed the fishing pressure to intensify so that in 1966 they were still able to harvest 2,000,000 tons of fish. By the late 1960s most of the fish caught were too immature to spawn and a total collapse of the stock was inevitable.

The role of over-fishing in this collapse is demonstrated by the recovery in the stock once the fishing pressure was appropriately regulated. As is frequently observed in over-fished populations, the fishery became dependent on a small number of strong year-classes produced in years especially favorable for recruitment. This herring population remained exceedingly depressed until the 1983 year-class entered the spawning stock. The stock has now recovered and is managed by the coastal states of the EU, Faroe Islands, Iceland, Norway, and Russia. These countries have adopted a long-term management strategy with the following aims:

- The spawning stock shall exceed 2.5 million tonnes.
- The fishing mortality of adults shall not exceed 0.125.
- The fishing mortality shall be reduced if the spawning stock falls below 5 million tonnes.

North Sea haddock (*Melanogrammus aeglefinus*) offers a second example of population decline linked to over-exploitation. In this case, the rapid recovery of a population when fishing pressure was reduced strongly suggests that the population

was being over-fished. Figure 8.3 shows that immediately after both the First and Second World Wars, when fishing was greatly reduced, the catch per unit effort in the North Sea of haddock increased only to subsequently decline.

A final example of over-exploitation of a North Sea fish is the cod (*Gadus morhua*). Cod are taken by all countries bordering the North Sea including, Denmark, Scotland, England, the Netherlands, Germany, Belgium, and Norway. At present, ICES (International Council for the Exploration of the Sea) considers the stock outside safe biological limits and concludes that it has been exploited at above sustainable levels since the early 1980s. Trends in spawning-stock biomass (SSB) show a decline from a peak of 250,000 t in the early 1970s to current levels of about 40,000 t (Figure 8.4). The stock is now well below the ICES limit level of 70,000 t, below which ICES considers productivity of the

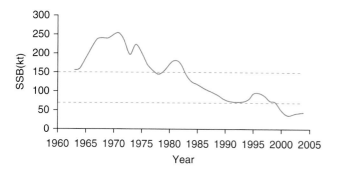

Figure 8.4

The spawning-stock size (SSB, '000 t) of North Sea cod. ICES limit (70,000 t) and precautionary (150 000 t) stock reference levels are also shown as dotted lines. (After Horwood et al 2006.)

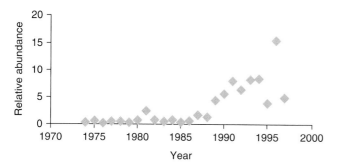

Figure 8.5

The change in abundance of young striped bass in the Hudson River, New York (1974–1997), following closure of the commercial fishery. (Data from the River Hudson Year-class Reports.)

stock to be impaired. The over-exploitation has long been acknowledged by fisheries scientists, however it has proved difficult to implement sufficiently stringent protective measures within the European Union to allow the stock to recover. Horwood et al (2006) concluded that by the beginning of 2005, restrictions on fishing had reduced fishing mortality rates by an estimated 37%, but this is insufficient to ensure recovery of North Sea cod within the next decade.

The final example shows the dramatic improvement in a fish population that can occur if fishing pressure is removed. Striped bass (*Morone saxatilis*) are an important commercial and sport fish in the USA. Over-exploitation caused a collapse in the Hudson River, New York, population so that by the 1970s abundance was only a fraction of that 100 years earlier. Subsequently, pollution of the estuary by PCBs (see the section below on the Hudson Estuary for more details) was found to have entered the food chain and the striped bass, forcing the closure of the commercial fishery. Contemporaneously with this closure, there were also improvements in water quality in the vicinity of New York increasing the availability of lower estuarine habitat. As shown in Figure 8.5, the closure of the fishery produced a

dramatic increase in young fish abundance as the spawning stock of adult fish increased. This is an odd example of a situation where industrial pollution has aided the recovery of a fish population by destroying the fishery.

The loss of top predators

In recent years, there have been a series of reports that document the loss of top predators through over-exploitation. These reports suggest that over-exploitation is not simply reducing the abundance of the few fish species which man particularly targets, but is producing a far more general degradation of marine ecosystems. It has long been recognized that fishing reduces the size of fish and thus will tend to reduce the average position in the food chain of fish such as cod. However, recent reports point to a far more general loss of predatory fish, termed the fishing-down of food webs (Pauly et al 1998; Jackson et al 2001). In the North Sea, for example, Hooper et al (2005) have reported that large, predatory fish have decreased in abundance as small fish have increased. Myers & Worm (2003) have documented the rapid worldwide depletion of large top-predator fish communities as shown by the reduction in catch per unit effort in all parts of the world (Figures 8.6 and 8.7). They concluded that:

- Declines in large predators in coastal regions are global.
- Industrialized fisheries typically reduced community biomass by 80% within 15 years of exploitation.
- Large predatory fish biomass is now at about 10% of pre-industrial levels.

Habitat damage and loss

Fish communities and fisheries have been damaged by habitat degradation and loss. Many of the most productive marine habitats are close inshore and these shallow habitats, frequently used by fish as nurseries, have been particularly badly damaged. Estuarine habitats are important nursery grounds which unfortunately also support some of the highest population densities of man. Great cities such as New York or London have greatly impoverished their local estuarine habitats. Historically, the most important cause of damage was usually waste water discharges. For example, in 1957 a 70-km stretch of the tidal Thames was fishless because of the lack of dissolved oxygen which had been consumed by bacteria which had reached high densities because of the nutrients supplied by the sewage discharges. Once the sewage treatment plants were improved and the discharges moved further offshore, oxygen levels increased and fish returned to the estuary. The re-establishment of fish since the 1960s has been carefully

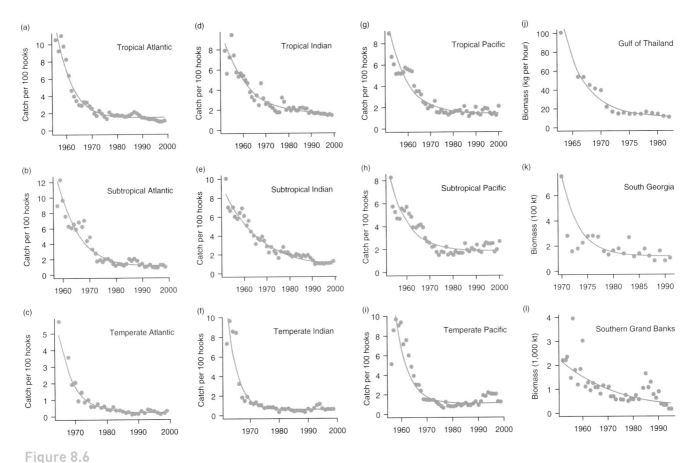

Figure 8.6

The change in the catch per unit effort of large predatory fish. (a–i) Oceanic and (j–l) shelf ecosystems. (After Myers & Worm 2003; reproduced with permission of *Nature*.)

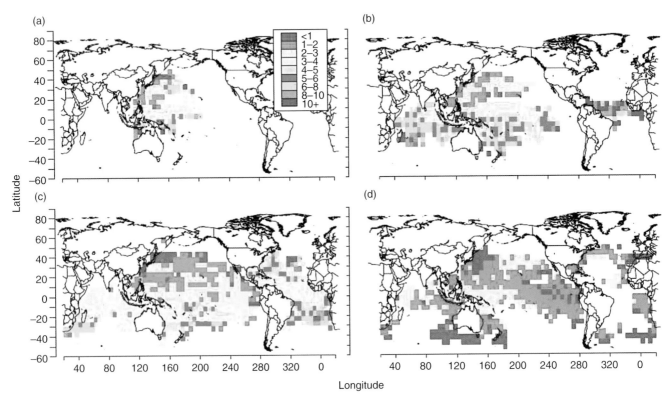

Figure 8.7

Relative predator biomass in (a) 1952, (b) 1958, (c) 1964 and (d) 1980. Color codes depict the number of fish caught per 100 hooks on pelagic longlines set by the Japanese fleet. (After Myers & Worm 2003; reproduced with permission of *Nature*.)

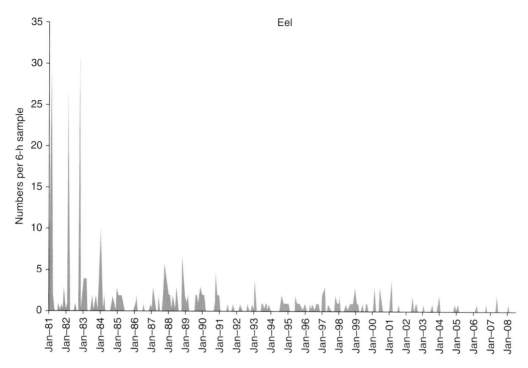

Figure 8.8

The dramatic almost exponential decline of eel, *Anguilla anguilla*, adults caught at Hinkley Point, Somerset, England between 1980 and 2008. Each monthly sample comprised the total capture over 6 hours from the Hinkley Point B nuclear power station cooling water intake filter screens. (Data courtesy of Peter Henderson, Pisces Conservation Ltd.)

monitored and is described in Attrill (1998). More than 80 species of fish have been recorded from the estuary since the 1970s.

Anoxia linked to eutrophication is not the only way in which man has reduced the fish communities of estuaries and shallow inshore waters. In many estuaries there has been extensive drainage of saltmarsh and other littoral habitat. This habitat loss further reduces fish abundance and diversity.

In the tropics, mangrove forests are important habitat for small and young fish. Around 50% of the world's total mangrove area is found in Indonesia, Australia, Brazil, Nigeria, and Mexico. In 2008 the UN Agriculture Organization (FAO) published a report estimating that 20% of the world's mangrove area had been destroyed since 1980. The total mangrove area has declined from 18.8 million ha in 1980 to 15.2 million ha in 2005. The main causes of this destruction include population pressure, conversion for shrimp and fish farming, agriculture, infrastructure development, and tourism, in addition to pollution and natural disasters (see Chapters 9 and 10).

Obstructions to movement have also had huge impacts on marine fish. Many marine fish including shads, lampreys, eels, salmon and flounders are migratory and spend part of their lifecycle in fresh waters. The construction of dams and weirs has often had a catastrophic effect on anadromous fish which move from the sea up rivers to spawn. For example, in the River Severn, England, it was noted in the nineteenth century that "Lampreys too, which were formerly considered of more importance than salmon, and were caught in the upper Severn, have altogether ceased to visit it since the erection of the first weir in 1843." The loss of habitat for anadromous fish has frequently been noted. For example in the Hudson River, New York, "the net effect of dams on most of the tributaries is the removal of a great number of miles of suitable anadromous fish spawning habitat from the system" (U.S. Fish and Wildlife Service).

The populations of fish with complex lifecycles can be laid low by the combined action of a number of factors. Populations of the catadromous eels of the genus *Anguilla* have recently collapsed in many parts of the world including Western Europe and the American east coast. A typical example of the approximately exponential decline observed since the 1980s is shown in Figure 8.8. The graph illustrates the change in the number of adult-sized eels caught in a lower estuarine habitat in the Bristol Channel England. The 28-year time series comprises monthly samples collected from the same volume of cooling water entering the nuclear power station at Hinkley Point. Thus, this time series shows the decline observed with a constant sampling effort, something which is rarely available for commercial landing data as commercial fishermen change their effort in response to economic conditions and available technology.

These animals would have recently migrated down the rivers and were about to start their migration across the Atlantic Ocean to their spawning grounds. The reasons for this global collapse are still unclear, in part because of the difficulty in observing and monitoring eels; no one has yet observed a mature eel on the spawning grounds in the Caribbean. Possible reasons for the crash include: interference with migration caused by dams; injury and death caused by the blades of pumps and other devices installed to drain low-lying areas; disease introduction caused by human movement of Asian eels into Europe; and finally over fishing – particularly for elvers which have a high commercial value. There may even be an element of climate change in the decline. All of these factors are likely to have contributed to the decline of survival from birth to adulthood of eel which is now expressed by the rapid decline in eel numbers.

Are marine fish populations ever stable?

All of the above examples have shown populations undergoing great changes in abundance and it is often argued that marine fish populations naturally show high population variability. This observation has then been frequently used, at least in the early stages of population decline, to

argue that the observed fall in abundance is within the natural, expected range of variability and therefore not linked to the activities of man. There is no evidence to suggest that natural fish populations are highly variable and as will be shown below support for the view that unexploited populations can be notably stable, showing little change in mean abundance over the long term. Fish populations are generally less stable in estuaries because the populations usually comprise the younger age classes. Studies over the last 30 years in Bridgwater Bay, a lower estuarine site in the Bristol Channel, have shown many fish populations are also observed to be fluctuating around reasonably constant levels. For example, Figure 8.9 shows the relative abundance of two of the commonest species, whiting, *Merlangius merlangus*, and flounder, *Platichthys flesus*. While both fish show regular seasonal changes in abundance, they both show little evidence for long-term trends. These examples lend support to the view that density-dependent control is operating to hold the population at an approximately constant long-term level. Neither of these species is under high fishing pressure in the Bristol Channel.

If density dependent processes are commonly operating, then it should be possible for populations to maintain population stability while they are being fished, provided the compensatory ability of the population is not exceeded. The shrimp population in Bridgwater Bay offers just such an example. Fixed-net fishermen in the Severn Estuary are

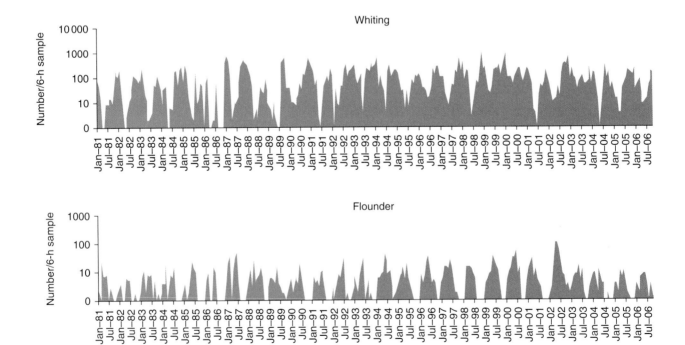

Figure 8.9

The relative abundance of whiting and flounder from 1980 to the present in Bridgwater Bay, England. This is a lower estuarine habitat used as a nursery by both these species. (Data courtesy of Peter Henderson, Pisces Conservation Ltd.)

continuing a fishing tradition which was recorded in the Domesday book in 1086. Their principal catch is shrimps and prawns with smaller catches of fish such as sprat and sole from the intertidal areas of the Bristol Channel (Figure 8.10). Figure 8.11 shows the remarkable stability of shrimp, *Crangon crangon*, in Bridgwater in an area where fixed-net fishermen work. The population fluctuates around a long-term, markedly constant, mean. This shrimp is still abundant after a history of exploitation by man extending over at least 1000 years.

While the above example shows that artisanal fishing can be maintained sustainably, there are fewer examples of large scale modern fisheries which have shown such stability.

Perhaps one of the best examples of stabilization for a large-scale commercial fishery is the Icelandic cod. As shown in Figure 8.12, over the last 50 years cod catches have decreased from above 500,000 tons in the 1950s to around 250,000 tons in 1998. Over this period the fishing mortality, *F*, also gradually increased. After the Second World War, Iceland gradually asserted rights to control the cod fishery and exclude foreign vessels principally from

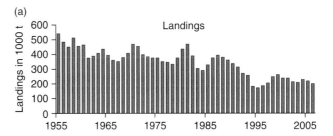

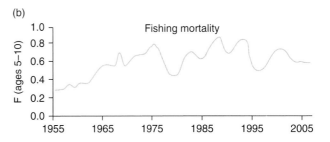

Figure 8.12

(a) The total landings of cod from Icelandic waters between 1955 and 2007. (b) Estimated fishing mortality. The fishing mortality is calculated for adult cod between 5 and 10 years of age.

Figure 8.10

Mr Sellick, a fixed net fishermen in Bridgwater bay with his traditional nets and mud horse. (Photograph Peter Henderson.)

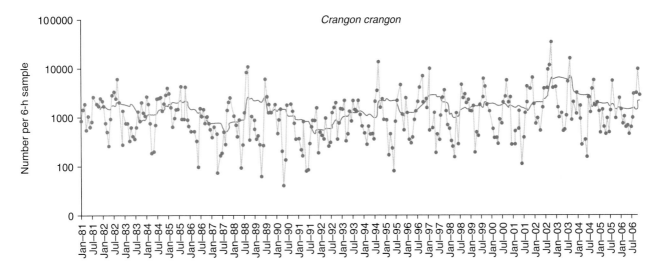

Figure 8.11

The relative abundance of shrimp, *Crangon crangon*, in Bridgwater Bay, Somerset. Over the 30-year period of this study this population was subject to commercial fishing pressure from the fished net fishermen. (Data courtesy of P. A. Henderson, Pisces Conservation Ltd.)

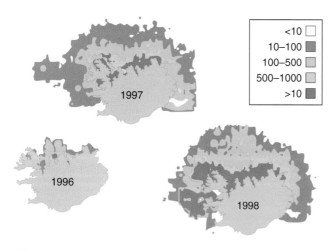

<10	☐
10–100	◼
100–500	◻
500–1000	◻
>10	◼

Figure 8.13

The extension in the range and abundance of juvenile cod in Icelandic waters following the introduction of a new catch control rule in 1995. From a speech delivered by the Iclandic fisheries minister Mr Árni M. Mathiesen, 2000.

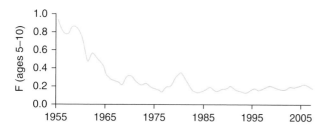

Figure 8.14

The change in the spawning stock biomass for Iceland cod.

Britain and Germany. Finally, in 1975 Iceland proclaimed authority for fishing in waters up to 200 miles (320 km) from the coast, triggering the third and final "cod war" between Great Britain and Iceland. An agreement was finally reached that limited British fishing within the 200-mile limit. In 1995, the Icelandic government, upon a recommendation of the Marine Research Institute and a wide consensus of other parties, adopted a long-term Catch Control Rule for cod, where a constant proportion of 25% of the fishable biomass was defined as the catch quota for any given year. Modeling had indicated this was a safe long-term limit. Subsequently the cod population has stabilized and shown slight recovery (Figure 8.14). The increase in abundance of juvenile cod, shown in Figure 8.13, was surprisingly rapid and the spawning stock biomass shown in Figure 8.14 has stabilized and recently risen slightly, indicating that a sustainable fishery is operating. However, this fishery is taking from a population with a biomass far below historical levels.

The Hudson Estuary: an example of multifactorial historical changes and fisheries collapse

In the above sections we have considered the various ways man affects fisheries in isolation. Further, we have only considered single species at a time. In reality, man exerts numerous different contemporaneous pressures on populations. Further, he changes many species within one community resulting in changes in community structure that also affect fisheries, occasionally in unforeseen ways. We can illustrate this multi-factorial situation by looking at one habitat in more detail. Because it is a well-studied estuary which has supported important fisheries, the Hudson Estuary (Figure 8.15) makes a particularly good example. The basic pattern of damage and change described is a typical experience for estuaries in all developed counties.

The virgin estuary

The Hudson Estuary stretches 153 miles from Troy Dam to New York Harbor, nearly half the river's 315-mile course between Lake Tear of the Clouds, its source in the Adirondacks, and the Battery at the tip of Manhattan. The estuary is tidal, or tidally influenced, as far upriver as Troy (Figure 8.15).

Before man's activities affected the river and its watershed, it was very different from that seen today, flowing from the highlands in the north through densely wooded lands until reaching the sea at Manhattan via salt marshes and islands. Estuaries are amongst the most productive of earth's ecosystems, and the Hudson's resouces have been known to man for thousands of years; evidence of early Native Americans' food-gathering remains in heaps of oyster shells on its shores. Henry Hudson and Dutch traders wrote of a river teeming with striped bass, herring, and giant sturgeon. More than 200 species of fish are found in the Hudson and its tributaries, and the estuary's productivity is ecologically and economically valuable to much of the Atlantic Coast; key commercial and recreational species like striped bass, bluefish, and blue crab depend on nursery habitat. Bald eagles, herons, waterfowl, and other birds fed from the river. Tidal marshes, mudflats, and other habitats in and along the estuary support a great diversity of life, including endangered species such as the short-nosed sturgeon.

The early influences of man

As populations of European settlers and their agriculture and trade increased, the population living around the

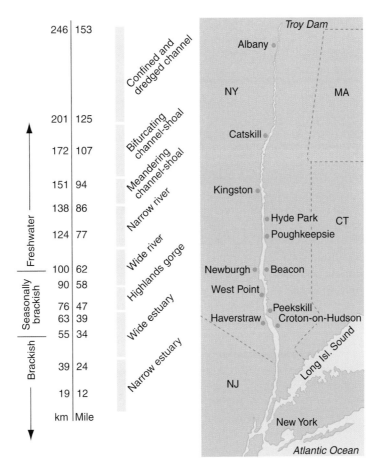

Figure 8.15

A diagrammatic map of the Hudson Estuary showing the change in salinity. (Courtesy of Mark B. Bain, NY Cooperative Fish & Wildlife Research Unit, Hudson River Sturgeon Study.)

Hudson started to deforest the countryside, in the process of clearing land and gathering wood for fuel and building materials. Initially minor, the effect of this on the Hudson was to increase fluctuations in the flow of the river; forests act as a sponge, holding back rainfall, and regulating its release into the river. The result of deforestation probably increased the periods of high flows and the severity of periods of low river flow. A secondary effect is the increase in the amount of silt washed into the river, which would clog the gravels, important to many bottom-spawning fish, and increase sedimentation in stretches of the river with low velocity. Increased sedimentation further downstream would also tend to favor the braiding of channels, the deposition of shallow marginal areas, the amount and variety of in-stream and marginal vegetation, and the deposition of saltmarsh at the river's delta. Thus a change in man's activities in the Hudson watershed would result in the alteration of the existing habitat, and the creation of (for the Hudson) new types of habitat. This first alteration of the habitat by man seems likely, then, to have facilitated many of the other changes which have since occurred.

Around 1760, Piermont Marsh (south of the Tappan Zee Bridge) began to be dominated by weedy, rather than arboreal, plant species. Core samples reveal an early dominance of hemlock, while about 3000 years ago the profiles became dominated by pine. Most recent sediments show anthropogenic influence with the dominance of giant reed (*Phragmites*) seeds replacing the cattail marsh (*Typha*) (Wong and Peteet 1998).

The first mile of Sparkhill Creek, running in to Piermont Marsh, was navigable to the flat-bottomed sloops of the eighteenth century. During this period, the creek was used as the principal entry port on the west side of the lower Hudson Valley. The port was the major access route to the fur and timber country lying to the west and a convenient shipping point for the agricultural industries (crops and milling) of the local area.

By the early nineteenth century development of larger sloops with deeper drafts made use of Sparkhill Creek impractical. To solve this problem the mile-long pier was constructed in 1839 at the terminus of the Erie railroad. The pier still stands today and is used by local residents for access to the Hudson River. The pier has likely had a significant impact on development of the marsh by accelerating deposition of sediments (Hudson River NERR – http://inlet.geol.sc.edu/HUD/piemont-marsh.html).

New species introductions

The development of man's activities around the estuary led, deliberately or accidentally, to the introduction of several species. In the early 1800s purple loose-strife and mute swan were introduced from Europe. Intentionally introduced fish species probably include Atlantic salmon, brown trout, rainbow trout, goldfish, carp, smallmouth bass, largemouth bass, white and black crappie. Each of these species is now well established in the Hudson catchment. Later in the 1820s the linking of the Hudson to St Lawrence by the opening of the Champlain–Hudson Canal resulted in several more species entering the river. These included the stonecat, white bass, and fantail darter. The establishment of these species is likely to have had a significant effect on the community as they began to compete for food and territory with the native species. It is also likely that their establishment was aided considerably by the changes wrought on the Hudson's ecosystem by man's influence, described above, which allowed them niches in which to thrive. The degree of perturbation caused by the introduction of these species is impossible to assess from the present, but evidence from modern introductions of exotic animals throughout the world suggests that the effects would be significant and wide-ranging.

Figures 8.16a & b represent a very simplified food web for the Hudson, before and after the increase in sedimentation, shift to dominance by marginal macrophyte species, and the introduction of carp. In Figure 8.16b, elements of the food web that increased are shown in shades of magenta; and elements suffering a decrease in abundance or strength in shades of light blue.

The Troy Dam

In 1825 the Troy Dam was built, stopping all migration of anadromous and catadromous fish beyond this point. The shad spawning once extended as far as Glens Falls. The full extent of the impact of this on the migratory fish species in the Hudson is impossible to quantify, since no reliable data exist from this period, but the loss of many miles of available spawning grounds for shad and other anadromous fish must have considerable effects on the populations. This dam and others built upstream of it have turned the Upper Hudson into a 40-mile chain of slow-flowing lakes. The native flowing water community in this region has presumably been replaced by one more adapted to the slow-flowing nature of the river.

Loss of foreshore

In the 1840s the railroads were started on both banks of the Hudson. The building of the railways produced several effects:

- They acted as a barrier to stop the free movement of people and animals from the surrounding area to the river.
- The construction of railway bridges across the bays of the Hudson created large areas of slow flowing water. This has developed over time into the reed-filled, marshy areas much in evidence today. This is, in effect, augmenting the changes caused by deforestation of the Hudson watershed, as described above. The gravel littoral and sublittoral habitats originally in many of these bays were lost to the aquatic community. Many species of crustacean live in the shallow waters of estuaries and these areas are often habitat for young fish. The loss and disturbance of the shallow-water environments in the river would have changed its characteristic fish fauna significantly, and with the appearance of new changed habitats, potentially allowed the proliferation of introduced species which otherwise would not have thrived.
- The clearance of vegetation for the railroad will have had an effect similar to the early forest clearances, though on a smaller scale, in increasing rainfall and silt flux into the river.
- In more modern times, herbicide has been regularly used to clear railroad tracksides. Where it is close to the estuary this may affect any littoral or marsh habitats it contacts.

Decline in water quality

As the population around the Hudson increased, the anthropomorphic impacts on the river also increased. Discharges of raw sewage from the many communities along the length of the Hudson led to high bacteria counts and low oxygen levels, and an increase in organic nutrients. In the 1850s manufacturing waste was reported to be reducing the oyster production in New York Harbor.

Ecologically valuable wetlands were destroyed by land filling and reclamation to build on the waterside. Many of the tributaries of the river were dammed and their water quality affected by the intensification of agriculture and manufacturing.

Dredging

Concern has been expressed from time to time over the effect of dredging upon the shad. Above the city of Hudson, a small channel was dredged between 1797 and 1831, and

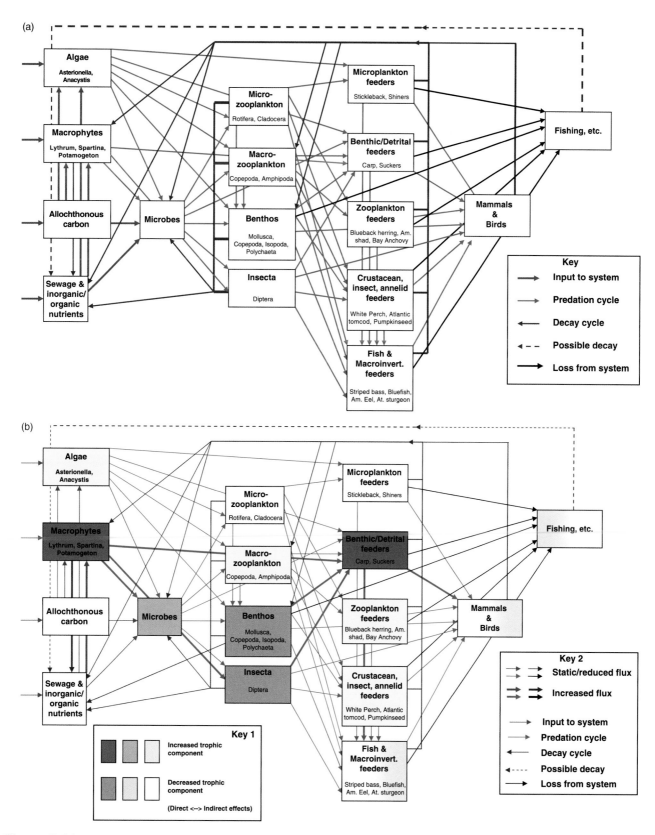

Figure 8.16

(a) Hypothetical Hudson River food chain, prior to the ecosystem changes and carp introduction. (b) Hypothetical Hudson River food chain, after increase in macrophytes and invasion by carp. ((a,b) Courtesy of P. A. Henderson, Pisces Conservation Ltd.)

this was improved upon in the years which followed. In 1899, a 12-foot channel existed from Coxsackie to Waterford. This was dredged to 27 feet from Albany to Hudson, and by 1931 this channel depth had been extended to New York City. These dredging operations destroyed many of the sandbars over which the shad spawned. Dredging reached its peak between 1926 and 1934, and during the years immediately following the abundance of shad increased. It seems obvious that dredging of the sandbars did not diminish the shad's potential to reach high population levels (from The Hudson – Fish and Wildlife (see Appendix, pp. 238–248)). The entire length of the navigation channel was deepened to 32 feet in the 1960s.

Intensification of fishing pressure, and cycles of abundance

The amount of fishing increased steadily throughout the eighteenth and nineteenth centuries, before key species began a series of cyclical peaks and troughs towards the end of the nineteenth century. Oyster landings were at more than 1.5 million bushels per year in the 1840s; by 1880 wild-caught oyster landings had decreased to below 1 million bushels per year, although the cultivation of oysters had by then begun to increase. In 1912 there were approximately 9000 acres of oyster bed under cultivation. By the end of the 1890s the shad fishing had peaked with catches as high as 1,162,000 lbs (a year (1887), falling to 18,200 lbs per year in 1916. In 1898 the sturgeon landings in the Hudson had peaked at 231,000 lbs; by 1925 they were between 1% and 20% of previous levels.

While fish populations were evidently subject to great fluctuation, both anthropogenic and natural, it should be noted that these cycles are not wholly attributable to peaks and crashes in population numbers. They may be equally dependent upon many variables, including:

- Changes in consumer and societal demand and trading practices.
- Invention or abandonment of fishing methods.
- Closure or opening of ports where fish landings were recorded.
- Changes in recording of catch data.
- Cessation of fishing in periods of war or instability.

Introduction of support for fisheries

As fishery catches in the Hudson began to decline, attempts were made to establish the cause of the declines:

- In 1850, it was reported that pollution by manufacturing wastes were responsible for marked decline in oyster productivity of the waters in and adjacent to New York Harbor.

- In 1870, pollution by raw sewage was reported as having an adverse effect on oyster fisheries.
- In 1877, disturbance of eggs and fry by ship traffic was first cited as a cause for the shad's decline
- In 1888, a report of the US Fish Commission stated that the caviar industry was responsible for the decline in the sturgeon fishery.

In response to the declines, various measures were used to support the fisheries, both by regulating fishing and by attempting to augment the wild populations of fish and shellfish. These continued through the latter part of the nineteenth and into the twentieth century:

- In 1868, the New York Fish Commission organized an artificial shad propagation operation with river fishermen; one of the earliest efforts on the Hudson at augmenting wild stocks.
- Between 1883 and 1895, NY State hatched and planted 33,322,500 shad fry in the Hudson, and the United States Fish Commission raised and planted 2000,000 eggs and 53,474,000 fry.
- In the early 1890s, cultivation of oysters increased greatly, aided by an 1887 law allowing the acquisition by perpetual franchise of subtidal areas for oyster farming.
- A 1908 State law required fixed nets to be lifted between sunset on Thursday and sunrise on Monday, in the hope of permitting greater shad egg production.
- In 1926, another state law introduced a closed season for both the short-nosed and Atlantic sturgeon, and set minimum landing sizes. However, there was little evidence of its efficacy.
- The minimum fork length for landing striped bass was set at 16 inches in 1938.

Continuing decline of fisheries – 1930s to 1960s

Through the 1930s and 1940s fish catches appeared to recover quite considerably, particularly alewife, blueback herring, striped bass, American eel, short-nosed and Atlantic sturgeon, white perch, Atlantic tomcod, and American smelt, which all peaked during the war years. In 1944, the last hatchery-raised shad fry planted in the Hudson, and the shad catch reached 1,731,545 lbs. However, by 1950 many of the fishery stocks were in decline again; the lowest landings of oysters and many fish species occurred in the 1960s. Many of the fishermen had given up the trade, and by 1964 there were only six regular commercial fishermen left on the Hudson Estuary.

PCB pollution

General Electric (GE) began using polychlorinated biphenyls (PCBs) in its manufacturing processes at the Fort

Edward and Hudson Falls plants in 1947 and 1952, respectively. The plants discharged manufacturing process wastewater containing PCBs directly into the Hudson River until 1977 when discharges to the environment were banned; residual contamination from both plant sites, and sediments along the length of the river continue to release PCBs into the environment. The Fort Edward dam was removed in 1973, which resulted in the downstream release of an estimated 1,300,000 cubic yards of PCB-laden sediment. PCB contamination now exists in all 200 miles of river sediment downstream of Fort Edward.

Elevated levels of PCBs were first discovered in Hudson River biota in 1969, but their importance was not recognized for several years. In the early 1970s, DEC began collecting limited data on PCBs in New York waters and fish. In 1973, the federal Food and Drug Administration (FDA) adopted a "tolerance" level for PCBs in food sold commercially, including fish, of 5 parts per million (ppm) in the edible portion, reduced to 2 ppm in 1984. At least 7 of the 11 species of Hudson River fish sampled between 1970 and 1972 had concentrations of PCBs which exceeded that level.

In line with the discovery of PCB contamination in fish came a decline in populations of top predators such as mink and river otter, potentially serious effects on fish, mammals, birds and other wildlife, and the necessity for measures to protect human health. In 1976 the first health advisory notice on the consumption of fish was issued; over the next few years a slew of restrictions and total bans were placed on commercial and recreational fishing and the sale of fish. These restrictions remain in place to date, to a greater or lesser degree, with a relaxation on recreational fishing allowing catch-and-release. Human consumption of fish and crabs was also regulated in a number of ways, ranging from advised maximum consumption to outright bans. Again, these restrictions mostly remain in force to date.

By the mid-1980s, PCB levels in fish from the Hudson had declined, although average levels in many species still exceeded the 2 ppm tolerance limit imposed in 1984. However, fish taken from the upper Hudson in 1992 and 1993 had PCB levels as high as those reported in the early 1980s. Additional releases from Allen Mill, at the Hudson Falls plant, may have contributed to the increased levels of PCBs detected in the fish. As a result, in 1994, advisories for the Troy Dam to Catskill reach of the lower river were revised from species-specific advice to *eat none* for all species except American shad. PCB concentrations in Hudson River fish gradually returned to pre-1992 levels as the Allen Mill release was brought under control. Despite these declines, the fish remain contaminated with PCBs. The effects of the almost complete cessation of fishing on the species' populations is unquantified, however, it must have allowed a considerable increase in fish populations, and consequential alterations in the food web.

Increase in power stations on the Hudson River

Through the twentieth century, the number of power stations along the whole length of the Hudson has increased dramatically, and with it the potentially harmful effects of impingement and entrainment of fish, mammals, and invertebrates, the alteration of river bank and bed, impacts of dredging, piling and other construction activities, thermal effluent effects and pollution. These factors have all augmented the degradation of the environment, and the often-dramatic alteration of natural habitats which has been the theme of man's interaction with the Hudson for the last four centuries.

The arrival of the zebra mussel

Prior to 1992, the nutrient-rich Hudson River estuary supported abundant phytoplankton populations that constituted a ready food supply for large populations of freshwater zooplankton, including rotifers, cladocerans, and copepods on a seasonal basis. The introduction and population explosion of zebra mussel (*Dreissena polymorpha*) has depleted the standing stock of phytoplankton and has impacted other components of the food chain. Benthic invertebrates are relatively abundant, but the species diversity is low, primarily oligochaetes and chironomids.

In 1986, the zebra mussel, an inhabitant of fresh and brackish Eurasian waters, arrived via the Great Lakes in the ballast water of ships (see also Chapter 9). First seen in the Hudson at Catskill in May 1991, zebra mussels now inhabit the Mohawk River and the Hudson River from Albany to Haverstraw Bay. Within little more than a year of their arrival the biomass of the mussels was greater than that of all other heterotrophic animals in the Hudson, and reached an estimated 550 billion individuals, at an average density of $4000\,m^{-2}$ over the freshwater tidal river (Figure 8.17). A secondary estimate was that, as filter feeders, the mussel population could filter the entire volume of the freshwater Hudson in 1–3 days. Their presence poses a number of very considerable threats to the ecosystem of the Hudson:

- Zebra mussels tend to colonize on rocky substrates in shoal areas, replacing or smothering any existing community that is in these habitats. Taxa of particular concern include Unionid and Sphaeriid clams. They also outcompete native mussel species for food and space, leading to a decline in native mussel populations.
- Phytoplankton and detritus are major food sources for lake and river food webs. Excessive removal of the phytoplankton by zebra mussels reduces the zooplankton species that feed upon them and can result in fisheries-related impacts.

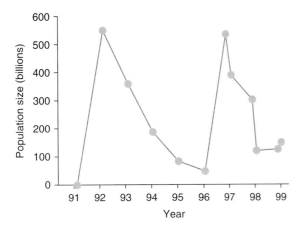

Figure 8.17

The estimated population of zebra mussels in the Hudson. (Data from http://www.ecostudies.org/research_hres.html)

- Mussels can filter large amounts of water and reduce the available food in the water column. Their filtering activity increases water clarity and hence light penetration. This, too, can dramatically change the benthic community structure.
- Zebra mussels cause significant biofouling in water intakes. This requires higher levels of biocide to combat the problem and this could lead to secondary effects in relation to the biocide chemical being released in to the environment.

Given their considerable numbers and their ecological effects (lakes and rivers colonized by the mussels often see 50–75% declines in phytoplankton and small zooplankton biomass, rises in water clarity of 50–100%, drops of more than 50% in filter-feeding zooplankton and native bivalves, and increases in macrophyte beds and animals associated with mussels), it is inevitable that their presence will have a profound effect on the food web of the Hudson. This is illustrated in Figure 8.16a and b, which represent a very simplified Hudson food web, before and after the introduction of Zebra mussels. In Figure 8.16b, elements of the food web increased by the changes are shown in shades of magenta, and elements suffering a decrease in abundance or strength by shades of light blue.

Long-term reduction of zebra mussels by natural predators has yet to be demonstrated, but at least 17 species of North American fish have been documented to consume attached zebra mussels and quagga mussels (*Dreissena bugenis*). Additional species are likely to consume zebra mussels (particularly fish in the sturgeon, sucker, and catfish families), but cases remain undocumented. Although numerous and widespread, the efficacy of molluscivorous fish as a control mechanism for zebra mussels is unclear. However, zebra mussels are more susceptible to fish predation than native unionids or *Corbicula* spp. because *Dreissena* shells

are weaker, adults are smaller in size, and most individuals are exposed to predators. (Kirk et al 2001)

Other invasive species

There are very many invasive species in the Hudson. In a paper on "Exotic species in the Hudson River basin – a history of invasions and introductions" (Estuaries 19:814–823), Mills et al (1996) state:

> We compiled information about the distribution of exotic organisms in the fresh waters of the Hudson River basin. At least 113 non-indigenous species of vertebrates, vascular plants, and large invertebrates have established populations in the basin. Too little was known about the past or present distributions of algae and most small invertebrates to identify exotic species in this group. Most established exotic species in the Hudson River basin originated from Eurasia or the Mississippi–Great Lakes basins, and were associated with vectors such as unintentional releases (especially escapes from cultivation), shipping activities (especially solid ballast or ballast water), canals, or intentional releases. Rates of species invasions of fresh and oligohaline waters in the basin have been high (ca. one new species per year) since about 1840. For many well-studied groups, introduced species constitute 4% to nearly 60% of the species now in the basin. Although the ecological impacts of the invaders in the Hudson River basin have not been well studied, we believe that about 10% of the exotic species have had major ecological impacts in the basin. Since the rates of entry and composition of exotic species in the Hudson basin are similar to those observed previously for the Laurentian Great Lakes, invasions tended to occur earlier in the Hudson basin, probably reflecting the earlier history of human commerce. While most exotics have had negative impacts on local flora and fauna, some fish species have provided unique angling opportunities and important economic benefits.

As this passage suggests, with a reasonably constant flux of new species arriving, the Hudson cannot be said to be at equilibrium. Many, possibly most, of these never become established, since conditions do not favor them. Of those that do establish themselves, some are quite innocuous, beneficial even, while some may never even be noticed. Nonetheless, it remains the case that a proportion of new plant and animal species go on, like the zebra mussel, to have profound effects on the ecosystem, and to require often drastic intervention by man.

A case in point is the water chestnut (*Trapa natans*). This Eurasian plant has in the last century become an aquatic nuisance species. Due to its dense, clonal, mat-forming, growths the species impedes navigation; its low food value for wildlife potentially can have a substantial impact on the use of the area by native species. The dense surface mats likely also reduce aquatic plant growth of other species beneath the shade of the floating canopy. The abundant detritus in the fall of each year and its decomposition could contribute towards lower oxygen levels in shallow waters and impact other aquatic organisms. The sharp spiny fruits can also be hazardous to bathers. First introduced to Collins Lake, NY in 1884, its infestation was reported at pest proportions in the upper Hudson in 1930. A 1936 survey reported only two plants downstream of the Troy Dam; by the mid-1940 it was widespread throughout the Hudson Estuary.

Measures to control the plant, consisting chiefly of harvesting the floating portion, and the use of the herbicide 2,4-D, were implemented from the 1950s onwards, but ceased in 1976, having achieved some considerable success in clearing the waterways. However, by the mid-1980s, populations were again at nuisance levels. Means of biological control, using species known to prey upon the plants, have been investigated since the early 1990s, and may offer another solution.

As related above, there are many other invasive species in the Hudson, including:

- *Anguillicola crassus*, an eel swimbladder nematode, a parasite, native to the Northwest Pacific–Asia (native host *Anguilla japonica*), spread to European *A. anguilla* in the 1980s, and was found in American eels in Texas and South Carolina in 1995, and in 1997 in Chesapeake Bay and the Hudson River Estuary. Early stages of this parasite infest copepods and can be spread by ballast water; later ones by natural or aquaculture movement of juvenile (yellow) eels. Severe effects on cultured eels, and detectable effects on wild eels.

- The Atlantic clam *Rangia cuneata* and American blenny *Hypsoblennius ionthas*, both introduced in the mid 1980s.

- *Botrylloides violaceus*, a colonial tunicate, native to Northwest Pacific, probably introduced in the 1970s (misidentified as *B. diegensis* or *B. leachii*), now found from Long Island Sound to Maine. Probably introduced by fouling. Now an abundant fouling organism.

- *Hemigrapsus sanguineus*, Asian shore crab, native to Northwest Pacific (Hong Kong–Vladivostok), discovered in New Jersey in 1987, now found from North Carolina to New Hampshire. Probably introduced by ballast water. Now the most abundant shore crab from New Jersey to Cape Cod, apparently competing with native crabs and affecting recruitment of intertidal biota.

- MSX oyster disease, first occurring in Long Island Sound in the 1980s, decimated oyster populations in 1997–98.

Current conditions in the estuary

Today, despite PCB contamination and the presence of invasive species, the Hudson River is one of the healthiest estuaries on the Atlantic Coast. Its rich history and striking environmental recovery have made it one of the nation's 14 American Heritage Rivers. As the Historical Timeline (in the Appendix, see p. 238) shows, the cycles of fish population strength have continued from the late 1960s to the present day:

- White perch populations increased until the mid 1980s, then they began a decline from which they have yet to recover.

- Atlantic and short-nosed sturgeon populations remain at low levels, and are now off-limits for fishing

- American shad abundance peaked in 1989, and then began a decline (to the point where full closure of the ocean harvest will be implemented in 2005), while other shad populations appeared stable.

- Conversely, after a serious decline in to the 1980s, striped bass populations have recovered to a remarkable extent, aided by compulsory and voluntary catch restrictions, and moratoria: almost inevitably, the outcome of the PCB contamination.

Concluding remarks

The performance of mankind in the protection and maintenance of fisheries is shameful, and while we now understand what is required there is still far too little political will to make the changes required. To achieve sustainable exploitation the following is required:

- The habitats needed by animals and plants must not be greatly damaged or reduced in extent. Particular care needs to be taken over spawning and nursery grounds. The introduction of alien species must be strictly controlled.

- Exploitation must be planned to allow the survival of sufficient adults to produce strong recruitment.

- Exploitation needs to be at a level that does not radically change the age structure and life history of the animal.

- The effects of climate need to be monitored and landings adjusted accordingly.

- A precautionary approach must be taken. Present practise is to believe in compensatory responses and aim for the maximum possible catch – this approach has failed repeatedly.

Chapter 9

Threats to marine ecosystems: the effects of man

Marine ecosystems must withstand damaging events that are both natural and man-made. In addition to the highly destructive, limited duration episodes such as great storms, tsunamis and oil releases from wrecked supertankers, there are also slower, long-term changes such as the increase in nutrients and sediment discharges from the land, or climate change. It is the shallow coastal and estuarine habitats that are most frequently impacted by both natural and man-made events, though some deep-water regions are subject to great changes when volcanoes erupt and hydrothermal vents open or shut down. Nor are impacts to marine ecosystems a new phenomenon. As Scheffers et al (2009) report, events such as tsunamis and hurricanes are shown to have been influencing marine ecosystems in, for example, the Caribbean, since prehistory.

In this chapter we will consider both natural and man-made threats and disturbances and, for convenience, classify both into sudden short-term impacts and longer-term slow changes. However, this distinction is not always easy to make as some slow changes, such as nutrient enrichment, can suddenly trigger a catastrophically damaging short-term event such as a toxic algal bloom. Figure 9.1 summarizes some of the major impacts to marine ecosystems as reported in the scientific literature, and also shows that many of these effects can be additive (Crain et al 2008). For example,

according to the number of published studies, combinations of stressors such as salinity and temperature, or sea level rise (SLR) and temperature with ultraviolet radiation (UV), can act together to make the impact even worse.

Recent problems

Threats to many natural habitats are often multifactorial and frequently involve man-made damage interacting with natural events. For example, anthropogenic damage to a barrier reef or mangrove may increase the exposure of a shore to wave action and lead to greater storm damage. There is also growing evidence that damage by man resulting in lower species richness has reduced the resilience of systems to natural variation. Halpern et al (2008) used models to map global and regional impacts from 17 separate datasets of human activities on 20 marine ecosystems (Figure 9.2) which show that almost all the world's oceans are affected in some way by human activities. These activities are itemized in Figure 9.3, which illustrates the diversity of problems which humans inflict on the sea, and the organisms that live in it. As can be seen, one of the most serious problems involves, predictably, climate change, a serious problem no doubt, but one that is one of the least tractable to fix. The most direct impact of man on the sea is through fishing and hunting (this topic was considered in Chapter 8), but pollution, physical damage such as shipping and tourism, and invasions of new species, are all having extremely serious effects on our oceans.

Let us take coral reefs as an example. In heavily populated parts of the world such as the Florida Keys, extensive areas of reef have been degraded since the 1970s. The areas of previously-healthy *Acropora palmata* dominated reef had been reduced by 1985 to dead coral rubble over-grown with algae (Figure 9.4). The result is a degraded species-poor

Marine Ecology: Concepts and Applications, 1st edition. By M. Speight and P. Henderson. Published 2010 by Blackwell Publishing Ltd.

	Salinity	Sediment	Nutrients	Toxins	Fishing	SLR	Temp	CO_2	UV	Invasives	Disease	Hypoxia	Disturbance	Total by stressor
Salinity	0	9	9	0	16	13	5	5	0	1	0	0		58
Sedimentation		5	0	0	0	2	1	0	1	0	0	0		9
Nutrients			17	3	7	8	2	5	0	2	0	0		58
Toxins				0	0	7	0	40	0	0	1	0		54
Fishing					0	1	0	0	0	0	0	0		4
SLR						0	1	0	0	0	0	3		27
Temperature							5	18	0	5	2	1		62
CO_2								3	0	1	0	0		18
UV									0	0	0	0		71
Invasives										0	0	0		1
Disease											0	3		12
Hypoxia												0		3
Disturbance														7
													Total studies = 202	

Figure 9.1

Stress factors (defined as any environmental or biotic factor impacting marine ecosystems that exceeds natural levels of variation) studied in the scientific literature as recorded by Web of Knowledge. Pairs of stressors are presented with the number of publications describing them. (From Crain et al 2008; reproduced with permission of Wylie-Blackwell.)

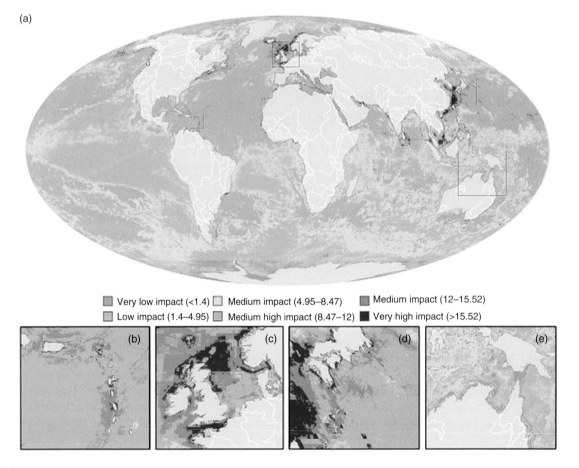

(a)

■ Very low impact (<1.4)   ☐ Medium impact (4.95–8.47)   ■ Medium impact (12–15.52)
☐ Low impact (1.4–4.95)   ■ Medium high impact (8.47–12)   ■ Very high impact (>15.52)

Figure 9.2

(a) Global map of cumulative human impact across 20 types of ocean ecosystems, with four highly impacted regions highlighted, (b) Eastern Caribbean, (c) North Sea, (d) Japanese waters, and one least impacted regions, (e) northern Australia and the Torres Strait. Data = predicted cumulative impact scores, I_c. (From Halpern et al 2008; reproduced with permission of *Science*, AAAS.)

(a)

(b)

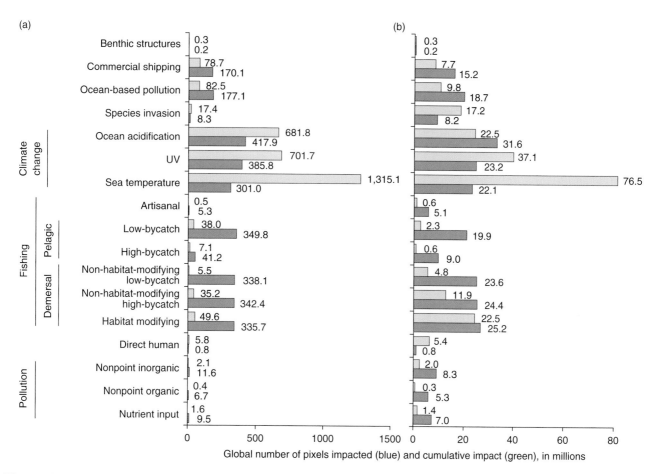

Global number of pixels impacted (blue) and cumulative impact (green), in millions

Figure 9.3

Anthropogenic impacts on marine ecosystems: Total area affected (sqaure kilometers; green bars) and summed threats (rescaled units; blue bars) for each factor (a) globally and (b) for all coastal regions <200 m depth. Values for each bar are in millions. (From Halpern et al 2008; reproduced with permission of *Science*, AAAS.)

Figure 9.4

Degraded coral reef in the Caribbean. (Photograph courtesy of Jonathan Shrives.)

ecosystem. In many regions there has been a progressive downward trend in coral cover. This loss of coral reef has occurred throughout the Caribbean (Figure 9.5) though some countries or regions are worse affected than others. In the Florida Keys, surveys of hard corals between 1996 and 2003 firstly revealed that differences between sites in the region were greater than within-site temporal variation. In general, it seems that declines in species such as *Acropora palmata*, *Millepora complanata*, and *Montastrea annularis* follow bleaching events, hurricanes, and algal blooms (Somerfield et al 2008). It also seems that Caribbean reefs are becoming flatter, or expressed in another way, less three-dimensional (rugose) (Alvarez-Filip et al 2009), with clear knock-on effects for reef diversity (see Chapter 2). At individual sites, the situation is often far from simple, and indeed, not always uniformly bleak. McGehee (2008) carried out a comprehensive study between 1995 and 2005 and, as some of her results show (Table 9.1), although many fish species have declined over the period, others have not changed in abundance. Further, some invertebrate species, such as certain gorgonians, have in fact increased.

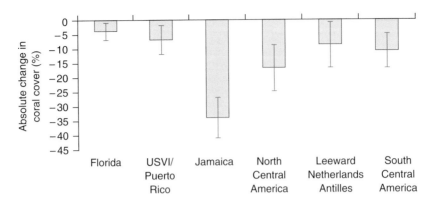

Figure 9.5

Changes in coral cover (± 95% C.I.) for the subregions of the Caribbean between 1975 and 2000. (From Gardner et al 2003; reproduced with permission of *Science*, AAAS.)

Table 9.1 Changes in abundance of species (assemblage 2) on Puerto Rican deep fore-reefs from 1995 to 2005. (From McGehee 2008; reproduced with permission of *Caribbean Journal of Science*.)

SPECIES	ASSEMBLAGE	CHANGE	*P* VALUE
FISHES			
Canthigaster rostrata (Bloch)	2	ns	0.124
Haemulon flavolineatum Desmarest	2	decrease	0.005
Holocentrus rufus (Walbaum)	2	decrease	0.002
Hypoplectrus chlorurus (Valenciennes)	2	decrease	0.016
Myripristis jacobus Cuvier	2	decrease	0.002
Sparisoma chrysopterum (Bloch & Schneider)	2	ns	0.704
Sparisoma viride (Bonnaterre)	2	ns	0.106
Stegastes partitus Poey	2	ns	0.841
Stegastes planifrons Cuvier	2	decrease	<0.001
SPONGES			
Aplysina cauliformis (Tsurumi & Reiswig)	2	increase	0.001
Chondrilla nucula Schmidt	2	ns	0.763
CORALS			
Agaricia agaricites (L.)	2	ns	0.879
Montastraea annularis (Ellis & Solander)	2	decrease	0.048
Porites porites (Pallas)	2	ns	0.702
Siderastrea radians (Pallas)	2	increase	<0.001

Table 9.1 (Cont'd)

SPECIES	ASSEMBLAGE	CHANGE	P VALUE
GORGONIANS			
Erythropodium caribaeorum (Duchassaing & Michelotti)	2	ns	0.342
Eunicea sp.			
Colonies (no.)		ns	0.053
Coverage (cm)		increase	<0.001
Gorgonia sp.	2		
Colonies (no.)		ns	0.096
Coverage (cm)		increase	<0.001
Pseudoplexaurella sp.	2		
Colonies (no.)		ns	0.115
Coverage (cm)		increase	<0.001
Pseudoplerogorgia sp.	2		
Colonies (no.)		ns	0.841
Coverage (cm)		increase	<0.001

Short-term anthropogenic impacts

Dumping at sea

Solid waste disposal presents all sorts of potential and actual environmental problems (Denton et al 2009), ranging from sediment loading to heavy metal pollution. The disposal of waste at sea is now strictly controlled in most parts of the world. In Europe, the dumping of wastes at sea is prohibited, except under licences issued under Part II of the Food and Environment Protection Act 1985 (FEPA II). The categories of licensed waste have included sewage sludge, solid industrial waste, and dredged materials. In the UK, under the OSPAR Convention, only dredged material, fish processing waste, inert materials of natural origin and vessels or aircraft may now be disposed of at sea. Dredged material now comprises virtually all of the material deposited at sea. Disposal of sewage sludge at sea ceased at the end of 1998. Solid industrial waste consisted mainly of waste rock from collieries, but also included small amounts of dumping of explosives and sediment processed in recovering coal. The dumping of colliery waste in the UK ceased in 1995.

In 1994 the UK accepted a global ban on the sea disposal of all radioactive wastes, under the London Convention 1972, superseding the ban which applied only to high level radioactive waste. Authorized discharges of low-level liquid radioactive waste into the sea are made, however, from a number of nuclear establishments.

Accidental acute release of pollutants

The greatest acute release of pollutants is caused by shipwreck, and crude oil is the cargo lost in greatest quantity. We will therefore focus on oil pollution, but other bulk cargoes of chemicals would also cause many similar types of impact. The effects of four crude oil releases are used to illustrate the types and extent of impact. It is also important

Figure 9.6

Oil tanker *Sea Empress* aground on rocks, Milford Haven, Wales. (Photograph courtesy of Julian Cremona.)

Figure 9.7

The brooding cushion-star, *Asterina phylactica*, adults and juveniles. (Photograph courtesy of Robin Crump.)

to note that not all oil spills are accidental; wars and revolutions tend to result in the inundative releases of oil and fuel supplies into the environment, often to thwart or deny the enemy the resource (Shaban et al 2009).

The *Sea Empress*

The *Sea Empress*, a 147,273 tonnes tanker, was wrecked at the entrance of the Cleddau Estuary while on her way to the Texaco refinery, Milford Haven, Pembrokeshire, South Wales in February 1996 (Figure 9.6). This resulted in one of the largest and most environmentally damaging oil spills observed in European waters. About 72,000 tonnes of crude oil were released into the sea along the coast of southwest Wales, a region renowned for the beauty and diversity of its coastline, and more than 100 km of coastline became seriously polluted by oil. By April, natural dispersion and a large cleaning operation had removed obvious signs of oil from the sea and shoreline. However, many shores were affected by residual oil pollution within the sediments during the summer of 1996, and these were remobilized by autumn storms.

Initial wildlife impacts were extensive. The plumage of birds at sea was contaminated with oil, resulting in many deaths. Where oil covered the shore, seaweeds, rock pool fish, and invertebrates were killed. Fishing was initially banned; however, following intensive Government testing of shellfish and finfish, this was lifted in stages.

It was concluded that long-term damage was more limited; the *Sea Empress* Environmental Evaluation Committee 1998 conclusions were: (a) there appeared to have been no impacts on mammals; (b) although tissue concentrations of oil components increased temporarily in some fish species, most fish were only affected to a small degree, if at all, and very few died; (c) several important populations of sea-

birds were not significantly affected, and there was no evidence of any effects on seabird breeding success; and (d) rare plants in the area were not significantly affected. However, the initial impact and the slow recovery of some organisms is illustrated by the population of *Asterina phylactica*, the brooding cushion star (Figure 9.7). By 1999 population numbers were only about 25% of those observed prior to the disaster, but recovery has subsequently been strong.

The Exxon Valdez

On March 24, 1989, the *Exxon Valdez* oil tanker ran aground in the Prince William Sound off Alaska, spilling 232,000 barrels (around 30,000 tonnes) of oil. The length of coastline impacted gradually increased, so that by 56 days after the release the oil extended 470 miles along the coast (Figure 9.8). The full effects of the incident are not really known, but only 25% of the migratory salmon population returned to the area the following season, and thousands of otters and birds died (Figure 9.9). The scale of the incident prompted the development of the Valdez Principles, which are slowly being adopted by industry. It also highlighted just how expensive a major oil spill can be – Exxon spent some US$2 billion cleaning up the spill, and a further US$1 billion to settle civil and criminal charges related to the case.

Figure 9.10 shows the likely impact on the long-term trajectory of the sea otter population (Garshelis & Johnson 2001). More otter pups were produced after the oil spill than immediately before, probably because the increase in clams increased food availability. There was massive clam mortality and habitat change following the 1964 earthquake and it is likely that otter populations began to increase appreciably after the mid-1980s. The oil spill

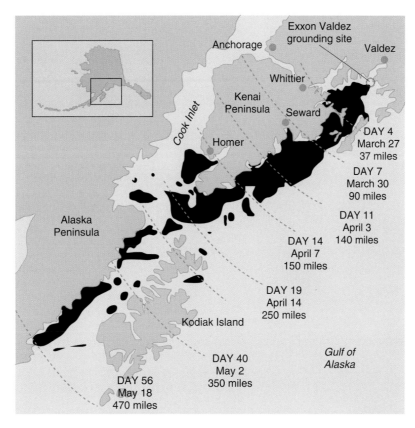

Figure 9.8

The dispersal of oil away from the release point following the *Exxon Valdez* grounding. (Reproduced with permission of the Exxon Valdez Oil Spill Trustee Council.)

Figure 9.9

(a) Oil lying on the intertidal beach following the *Exxon Valdez* grounding. (b) Dead otter covered in oil following the *Exxon Valdez* grounding. ((a,b) Reproduced with permission of the Exxon Valdez Oil Spill Trustee Council.)

therefore caused only a slight fluctuation in the population dynamics of the otters.

Dean & Jewett (2001) report hydrocarbon residues in kelp and eelgrass habitats in Prince William Sound. They found that much higher levels of hydrocarbons were deposited and retained in eelgrass habitats than in kelp beds. This was probably linked to the nature of the substrates and the different wave action in the two habitats. However, various species of marine animal showed the different responses to the oil release. The abundance of *Pycnopodia*

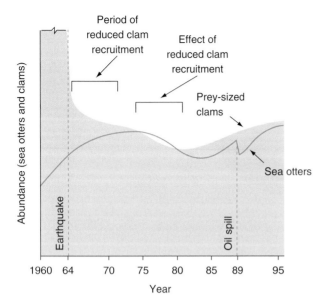

Figure 9.10

Hypothesized "trajectory" of the sea otter population in western Prince William Sound, Alaska, showing the impact of the *Exxon Valdez* grounding and oil release. (From Garshelis & Johnson 2001; reproduced with permission of Wiley Interscience.)

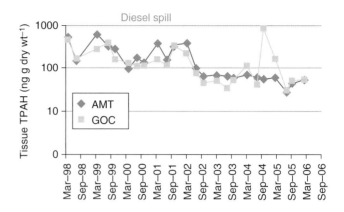

Figure 9.11

Concentrations of total polycyclic hydrocarbons (TPAH) in tissues of the filter feeding mussel, *Mytilus trossulus*, at the Alyeska Marine Terminal (AMT), the terminus of the trans-Alaskan oil pipeline, and Gold Creek (GOC) 6 km from Port Valdez where the *Exxon Valdez* first went down in 1989. (From Payne et al 2008; reproduced with permission of Elsevier.)

helianthoides, sunflower stars, for example, increased slightly, and juvenile pacific cod, *Gadus macrocephalus*, showed a great increase. In contrast, *Dermasterias imbricate*, the leather star, greatly decreased and did not recover until 1993.

Long-term monitoring of the *Exxon Valdez* disaster has now been carried out for many years (Payne et al 2008). As can be seen in Figure 9.11, there is a very significant decreasing trend (note the logarithmic *y*-axis) in the concentration of polycyclic aromatic hydrocarbons (TPAH) accumulating in the tissues of the filter-feeding mussel, *Mytilus trossulus*. By the end of the first decade of the twenty-first century, it appears that the Prince William Sound and its environs have reverted to a stable system with low levels of contaminants from the oil spillage. It has taken perhaps 20 years, but in this instance at least nature has repaired the damage done by man.

The *Amoco Cadiz*

Amoco Cadiz was wrecked on the coast of Brittany in spring 1978, spilling 220,000 tonnes of crude oil into the sea, of which between 10,000 and 90,000 tonnes became trapped in subtidal sediments. The effects of this release on small benthic organisms living in these sediments has been assessed by studying amphipods (Figure 9.12) of the genus *Ampelisca*, which inhabit fine sand sediments in huge numbers, feeding on organic matter. Because amphipods have low fecundities and do not have pelagic larvae, re-colonization from populations living in non-polluted sites is slow. As shown in Figure 9.12, amphipod populations took

12–15 years to recover from the damage caused by the oil, which is one of the slowest recoveries known for any affected organism (Poggiale & Dauvin 2001).

Studies on the impacts of the numerous crude oil releases observed since the 1960s lead to some general conclusions about their effects on marine communities. Life in areas covered with oil is initially greatly impacted so that succession must start again with empty niches. Most changes are unpredictable but temporary, and the most visible impacts are quickly repaired. The lowest impacts are experienced by the communities of hard substrates on highly-exposed sites (rocks and cliffs), conversely the greatest impacts are on soft substrates in low exposure sites (sands and mud). In the past, the worst problems derived not from oil or tar but from clean-up methods. For example, in response to a relatively small oil release in Southampton Water in the 1980s, Exxon used a team of cleaners to dig up and bag the oiled intertidal gravel. The result was the complete destruction of the intertidal pools and the loss of the physical habitat that would likely have been re-colonized if left in place. The longest recovery times are likely to be for affected species with no planktonic dispersal stage, however even these eventually recover. The inshore community has great powers of recovery so that short-term oil release catastrophes do not necessarily cause long-term ecological disasters.

Sudden man-made pressure changes; explosions and noise pollution

Explosives have the capability of producing the greatest and most damaging anthropogenic pressure shock waves. Chemical explosives for construction work are mainly used close to the coast, and their use is limited in area so that their impact on marine life is not generally great. The military use of

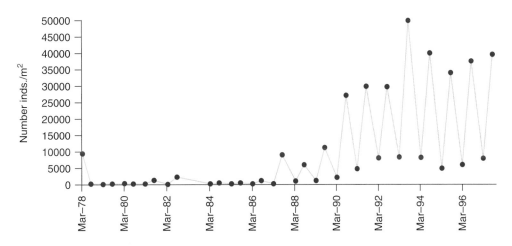

Figure 9.12

The recovery of *Ampelisca* populations in Brittany after the *Amoco Cadiz* oil release in 1978. (After Poggiale & Dauvin 2001; reproduced with permission of InterResearch.)

Figure 9.13

Blast fishing. (http://www.reefball.com)

explosives is potentially far greater and can extend far out to sea. The impacts of military explosives have not been quantified, but are only likely to be great in restricted areas during war. Explosives are also used by fishermen, and have been discharged on coral reefs to capture fish for the aquarium trade, or larger fish to eat, by stunning (Figure 9.13) (see Chapter 10). The indiscriminate destruction caused by explosive fishing is locally devastating.

The explorations for oil and gas deposits beneath the seabed use seismic survey techniques in which high-level, low-frequency sound is generated in the water column and the reflections back from the underlying geological strata detected. Marine seismic surveys have been performed since the 1950s, and in the beginning chemical explosives were used as the sound source. During the 1960s airguns were developed which have a lower environmental impact, and these are now used for nearly all seismic surveys. A typical airgun array can generate sound with total amplitude (peak to peak) of around 260 dB (relative to 1 μPa at 1 m). High levels of mortality have been found in fish exposed to 177 dB of sound and the threshold for internal injures to fish is around 160 dB. Therefore fish swimming close to an airgun when fired are likely to die, whilst others may change their behavior at considerable distances from the discharge. A recent major causeway project in California uses 150 dB (relative to 1 μPa) as a safe upper limit to avoid harm to fish. In one study by Engas et al (1996), cod and haddock catch rates declined by half over an area of 74 × 74 km during a seismic survey, and catch rates had still not returned to normal 5 days after the end of noise generation. Cetaceans frequently use sound for orientation, hunting, and communication, and so a strong response to sound would therefore be anticipated. North American studies have shown that whales, dolphins, and porpoises have changed their behavior at threshold distances ranging between 400 and 7000 m from a seismic survey vessel (Turnpenny & Nedwell 1994).

Pile driving is a common construction activity in coastal zones. Studies by Nedwell et al (2003) in Southampton Water, UK give insight into the potential for harm from percussive piling, which is the term used when piles are simply hammered into the substrate. The observed change in water pressure at a distance of 96.3 m from the pile driver is shown in Figure 9.14. The variation in peak to peak pressure declined approximately linearly from about 195 dB (relative to 1 μPa) at the pile driver, to about 152 dB at a distance of about 240 m. Thus a source level of 194 dB had declined to 150 dB (the safe threshold for no physical effects for fish) within 295 m.

The sound levels reported by Nedwell et al (2003) are lower than the maximum of 246 dB which have been reported for impact piling in deep water. Further, in deep, open waters the decline in pressure with distance is usually modeled as a geometric loss because the decline is primarily the result of dispersion over an ever-increasing area. Within restricted waters and harbors it would seem the transmission loss is primarily caused by absorption, resulting in a linear decline with distance.

Nedwell et al (2003) also recorded the change in sound pressure through time at Town Quay, Southampton, approximately 417.4 m from vibratory piling. The results showing the total sound pressure variation and the sound pressure frequency weighted for the hearing threshold of salmon are given in Figure 9.15. The figure clearly demonstrates the considerable variation in sound linked to passing vessels which swamps the variation in sound caused by vibratory

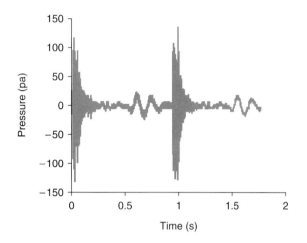

Figure 9.14

The change in water pressure through time during impact piling at Southampton. (After Nedwell et al 2003.)

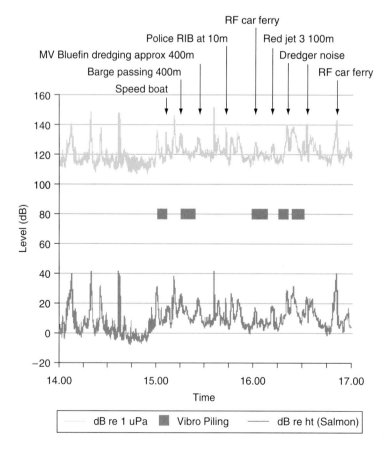

Figure 9.15

The variation in sound pressure at Town Quay, Southampton, about 417 m from vibratory pile driving. The upper plot gives the sound level in decibels relative (re) to 1 micro Pascal (μP). The lower plot gives the sound level relative to the hearing sensitivity of salmon. (After Nedwell et al 2003.)

piling which could not be detected in the signal. This study clearly demonstrated that salmon could not detect vibratory piling activity 400 m from the source.

Noise generated by dredging in the vicinity of ports and during the extraction of aggregates will depend upon the type and size of the dredger, and the material that is being dredged. Underwater sound measurements taken near the dredger *Queen of the Netherlands* indicated that the sound power emission was around 180 dB (relative to 1 µPa at 1 m) reducing to 140–147 dB (relative to 1 µPa at 200 m). This sound is of low frequency, in the range 10–1000 Hz (Malme et al 1989, cited in Richardson et al 1995). In enclosed, shallow waters, with a source sound level of 180 dB, a sound level below 160 dB would be reached within 133 m of the source. As fish would avoid moving close to a working dredger, since the sound would cause an avoidance response, acute damage would only occur if they were present within 133 m of the dredger suction head.

Ship engines also generate high levels of noise pollution, especially in estuaries and shallow seas. The noise produced by ships represents one of the most pervasive forms of man-made noise in the ocean (McCauley 1994), and in areas of high shipping density (such as the central and southern North Sea) produces a nondescript low-frequency noise (<500 Hz). This low-frequency noise propagates extremely well in deep water. Broadband source levels of ships between 55 and 85 m in length are around 170–180 dB relative to 1 µPa (Richardson et al 1995), with most energy below 1 kHz. The potential for ships to disturb the behavior of fish and cetaceans is great, and in estuaries is probably always occurring. During shipping movements, the noise levels will be sufficient to cause alarm reactions in fish within at least 200 m of any large ship greater than 55 m in length.

Dredging

Dredging (Figure 9.16) is a regular activity in almost all busy ports and has become considerably more extensive as ships have increased in size. In addition, man also dredges to replenish beaches and to mine aggregates for the construction industry. The potential effects of dredging are manifold. It may increase suspended solids, produce noise (see above), affect water quality and reintroduce into solution chemicals that have settled out and become attached to the sediment on the seabed. Fish and other organisms can be directly sucked into the dredger and killed, and habitat is lost or damaged. Undoubtedly, dredging, whether to deepen harbors or to extract aggregates, can have a serious influence on marine benthic communities. Cooper et al (2008) investigated the effect in the eastern English Channel of dredging for gravel aggregates to be used by the construction industry. They compared control (reference) sites where no dredging was carried out with low and high intensity dredged areas. Dredging in the low area ceased in 1996,

Figure 9.16

A cutter suction dredger operating in Felixstow harbour, England. (Photograph Peter Henderson.)

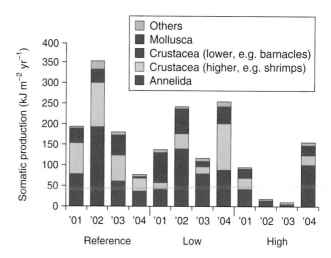

Figure 9.17

Relative contribution by major phyla to the total productivity (kJ /m^{-2} yr^{-1}) at high and low dredging intensities, compared with a reference (no dredging) site from 2001 to 2004. (From Cooper et al 2008; reproduced with permission of Elsevier.)

but resumed briefly in 2002 and 2003 in the high area. Figure 9.17 shows some of the results of benthic surveys carried out on the sites between 2001 and 2005. There is a clear direct effect of dredging in the high area in 2002 and 2003; almost all productivity by the phyla studied ceased, with the possible exception of a few crustacea and annelids. Most interesting however was the rapid rate of recovery after dredging stopped. By 2004, the high site showed equivalent productivity to the control site in the same year, though with a shifted distribution of phyla where annelids dominated.

Dredging is also carried out as a means of collecting bottom-dwelling bivalve molluscs such as scallops (*Pecten maximus*; Figure 9.18) and various species of clams. In the

Figure 9.18
Scallop *Pecten maximus*. (Photograph courtesy of Paul Naylor.)

UK, scallop dredges towed behind boats traditionally dig through the soft sediments to extract the molluscs, causing damage and death, not only to a fair proportion of scallops themselves, but also to any by-catch (all the other organisms caught by accident during the dredging process). New types of dredge are now being tested which use water pressure to lift scallops off the bottom and into collecting devices which reduce both types of damage (Shephard et al 2009). Dredging may not always have a significant impact, however. Constantino et al (2009) surveyed the effect of dredging for clams in southern Portugal, and detected no serious differences in benthic communities between dredged and control sites.

Given this extensive list of potential effects of dredging, mitigation of harm is a constant concern. In the USA, for several decades, State and Federal resource agencies have routinely requested that dredging activities are restricted to specified time periods known as environmental windows. These are defined time periods when sensitive species such as salmon, or life stages such as recently-metamorphosed young are not present in the dredging zone. Presently about 80% of all civil works operations and maintenance dredging are undertaken during environmental windows. Other major effects of dredging are briefly reviewed in turn below.

Increased suspended sediments

Dredging produces suspended sediment plumes that can differ in timing, scope intensity and duration from that found naturally. These may be outside the normal tolerances for the fish and other organisms living within or passing through the area affected. Plumes of high suspended sediment concentration (SSC; also known as suspended particulate matter – SPM) water are formed during dredging operations. Once the dredging has stopped, the time

taken for the sediment to settle is related to several factors, the most important being the size of the particles released. Small, light particles take much longer to settle than heavy ones. Tidal streams cause turbulence in the water column and will increase the time it takes for sedimentation. The strength of tidal streams varies both daily, with the strongest currents occurring in the middle of the tide, and monthly in relation to the spring–neap tidal cycle. Large waves can also have an effect.

It appears that the total amount of sediment lost to re-suspension is 2–5% of the extracted substrate. The exact proportion will vary with the type of sediment dredged, the dredging method, and the extent to which the excess water is allowed to drain off the dredger back into the water. The presence of debris, ranging from household garbage to logs and automobiles, in the sediments can increase re-suspension by interfering with the dredging process. Dredges have been designed to reduce re-suspension as they are effective in removing sediment with a minimum of re-suspension. However, field tests indicate that conventional dredges, if operated with care, can also remove sediment with low levels of re-suspension (Hayes et al 1988).

In the NRC report (1989), field and laboratory studies quantifying the extent and mechanisms of sediment re-suspension were summarized. The studies show that re-suspended sediment concentrations are generally less than 100 mg l^{-1}, except in the immediate vicinity of the dredging operation. In most of the field studies, re-suspended sediment concentrations were less than 10 mg l^{-1} at distances of the order of 100 m from the dredge. These results contrast sharply with some early assertions of re-suspended sediment concentrations of more than 1000 mg l^{-1}. Very high concentrations were observed in some laboratory studies (Herbich and DeVries 1986), but they probably reflect scaling difficulties between hydraulic parameters and sediment settling rates. None of the field studies conducted in the USA has revealed such high suspended sediment concentrations (McLellan et al 1989; Collins 1995).

Increasing dredging, or the widespread problems of river runoff after storms and floods, and soil erosion (Figure 9.19) increase the SSC, which then reduces the visibility in the water body. Since many estuarine fish are visual feeders, a reduction in visibility could have the effect of reducing their hunting success. In addition, fish obtain their oxygen by passing water over their gills. Gills are very finely divided and are very easily damaged. Increased SSC can cause gills to become blocked. In extreme cases this can lead to suffocation of the fish. The levels of SSC likely to be produced during most dredging operations are not likely to be high enough to completely block the gills of fish. However, the increased levels of sediment in the water may lead to an increase in mucus production on the gills, and an increase in gill clearing. Both of these have a metabolic cost to an animal and might, if the high SSC is long-lasting, affect the energy budget of the fish. Richie (1970) found no evidence of gill pathology in

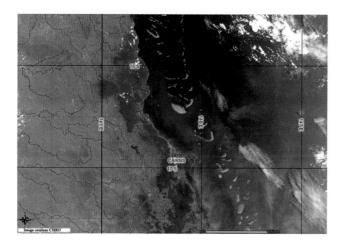

Figure 9.19
Satellite image of sediments in river runoff on part of the Great Barrier Reef.
(Photograph courtesy of CSIRO, Australia.)

11 species of estuarine fish exposed to conditions found in sediment plumes from dredging. However O'Connor and Neumann (1975) found disrupted gill tissues and increased mucus production in the white perch exposed to sublethal SSC of 650 mg l^{-1}.

Habitat damage and loss of feeding grounds

Seabed spawning fish and benthic invertebrates require particular sediment types to successfully bury or attach their eggs. Changes in the proportion of the different types of sediment found on the surface of the seabed due to dredging or sedimentation might make areas unsuitable for breeding (Courrat et al 2009).

High SSC has been observed to decrease feeding in several species of estuarine fish, including the Atlantic croaker, *Micropogonias undulatus*, and pinfish, *Lagodon rhomboides* (Minello et al 1987). For the silverside, *Atherina breviceps*, even quite small increases in turbidity have been shown to reduce feeding. The silverside is a visual predator hunting small planktonic prey. It is hypothesized that the reduction in the reactive distance leads to a decrease in their ability to forage.

Physical harm and smothering

Settling suspended solids can smother bottom-living organisms, and sedentary molluscs, barnacles and other filter feeders are particularly vulnerable. Large scale kills of mussels and other molluscs caused by sediment settlement have been reported.

Clarke and Wilber (2000) reviewed the known sublethal effects of high sediment loads on American fish. Sublethal responses observed include increased red cell counts, hematocrit levels, and hemoglobin concentrations in the peripheral blood. These are all consistent with the responses of fish deprived of oxygen. Eggs and larvae of fish are vulnerable to physical abrasion by suspended solids (Cain 1968). The settling out of suspended solids can lead to smothering of benthic organisms and eggs. Species of fish, such as the herring with demersal eggs are particularly vulnerable to this impact. Messieh et al (1981) showed that the burial of Atlantic herring eggs under even a thin veneer of sediment caused substantial mortality. Smothering of white perch eggs to a depth of 0.45 mm had no effect. Once the depth had increased to between 0.5 and 1 mm then 50% mortality was observed. Sediment layers of 2 mm resulted in 100% death (Morgan et al 1983, in LaSalle et al 1991). Adult and juvenile fish can generally avoid burial by moving, indeed they are often found returning very quickly after the disturbance ceases. The greater loss is probably the destruction of their food resource. Sessile species are most likely to be impacted by sediments for the simple reason that they are unable to get out of the way, and corals all over the world are experiencing increasing smothering or burying problems. *Siderastrea radians* is a common hard coral in the coastal lagoons of southern Florida, where Lirman & Manzello (2009) carried out experiments on the effects of chronic sediment burial on the ability of the coral (or to be precise, its zooxanthellae) to photosynthesize. *S. radians* is able to resist short-term (up to 2 hours) burial with few ill effects, and indeed is able to clear sediments off its surface within an hour if they are not too extensive, enabling it to survive in fairly high turbidity areas of a reef. However after a day or two at the most under sediment, primary production stops, and recovery is difficult. An added complication is that although this coral can clear light loads in normal seawater at 35 ppt (psu), this ability declines markedly as salinity drops to 20 or 25 ppt. A heavy rain storm inland is a double blow, since *S. radians* would be exposed not only to high sediment loads via erosion, but to lowered salinity from rainwater runoff.

Marine organisms other than animals can also be buried by sediment; seagrass is a case in point. Just like any other angiosperm plant, seagrass which becomes buried by sediment, whether deriving from marine or terrestrial activities, will suffer varying degrees of mortality depending on the depth of burial and the time under the sediment. Figure 9.20 shows the shoot mortalities of six species of seagrass from various parts of the world in experimental burials (Cabaço et al 2008). Most show a similar trend; when high sediment loads settle on seagrass beds there are likely to be significant impacts to these important and fragile ecosystems. Note of course that the area affected by sedimentation is related to the amount of sediment released and the particle size of that sediment. Larger sediments tend to fall quicker but will cause a greater degree of smothering in the areas on which they fall. Seagrass beds by their very nature are quiet and sheltered areas, where even small particles will tend to settle out and smother anything beneath them.

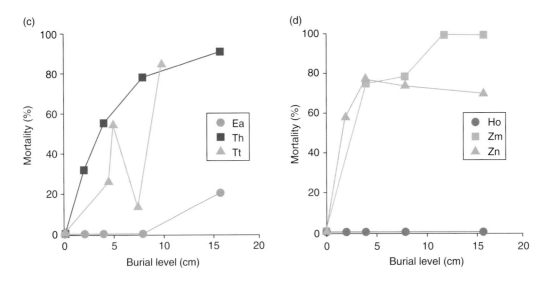

Figure 9.20

The relationship between shoot mortality and burial levels in seagrass species subject to experimental burial. Ea = *Enhalus acoroides*; Th = *Thalassia hemprichii*; Tt = *Thalassia tesdudinum*; Ho = *Halophila ovalis*; Zm = *Zostera marina*; Zn = *Zostera noltii*. (From Cabaço et al 2008; reproduced with permission of Elsevier.)

Entrainment

Entrainment is the term used for the capture of organisms by becoming sucked into the dredging machinery with the water and sediment. The physical abrasion with the silt will certainly cause a high level of damage to any entrained organism and will typically result in death.

Changes in salinity and tidal regimes

Large changes in channel morphology in estuaries can affect the salinity and tidal regime. If the dredged area occurs in the mixing zone between fresh and seawater the increased depth can cause the site to become more saline. This can have long-term effects on the organisms living in the area. Changes in tidal heights will affect the type and amount of intertidal resources available to the fish community.

Contaminated sediments

While only 2–5% of the sediment may be introduced into the water column, even 1% could be a substantial problem if these sediments hold toxic contaminants. Toxic compounds such as heavy metals, anti-fouling compounds, and aromatic hydrocarbons are common reported in high concentrations in estuaries with a long history of industrial activity. The sediments in estuaries are often contaminated with chemicals released from industrial process, water run-off, and sewage waste (see later). Many chemicals have an affinity for the fine particles found in marine sediment, to which they are often tightly bound. They can become concentrated in the sediments and get re-suspended during

dredging operations. However, the strong hydrophobic nature of most contaminants associated with sediments results in minimal release of dissolved contaminants into the water column (Digiano et al 1993).

An example of a successful dredge of highly contaminated sediments is the New Bedford Harbor Superfund site. Otis ran a pilot study using three types of dredges which verified that releases of dissolved contaminants were rather small (Otis 1992). Reporting on the dredging of the PCB-contaminated hot spot, Otis (1994) reported "no problems with sediment re-suspension or contaminant release in the water column using an extremely slow production rate." This result was upheld by laboratory studies (Digiano et al 1995) examining the partitioning of contaminants to estimate the potential release of PCBs during dredging operations and comparing the results to those of the pilot study. The microbes found in the water and on the sediment surface can take up chemicals from re-suspended contaminated sediments. These can then pass along the food chain and become concentrated, a process termed bioaccumulation. As fish are generally towards the top of the food chain the degree of bioaccumulation in their flesh can be appreciable.

Natural short-term impacts

Probably the commonest natural event that disrupts and changes natural coastal marine ecosystems is a great storm. The most intense storms can not only kill animals and plants but also change the form of beaches and shores, destroy coral reefs, and bring great quantities of sediment into

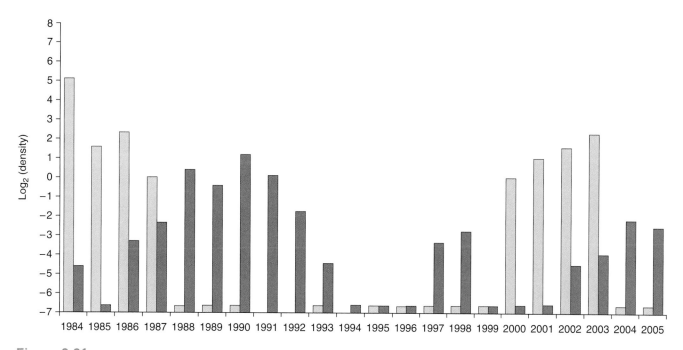

Figure 9.21

Population density fluctuations in crown of thorns starfish on two reefs on the Great Barrier Reef. Numbers are averages of per 2-minute Manta tow, $\log_2$-transformed, so that an increase of one unit on the y-axis is equivalent to a doubling of the population density. (From Uthicke et al 2009; reproduced with permission of *Ecological Monographs*, ESA.)

suspension. Not all the effects are destructive; storms can create new habitat and release and distribute nutrients creating conditions for future plankton blooms. In the temperate parts of the planet unusually cold winters can also have a large impact in shallow waters if they freeze; few littoral species can withstand freezing. However, it is clear that such natural events produce little long-term impact. Even the harshest winters leave almost no discernable impact after 10 years and recovery from even the most powerful storms is often more rapid.

Long-term and continuous impacts

Population explosions

There have been numerous reports of large-scale population blooms and outbreaks that have had major effects on other members of marine communities. In general, the causes for these outbreaks are difficult to identify and may be multifactorial. A well-known example is the population explosions in the Crown of Thorns starfish (known as COTS). This species complex occurs all over the tropical regions of the world (Vogler et al 2008), where it is an important and destructive predator of hard corals. Population explosions are a feature of COTS ecology, as

exemplified from the Great Barrier Reef in Australia in Figure 9.21 (Uthicke et al 2009). The periodicity of these large fluctuations in starfish density appears to be in the region of 10–15 years, and whenever the species breaks out coral can suffer significantly, as shown earlier in Chapter 2 (Wakeford et al 2008). Quite what drives these booms and busts is not clear (Pratchett 2005b). Booms may be caused by weather variations on land, linked to nutrient runoff and increased algal blooms, leading to more food for starfish larvae, and hence elevated juvenile recruitment may be responsible. Alternatively, it may have been initiated by the removal of tritons and other predators on starfish by man. Busts may result from one of several classic ecological factors, such as food depletion (the live coral runs out), delayed density dependent influences of predator fish or gastropods (see Chapter 10), or possibly diseases. Whatever the actual cause or causes, it is hard to distinguish between natural and anthropogenic sources.

Tidal power generation and other obstructions to movement

The development of dams and weirs in rivers has long affected the movement of anadromous and catadromous fish. The sea lamprey, *Petromyzon marinus* (Figure 9.22) in the River Severn in England gives a good example of the fate of a species unable to move successfully to their upriver spawning

Figure 9.22

A lamprey attached to the side of a fish. (Photograph Peter Henderson.)

grounds. Prior to the erection of weirs in the nineteenth century lamprey were clearly abundant and supported a valued fishery within the River Severn. From the earliest recorded times lamprey from the Severn were prized, and until 1830 it was tradition for the Corporation of Gloucester to send the sovereign a lamprey pie for Christmas. The high esteem that lamprey were held in is borne witness by the reports that Henry I died at Rouen in 1135 from a "surfeit of lamprey." The great decline in abundance was recorded as it occurred. Randell in his book *The Severn Valley* published in 1882 wrote "Lampreys too, which were formerly considered of more importance than salmon, and were caught in the upper Severn, have altogether ceased to visit it since the erection of the first weir in 1843." A few years later Day (1890) wrote "These fish up to within recent years were pretty abundant in the upper portion of the Severn, decreased very perceptibly from twelve or fourteen years since." Thus Day indicates that the decline occurred around 1865, while Randell points to immediately after 1843. It would seem that a great decline occurred between these dates and it is likely that lamprey disappeared first from the upper reaches of the river. In the first half of the twentieth century Lloyd (1941) states that they still ascend the Severn during the spring and have been taken in small number only in the Berkeley kypes. The species is now rarely caught in the Severn Estuary and Bristol Channel.

More recently, there has been growing interest in tidal power stations. There are a wide variety of designs, from small-scale units to large dams holding multiple turbines. The principle direct potential impact of all tidal generators is the death or injury to organisms passing close to rotating turbines and other machinery. Most of the available data on turbine damage have been obtained from hydroelectric schemes, with more limited information from low-head tidal generation. We therefore do not have a large body of experience upon which to predict the impact on fish. However, the available data do indicate the potential problems.

Observations of fish damage in turbines suggest that mortality rates vary from zero to over 60% of the number of fish passing through. The difference is due in part to the species involved and also to turbine type, design, and method of use. The three main sources of damage commonly reported are:

- Physical impact by fixed and moving components of the turbines.
- Pressure effects, including cavitation, due to the gradients required for machine operation.
- Shear or turbulence effects, especially immediately downstream of the turbine blades.

The types of damage attributed to these three causes include gashes, severed heads, removal of scales, various forms of eye damage, fractured backbone, and external and internal bruising. The proportion of each type of damage recorded at each scheme, and the percentage of the total throughput population affected, vary greatly, for reasons which are not clear. Much of the evidence comes from experiments in which fish were introduced to and collected from the turbines by special means. The fact that large proportions of fish apparently pass through most if not all turbines without damage adds to the difficulty of interpreting the problems reported.

Cavitation, a hydrodynamic condition responsible for the severe erosion of metal from turbine blades, is well understood and countered successfully as a routine part of the process of turbine design. Cavitation appears in areas of sufficiently low pressure for the water to vaporize, and it is the subsequent sharp collapse of these bubbles when they pass into areas of higher pressure that creates the very localized and concentrated attack, which over time leads to metal erosion. Some observers have suggested that in some turbines the bruising and pulping of fish tissues may be caused by this process, though this seems to have been concluded without reference to whether the process could have been occurring at the time of fish passage; it is very much more the exception than the rule that it happens at all, being confined to extreme rather than normal operating conditions.

In addition to any acute physical harm following passage across a turbine, fish can also be temporarily stunned or disorientated. Such effects are more likely to occur with small shoaling species such as anchovy or young herring. Such temporary effects may make them more vulnerable to predation from both birds and large predatory fish.

Tidal power generation can also have indirect effects on organisms by changing, for example, salinity, patterns of sedimentation, and the amount of intertidal habitat.

Use of water; heated and high salinity discharges

Power stations, chemical works, refineries, and desalination plants all take large volumes of seawater and discharge it

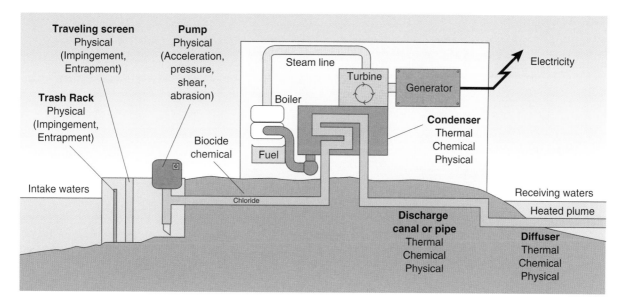

Figure 9.23

Schematic representation of a direct once-through cooling system showing the impacts different parts of the cooling water system can have on marine organisms which are sucked into the intake. (Reproduced courtesy of Pisces Conservation Ltd.)

back into the sea physically and chemically modified. At present, the greatest single users of seawater are coastal power stations using once-through cooling. The impacts of these plants will be considered in more detail below as illustrative of the impacts associated with the use of seawater for cooling. In recent years there has been a considerable growth in the number of coastal desalination plants. This activity typically releases hypersaline heated effluents, the impacts of which are also considered in more detail below.

The impacts of once-through cooling water systems

In direct once-through systems, water is pumped from the sea or estuary via large water inlet channels, it passes through the plant via heat exchangers or condensers, and the heated water is discharged directly back into the surface water. A diagram of a typical power station system is shown in Figure 9.23. Once-through systems are mostly used when large cooling capacities are required ($>1000\,\mathrm{MWth}$). Typical water flows for large power plants to cool $1\,\mathrm{MWth}$ are in the range $0.02\,\mathrm{m^3\,s^{-1}}$ (T = 12K) to $0.034\,\mathrm{m^3\,s^{-1}}$ (T = 7K). The typical water use by a 1200-MW nuclear power station would be about $30\,\mathrm{m^3\,s^{-1}}$ and this would be discharged back to the sea at about 10°C warmer than the ambient temperature.

For once-through systems the major environmental issues are as follows:

- Those associated with the use of large amounts of water. These include impingement and entrainment of fish and other aquatic life.
- The discharge of heated water.

- Sensitivity to biofouling and the need to add antifouling agents.
- Corrosion and scaling problems.
- The release of heavy metals.
- The use of additives and the resulting emissions to water.
- Construction of intake structures, intake canals, etc.
- Changes in water flow and bed scour

These are considered in turn below.

Entrainment is the term used to describe the process by which organisms get drawn into the water extraction system with the flow of water. In power stations entrained organisms pass through the condenser circuit and are discharged back to the environment. Numbers of entrained organisms are very large. For example, Sizewell A, on the Suffolk coast of England, with a pumping rate of $34\,\mathrm{m^3\,s^{-1}}$ was estimated to pass through the condenser circuit and likely kill 2×10^{10} fish eggs and 4.9×10^7 fish larvae per year (unpublished survey by Fawley Aquatic Research Laboratory, 1992–93). Entrainment in estuaries can also result in huge losses. The impacts of the power plants taking water from the Hudson Estuary, New York, have been studied for many years.

Table 9.2 uses data from the Draft Environmental Impact Statement for the Roseton, Indian Point, and Bowline power stations to calculate the average annual number of individuals of six of the commonest fish species entrained by these three power stations. The total number of eggs, larvae and post-larvae of five species of fish entrained and likely killed by these three power stations alone is estimated to be about 2×10^9. The impact of the size of these losses has been a

Table 9.2 Estimated average numbers of selected fish species entrained annually at Roseton, Indian Point, and Bowline stations; based on in-plant abundance sampling, 1981–87. (Data presented by the Department of Environmental Conservation of New York State, http://www.dec.ny.gov/docs/permits_ej_operations_pdf/ FEISHRPP1.pdf)

SPECIES	POWER STATION			TOTAL
	ROSETON	INDIAN POINT	BOWLINE	
American shad	3,128,571	13,380,000	346,667	16,855,238
Bay anchovy	1,892,500	326,666,667	81,000,000	409,559,167
River herring	345,714,286	466,666,667	13,814,286	826,195,238
Striped bass	129,857,143	158,000,000	15,571,429	303,428,571
White perch	211,428,571	243,333,333	13,257,143	468,019,048
Total	692,021,071	1,208,046,66	123,989,524	2,024,057,262

continual source of contention for many years, in part because of the naturally high mortality rates of young fish and in part because some claim that not all the entrained fish are killed. However, in the USA this loss is no longer viewed as acceptable and future power station cooling systems will be required to meet much reduced entrainment levels.

Impingement is the general term used for collision of organisms with filter screens protecting the water intakes. Typically fine screens have a mesh size of about 1 cm, and will retain most adult fish and crustaceans. In many direct-cooled power stations impinged animals are killed in large numbers. Fish kills are typically in the hundreds of thousands to millions per year, while shrimps and prawns are impinged in millions of individuals per year. For fish >3 cm in length, the impingement mortalities per annum at Sizewell A and B, Hinkley B, and Dungeness B in the UK are all in the order of 10^6 individuals.

Once-through cooling water systems are vulnerable to biofouling from a variety of organisms, and it has been standard practice to inject chlorine into the system as a biocide. Other biocides can and have been used. The effects of biocides, and chlorine in particular, released in the effluent discharge on the natural environment are poorly understood. A chlorine or bromine-based biocide typically has residual levels of up 0.2 mg l^{-1} at the discharge. It is quite difficult to measure accurately the biologically active chlorine in seawater, and chlorine is toxic to aquatic life at very low levels. For some micro-organisms inhibition of normal activity has been reported at levels that cannot be measured. Davis and Coughlan (1978) showed that bacterial activity was suppressed at chlorine levels below those that were chemically detectable. Zooplankton show severe metabolic and reproductive suppression after exposure to levels as low as 0.01 mg l^{-1}

in seawater (Goldman et al 1978). Snails exposed to cooling tower effluent at the Appalachian Carbo Power Plant suffered 50% mortality at 0.04 mg l^{-1} chlorine and 80 mg l^{-1} copper (Dickson et al 1974). Chlorine concentrations of 0.05 mg l^{-1} caused about 50% of Pacific oyster, *Crassostrea gigas*, larvae to develop abnormally (Bamber & Seaby 1997). The larvae of American oysters have a 48-hour LC$_{50}$ of less than 0.005 mg l^{-1} (Mattice & Zittel 1976). Lethal concentrations for fish occur in the concentration range 0.008–0.01 mg l^{-1} (Langford 1983). Fish show many sublethal responses. Fish will avoid chlorinated water; the avoidance threshold for trout may be as low as 0.001 mg l^{-1} (Sprague & Dury 1969; White 1972).

Thermal pollution caused by the release of cooling water from direct cooled power stations has been subject to investigation for more than 50 years. Each power station has a temperature of discharge designed to be a set number of degrees above ambient. This is often an increase (ΔT) of about 10°C. Thus, during the summer, when intake temperature in a temperate sea might exceed 20°C, the discharge can exceed 30°C. When released into tidal waters heated water typically floats over the cooler receiving waters, and this buoyant plume swings with the tide and continually changes shape. Because the heated water floats above the seabed there is often little direct thermal impact on the benthic community (Figure 9.24). An example of the extent of the surface thermal plume is given in Figure 9.25 for the 30.4 m^3 s^{-1} discharge from Sizewell A nuclear power station. This shows the extent of the plume close to low water slack tide. Note that a temperature >9°C above the ambient of about 9°C only occurs within about 100 m of the discharge point.

All aquatic life is influenced by temperature and thus thermal pollution has a high potential to change impacted aquatic communities.

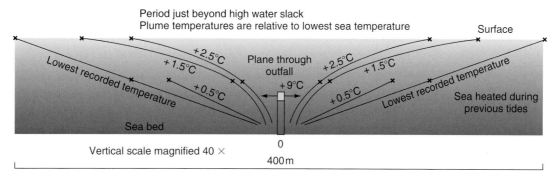

Figure 9.24

Cross-section of the Sizewell A discharge plume showing the general change in temperature with depth. Note that in the vicinity of the discharge point the warm water does not impinge the seabed. (After Parker 1977; reproduced with permission of Pisces Conservation Ltd.)

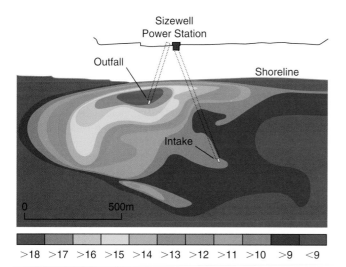

Figure 9.25

The Sizewell A cooling water discharge plume. Surface temperature (°C) contours for the discharge are shown for the point just beyond low water in the tidal cycle. (Based on data in the CERL 1975 Hydrographic survey; Pisces Conservation Ltd.)

Planktonic life

In most cases where measurements have been made in receiving waters the effects of the discharge on phytoplankton have been restricted to the near-field plume, usually within a very short distance of the outfall (Langford 1990). Generally it has been difficult to separate chlorination from thermal impacts. For example, intermittent chlorination caused reductions of 80—90% of photosynthesis measured some 50 m from the vertical discharge "boil" at the San Onofre nuclear power station on the California coast (Eppley et al 1976). The thermal plume close to the outfall typically remains discrete and unless the outfall is fitted with diffusers there is typically little mixing with the receiving water. Thus, low levels of photosynthesis in heated water close to

an outfall are related to damage during passage through the condenser circuit.

Few studies have been undertaken on the effects of the plume after appreciable mixing with the receiving water. A notable exception is the study by Smith et al (1974) at the estuarine Indian River power station, USA. A regression model was used to estimate rates of photosynthesis at points 2.5 km upstream and downstream of the discharge canal mouth, mostly within the areas bounded by the 2°C isotherm. When natural water temperatures were around 5°C, and in the absence of chlorination, the increase in the rate of photosynthesis ranged from 3% to 260% over a ΔT range of 2–8°C. Elevated rates were predicted over areas of 2 km upstream and 2–3 km downstream. In summer, when natural temperatures were around 25°C, the predicted changes in photosynthesis ranged from −7% in the discharge canal to +3% at the 2°C isotherm.

Close to the discharge point where temperatures are most elevated, impacts on planktonic crustaceans have been noted. For example, Brylinski (1981) in a study at Dunkerque power station found a close correlation between temperature and the rate of development of the copepod *Temora longicornis*.

While changes in growth and development can be anticipated, within the area with temperatures elevated by 2°C or more, there is little evidence that thermal discharges increase mortality. Recent studies by Bamber & Seaby (2004) found that an increase (ΔT) of between 8.3 and 10°C had no significant effect on the copepod *Acartia tonsa*. These authors do note that chlorination can cause high mortalities to planktonic animals. Similarly, Bamber & Seaby (2004) showed that larval common shrimp, *Crangon crangon*, exposed to a ΔT range of about 8°C or more had a significantly higher 48-hour mortality rate than that of the control.

Power stations may also be able to increase seawater temperatures in subtropical areas as well as temperate regions. Jiang et al (2009) investigated the critical thermal maximum (CTM) temperature of calanoid copepods in relation to body size, and found that large species were more sensitive to relatively hot water. The authors concluded that the composition of copepod communities close to power stations might show reductions in species diversity and a trend towards smaller body sizes, with potential consequences for marine food chains in the locality.

Fish

As early as the 1930s, Bull (1936) demonstrated, from a range of marine species covering a number of taxa and ecotypes, that fish could detect and respond to a temperature front of 0.03–0.07°C. Fish will therefore attempt to avoid stressful temperatures by actively seeking water at the preferred temperature, but this becomes increasingly a matter of chance once coordination begins to break down. If an uncoordinated fish is moved to cooler water it may recover, but the chances of recovery decrease with duration of exposure.

At less than stressful levels, increasing temperatures allow increased rates of metabolism, and (notably with regard to migratory activity) increased swimming speeds but decreased endurance (Turnpenny & Bamber 1983; Beach 1984). The temperature at which locomotory activity becomes disorganized, and thus the fish loses its ability to escape from adverse conditions, has been termed the critical thermal maximum (CTM). Once temperatures exceed 40°C, heat death ensues: enzymes are inactivated, proteins denature or coagulate, and fats melt. The last comprehensive review of this subject, from the molecular to whole organism level, was that of Rose (1967).

Some species are better than others at adapting their physiology or behavior: in general, estuarine species are the most resilient, since they are subject to regular environmental fluctuation. For any fish, there are temperatures that it prefers; temperatures to which it can acclimate; temperatures that it would seek to avoid but at which it can survive for various periods of time; and temperatures that are lethal. Moreover the ability of individuals to survive is not the same as the ability of the species to continue to prosper; increased temperatures may advance or delay breeding seasons, encourage breeding in the wrong place, or inhibit fish migration.

Indirect effects of temperature on fish include responses to the reduced solubility of gases, particularly of oxygen. This change in availability can be exacerbated by the elevated temperature simultaneously increasing the rate of oxygen removal by bacteria in the presence of pollutants such as sewage. The sort of temperature elevations that are encountered outside the immediate vicinity of a power station discharge are between 1 and 3°C, which would decrease the solubility of oxygen by only about 0.5 ppm. Were the water to be 100% saturated with oxygen then this reduction in solubility would lead to out-gassing. Gas-bubble disease, caused by nitrogen release within fish exposed to raised temperatures, has been recorded in *Oncorhynchus* young. These died in 10–30 seconds in air-saturated water with a ⊗T of 1.7°C (Snydes 1969). A similar effect was found by Coughlan (1970) in *Salmo salar* kelts when maintained in supersaturated, heated seawater (c. 13°C) for 2 months.

The effects of temperature on the biology and ecological requirements of fish have been extensively studied and reviewed. Temperature can affect survival, growth and metabolism, activity, swimming performance and behavior, reproductive timing and rates of gonad development, egg development, hatching success, and morphology. Temperature also influences the survival of fishes stressed by other factors such as toxins, disease, or parasites. Several reviewers have focused on thermal biology, specifically: lethal and/ or preference temperatures (e.g. Cherry et al 1977; Coutant 1977a). Others have widened their reviews to include data on growth, preference and lethal temperatures (e.g. Jobling 1981). Much of the literature is summarized by Langford (1990). More recently, Laurel et al (2008) studied the role of water temperature on the growth and survival of the larvae of Pacific cod, *Gadus macrocephalus*. The first thing to notice about the results shown in Figure 9.26 is that cod larvae have a working temperature range between 0 and 8°C. However, colder water appears better for their growth rate (size reached days post hatch – DPH) and for their condition. It is clear that for this species of a cold-adapted fish (see also Chapter 8), any artificial increases in seawater temperature (either locally from a power station or regionally because of global warming) may have profound effects on distribution and survival.

In open-water discharges, fish deaths caused by the temperature of the discharge are unknown. This is almost certainly because of their mobility. There are extensive observations that show that discharges change the distribution of fish. The majority of these indicate that fish are attracted into thermal discharges, although this may not be because of the temperature but rather because of the opportunities to scavenge dead animals released with the discharge.

Benthic life

If bottom substrates close to a thermal discharge are impacted by the plume, larger algal forms such as *Ascophyllum* spp. and *Fucus* spp. tend to be replaced by *Enteromorpha* and *Cladophora*. At the Maine Yankee power station *Ascophyllum nodosum* and *Fucus vesiculosus* was found to decline in regions where the effluent elevated the temperature between 5 and 7°C above ambient. Where temperatures exceed 5°C above ambient large algal species will be reduced or eliminated and at temperatures of 10°C or greater above ambient only simple temperature-tolerant forms will be present.

An important early study on the effects of a power station discharge on a sandy beach community was undertaken by Barnett (1971) near Hunterston generating station,

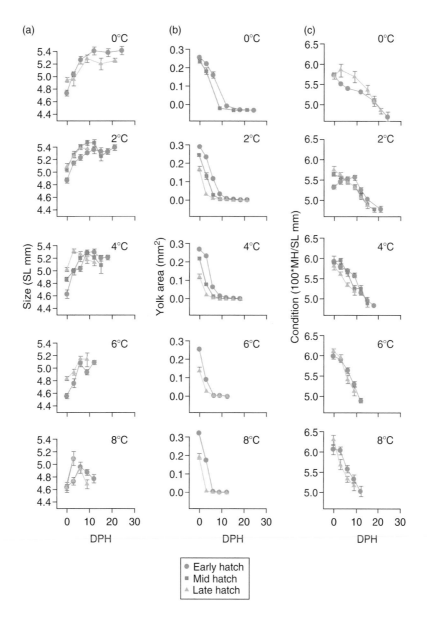

Figure 9.26

Post-hatch development changes in morphometrics of early-, mid- and late-hatch Pacific cod larvae across various temperatures. a = changes in standard length; b = changes in yolk area; c = changes in larval condition measured as the ratio of length to myotome height. (From Laurel et al 2008; reproduced with permission of Oxford University Press.)

Scotland. He concluded that on the beach adjacent to the outfall, where the temperature was 2–4°C above ambient, the growth rate of the bivalue *Tellina tenuis* increased, but population density change could not be linked to the effluent. The growth and seasonality of the amphipod *Urothoe brevicornis* was also changed by the effluent. At Kingnorth power station in the Thames Estuary, Bamber & Spencer (1984) found that tidal temperature fluctuations of 10°C caused by the effluent plume stressed the benthic community and resulted in the loss of about 50% of the expected species. Almost all the species present in the plume were littoral species that were adapted to withstand high summer temperatures when exposed to the sun on beaches (Bamber 1990). It seems likely that changes in macroinvertebrate growth and seasonality will occur when temperatures exceed 2°C above ambient, and large changes in community structure including significant loss in diversity are likely to occur when temperatures exceed 8°C above ambient.

Desalination plants

Growing water demand, especially in arid parts of the world such as the Middle East, has greatly increased the number

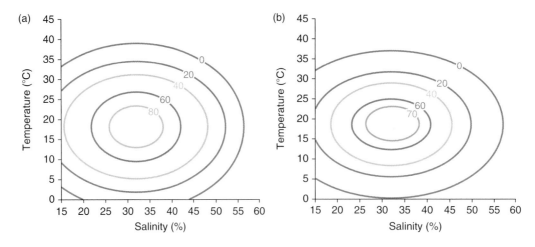

Figure 9.27

Estimated response surface curves of percentage survival for 12-day-old (a) and 18-day-old (b) *Pagurus criniticornus* larvae. (From Blaszkowski and Moreira 1986; reproduced with permission of Elsevier.)

of desalination plants. A large reverse osmosis (RO) plant extracts 100 million gallons per day (MGD) of seawater, and following passage through the reverse osmosis (RO) plant, discharges back to the environment 50 MGD of water with a salinity twice that of ocean water (about 67 ppt). This effluent may also be warmer than the ambient temperature. There are a number of ecological impacts of desalination plants, the most important of which are listed below.

- The effects of a denser than seawater hypersaline effluent on the local benthic community.
- Secondary impacts linked to seawater stratification caused by the salinity, in particular hypoxia (the reduction in dissolved oxygen) close to the seabed.
- Entrained plankton that enters the RO plant.
- The effects on planktonic life in the ocean following contact with the hypersaline effluent.
- The impact of chemical treatments added to the effluent as a result of RO operation.

Each of these potential issues is examined in more detail below.

The increase in salinity of the discharge increases the density of the water, and therefore the discharge water will sink to the seabed and impact the bottom-living fauna. The ability to withstand hypersaline conditions varies greatly between marine organisms. The ability to handle salinity variation has been most extensively studied with respect to littoral and estuarine biology, because in these habitats salinity acts as a physiological barrier for species unable to adapt. Euryhaline animals, capable of tolerating a range of salt concentrations, exhibit several patterns of physiological adaptation to salinity change. Among the vertebrates, blood osmotic concentrations are regulated within a rather narrow range by controlling ion fluxes and the accumulation of organic osmolytes. Adult fish and other marine vertebrates are unlikely to suffer acute harm from the hypersaline discharge, both because of their ability to regulate their internal salinity and, secondly, because their mobility allows them to rapidly leave unfavorable waters. Invertebrates show several adaptive patterns, but they can be roughly classified as conformers, regulators, or a mixture of the two. The common shore crab *Carcinus maenas* demonstrates both invertebrate approaches. At salinities above 25 ppt, the blood osmotic concentration tracks that of the ambient water; hence it is a conformer. At salinities below 25 ppt, it uses physiological mechanisms to regulate blood salt levels. This regulation can be maintained down to salinities of 8 ppt, so it cannot survive in freshwaters. Most fully marine subtidal invertebrates are likely to be conformers under hypersaline conditions, so that their internal salinity will reflect that of their environment. Their ability to withstand and live in hypersaline conditions will therefore be limited by the maximum salinity under which their cells and physiology can function.

The effects of salinity vary with ambient temperature, and a prediction of the ability to survive the physical conditions created by the plume must consider both the temperature and salinity effects in combination. A clear example of the synergistic interaction of temperature and salinity comes from the study of Blaszkowski & Moreira (1986) on the larval stages of the hermit crab *Pagurus criniticornus*. The response surfaces showing the change in survival for larvae of different ages are shown in Figure 9.27. As the authors state in their abstract: "Salinity also influenced temperature tolerance with thermal limits being wider at 25 and 35 than at 45 ppt salinity." In other words this hermit crab could not withstand such a high range of temperature when exposed to hypersaline water.

Stratification and hypoxia

Hypersaline discharges typically produce stratification between the surface and bottom waters, and this causes poor or incomplete mixing throughout the water column. Good mixing is an important feature of marine systems since almost all the dissolved oxygen present close to the seabed is derived from the surface. When vertical mixing is poor the bottom layers become oxygen-deficient, a condition termed hypoxia. Hypoxia is typically caused by the respiration of the bottom-living community, but, as will be described below, can be made more severe if organic material is introduced. The powerful role of salinity stratification in determining dissolved oxygen concentration has frequently been noted. As Rabalais (2005) states, "the existence of a strong pycnocline, usually controlled by salinity differences is a necessary condition for the occurrence of hypoxia." In addition, low levels of dissolved oxygen in surface water can come about by oxygen-poor subsurface water mixing at the surface.

The effects on planktonic life exposed to hypersaline effluent

The effluent discharge gradually mixes with ocean water that holds planktonic organisms, including the early life-stages of fish. This plankton is exposed to sudden changes in temperature and salinity that are potentially deadly or teratogenic. We still know insufficient detail about the effects of hypersaline water. Iso et al (1994) carried out preliminary experiments on the incipient lethal high salinity (ILHS) of fertilized eggs, larvae, juveniles and adults, of two species of fish and a bivalve. They found that the ILHS was about 50 ppt or even higher (range 50–70 ppt) and that high salinity caused delays in embryonic development. Ottesen and Bolla (1998) found that for Atlantic halibut larvae, abnormal development of the caudal notochord, sometimes resulting in a 90-degree bend of the tail, occurred during incubation of early yolk sac larvae in high-salinity water.

Some effects are less obvious; Swanson (1996) found that the rate at which the yolk sac of larval milkfish (*Chanos chanos*) was utilized varied with salinity. Swanson found that at 50 ppt salinity the yolk was utilized less efficiently than at 35 ppt and she suggested that the high salinity reduced conversion efficiency. A similar, though less pronounced, effect was also found in *Bairdiella icistia* (May 1974), at 40 ppt salinity the yolk was used more rapidly than at 35 ppt. Effects are not restricted to fish. A study by Uye & Fleminger (1976) of several *Acartia* copepod species from the Californian coast found the hatching time increased at both hypersaline and low saline environments. Fewer than 50% of eggs hatched at salinities above 50‰ and some of these had deformities. It is well known that the development of early stages of multicellular organisms can be damaged by such sudden shocks, producing malformed embryos which cannot develop successfully (Brown and

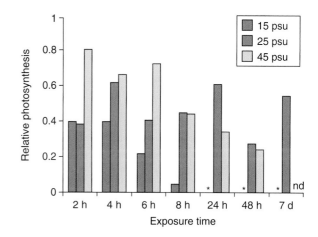

Figure 9.28

Relative photosynthesis of *Siderastrea radians* exposed to different salinity regimes. * = respiration exceeds photosynthesis. (From Lirman & Manzello 2009; reproduced with permission of Elsevier.)

Nunez 1998). A commonly observed teratogenic effect on fish embryos is the malformation of the vertebral column, and there are indications that both salinity and temperature shock can induce such damage.

Adult animals (or colonies thereof) show all sorts of reactions to different salinity levels. Figure 9.28 shows just such an example (Lirman & Manzello 2009). Here, the hard coral *Siderastrea radians* (or, to be precise, its zooxanthellae symbionts) shows varying rates of photosynthesis under equal light regimes according to the salinity of seawater. It is clear from the figure that hypersaline seawater poses no real obstacle to primary productivity at least over a few hours, whereas very dilute (brackish) water significantly influences photosynthesis. The implications for corals exposed to varying salinities (high and low) are thus rather predictable.

The impact of chemical treatments added to the effluent

It is common practice to use a variety of chemicals in RO plants to maintain the RO membranes. Any chemical used to run or clean the RO plant may be discharged to the sea. Einav et al (2002) suggest the following chemicals are commonly used in desalination plants:

- NaOCl or free chlorine, used for chlorination, preventing biological growth (antifouling).
- $FeCl_3$ or $AlCl_3$, used for the flocculation and removal of suspended matter from the water.
- H_2SO_4 or HCl, used for pH adjustment.
- SHMP $(NaPO_3)_6$, and similar materials, prevent scale formation on the pipes and on the membranes.
- $NaHSO_3$, used in order to neutralize any remains of chlorine in the feed water.

Chemical pollution and antifouling compounds

From the beginning of the industrial revolution and the subsequent development of huge cities there has been an accelerating increase in the release of waste, rubbish, and chemical pollutants into the sea. In fact, it is likely that some of the most serious factors influencing trophic diversity and structure of marine communities involve not just physical disturbance, but also pollutants such as metal contamination (Schratzberger et al 2009). While some other compounds such as phosphates are not of themselves harmful, they are nutrients and can cause excessive blooms of lower organisms such as bacteria and dinoflagellates, resulting in anoxia and in extreme cases the loss of all higher life. In the twentieth century estuarine rivers near major cities such as London and New York held anoxic rivers that were almost fishless. Recent improvements in sewage treatment and disposal have allowed fish to return to both the Thames and the lower Hudson estuary. The effects of sewage are considered in more detail below.

In addition to excessive nutrients man has released numerous damaging compounds, and even a general introduction to the main groups of these compounds and their effects would be beyond the scope of this book. We therefore illustrate the types of effects observed using two compounds of global importance. Copper pollution is a typical product of human activity and is one of a number of harmful heavy metals frequently encountered in elevated concentration in estuarine and inshore sediments. Tributyltin is a man-made organic compound and is one of many harmful synthetic organic compounds that have polluted the seas during the twentieth century.

The effects of metals

The polluting effects of heavy metals in the sea have been documented for a long time. One classic example concerns mercury which was deposited in industrial wastewater into Minamata Bay on the Japanese island of Kyushu in the 1950s and 1960s. The poison was accumulated by filter feeding bivalves, and eventually was thought to be responsible for the deaths of over 10,000 local people. Mercury still enters marine systems today (Muenhor et al 2009) but monitoring will hopefully keep levels tolerable. There are a variety of sources for elevated copper concentrations present in the sea, which include sewage discharges, fungicides and herbicides, and antifouling paints. The use of copper as an antifouling compound shows the potential for high concentrations to damage ecosystems. The effects of copper on the colonization rate of sedentary organisms are illustrated for the coral *Acropora tenuis* in the work of Reichelt-Brushett & Harrison (2000). Up to around 20 μg per litre copper, coral larvae appear to settle normally, but as concentrations in seawater increase beyond this level,

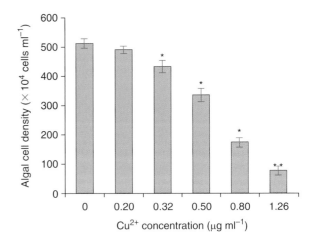

Figure 9.29

Effects of Cu^{2+} at different concentrations on growth of the diatom *Phaeodactylum tricornutum* after 72 hours exposure (stars indicate levels of significant differences). (From Wang & Zheng 2008; reproduced with permission of Elsevier.)

settlement is significantly reduced, so that by ≥50 μg per litre copper, larvae do not settle. Thus, in water with relatively high concentrations of copper, coral regeneration simply will not happen. Once organisms have settled or metamorphosed, copper can still influence their growth and/or survival. Diatoms are obviously extremely different creatures to corals, but as Figure 9.29 illustrates, their growth rate, measured as cell replication, significantly declines beyond the 20 μg per litre copper level (Wang & Zheng 2008).

Ship antifouling and tributyltin

Tributyltin (TBT) is a man-made organic compound containing tin, which enters the sea because it is present in wood preservatives and used in marine antifouling paints on ships and other objects placed in the sea. It is toxic to fish and other marine life such as gastropod molluscs. Marine life is known to bioaccumulate TBT compounds, which also bind well to sediment where they can persist for long periods. High levels of TBT have been found in water, sediment and biota near to centres for pleasure boating activity, especially in or near marinas, boat yards, and dry docks. Controls on the use of TBT in antifouling paints were introduced in many parts of Europe in 1986, when the sale of TBT-based paints was banned. In 1987, the use of TBT-based paints on boats under 25 m and mariculture equipment was also prohibited. Even so, TBT levels around parts of the Mediterranean were still very high by 2003 (Figure 9.30) (Diez & Bayona 2009). The major problem or ecological impact of TBT is imposex in molluscs (de Castro et al 2008), where females develop at least partial male reproductive organs and become effectively

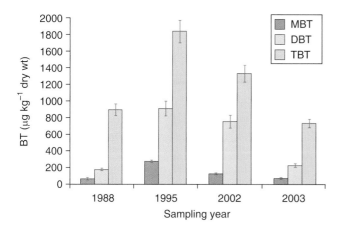

Figure 9.30

Mean concentration of butyltin from surveys in Barcelona harbor. (From Diez & Bayona 2009; reproduced with permission of Elsevier.)

sterile. The degree of harm is measured by the relative penis size index (rpsi), calculated as

$$\frac{(\text{Mean length of female penis})^3}{(\text{Mean length of male penis})^3} \times 100$$

Sites with many boats and hence high concentrations of TBT could be identified by examining the age structure of dogwhelks, *Nucella lapillus*. In the UK, whelk size frequency diagrams show very few if any juveniles (indicted by their small size) in sites with large numbers of small boats, such as Brighton on the south coast of England. Once TBT is removed the system recovers, as was reported by Svavarsson (2000) for Reykjavik harbor, Iceland. In 1992–93, there was a high incidence of imposex (over 60%), but within 5 or 6 years, the problem had diminished to 25% or less.

Sewage

Sewage is defined as waste matter from domestic or industrial establishments that is carried away in sewers or drains. It is impossible to estimate just how much sewage is pumped untreated into our oceans, but no country, developed or developing, can honestly be said to be free from guilt. The problem is as usual a combination of technology and economics; it is clearly much simpler and cheaper to run a long pipe out to sea where untreated, raw sewage can be dumped, hopefully "out of sight – out of mind." It is often very difficult to accelerate the construction of complex sewage treatment facilities to keep up with the rate of coastal colonization and urban development (Thronson & Quigg 2008), so the problems increase. Indeed, Kocasoy et al (2008) suggest that the incidence of water-borne diseases which can be related to lack of sewage treatment is actually increasing world-wide. Even small boats contribute to the levels of raw sewage in the sea, especially in the tourist season (Leon & Warnken 2008).

Sewage is a major ecological problem for a number of reasons. First, it causes direct ecological damage. Suspended solids may blanket river and seabeds, preventing respiration of benthic flora and fauna (see earlier in this chapter). There are many constituents of sewage, both from domestic and industrial sources, which have an effect on aquatic life. Faecal material, both human and agricultural, is frequently found on coral reefs for example (Bonkosky et al 2009), with many associated impacts on human health and tourism. The breakdown of proteins and other nitrogenous compounds from these sources releases ammonia, potentially toxic to fish in low concentrations. Figure 9.31 shows the results of seawater samples taken at different times of year in the Gulf of Trieste in the northern Mediterranean (Mozetic et al 2008). There are clear associations between ammonium concentrations and faecal coliform bacteria levels, the highest of both being close to the surface of the sea. Second, sewage poses considerable health risks. Bathers are at increased risk of contracting illnesses in sewage-contaminated waters, due to the bacteria and viruses present in sewage effluent. In particular, gastrointestinal disorders have been linked to sewage pollution, with viruses implicated as the cause. Shellfish grown in sewage-contaminated waters may cause food poisoning, as mussels and oysters can accumulate human pathogens through their filter feeding apparatus. It is not uncommon for shellfisheries to be closed because of contamination. Finally, sewage pollution also results in economic loss. Debris associated with sewage probably has the highest monetary cost associated with its presence on our beaches, due to loss of tourism and blockage removal. Closure of shellfisheries due to sewage contamination can lead to high loss of income and closure if breaches of consent occur frequently.

Actually tracing or identifying the point source of marine sewage pollution can be difficult. Molecular markers have recently been used in Puerto Rico to study non-point source sewage contamination in coastal waters (Bonkosky et al 2009), whilst Luo et al (2008) employed linear alkylbenzenes (LABs) in sediments as markers for sewage contamination in the coastal waters and estuaries of the Pearl River Delta in the vicinity of Hong Kong, China (Figure 9.32). Notice that sites close to urban and industrial areas and/or up estuaries with limited flushing by the sea (ZJ, DJ, and XJ) showed the highest levels of LABs (though not as high as actual raw sewage sludge – SW), whilst open water sites in the estuary and the sea itself (PRE and SCS) were the cleanest. Clearly, this type of pollution becomes diluted the further away from the point source, but nevertheless the ability of the sea to remove any significant problems is limited.

Land reclaim

Coastal wetlands have been converted into agricultural land for many thousands of years. In the last century this

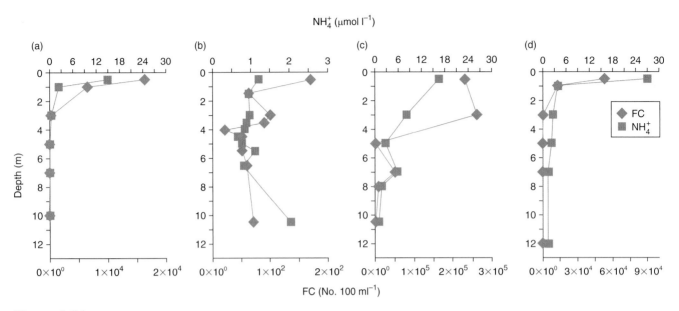

Figure 9.31

Vertical distribution of ammonia (NH_4^+) and faecal coliform bacteria (FC) in March, May, August and October in the Gulf of Trieste. (From Mozetic et al 2008; reproduced with permission of Wiley InterScience.)

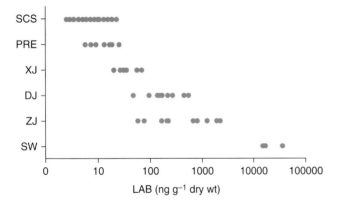

Figure 9.32

Geographical distrbution of total LAB concentrations in the study areas. SW = sewage sludge; ZJ = Zhujiang; DJ = Dongjiang River; XJ = Xijiang River; PRE = Pearl River Estuary; SCE = South China Sea. (From Luo et al 2008; reproduced with permission of Elsevier.)

loss has accelerated as industrial uses for airports, power generation, oil refining, and other chemical processes have taken large areas of salt marsh and beach habitat. In many parts of the world more than 50% of the natural saltmarsh habitat has been lost. These habitats are important to coastal ecosystems as they transfer energy and resources from the land. They are also important nurseries for many marine animals, including fish. One very good example comes from Holland (the Netherlands). The port of Rotterdam, one of the most important in Western Europe,

if not the world, has been in existence for many hundreds of years, gradually expanding onto land reclaimed from the North Sea (Hommes et al 2009). Close by however is the Wadden Zee (Wadden Sea), an internationally important area of wetlands famous for birdlife, and a vital breeding ground for various commercial fish stocks in the North Sea (see Chapter 8). More land reclamation, vigorously opposed by environmentalists of course, has to assess the impact on these natural habitats, and the advantages gained by expanding Mainport Rotterdam weighed against the possible destruction of habitats. Compensation in the form of new Marine Protected Areas (see Chapter 10) in the locality may be the way forward. Even if land is not reclaimed, many projects these days convert coastal seawaters and estuaries into brackish or freshwater lakes for tidal power, recreation, and other uses. This process can seriously influence the chemistry and ecology of the waters, especially in terms of salinity changes and eutrophication (Hodoki & Murakami 2006), with fundamental consequences for the natural ecology.

We discuss the vital and multivarious importance of mangroves in several parts of this book, and as mentioned elsewhere, probably the biggest threat to them and their associated species and ecosystem services is their conversion to aquaculture, in particular for shrimps and prawns as well as various species of brackish water fish. Huge areas of mangroves are being destroyed all over the world to turn into lakes and ponds for the intensive production of shrimp (Figure 9.33). In the Philippines, for example, Primavera (2005) reports that only a fifth of the

Figure 9.33

Shrimp ponds built from mangrove forests in Indonesia. (Photograph Martin Speight.)

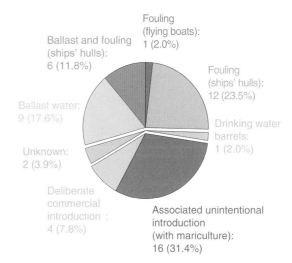

Figure 9.35

Probable primary methods of introduction of non-native British marine flora and fauna. Numerals indicate the number of species and the percentage of total introductions. (Eno et al 1997; from *Non-native Marine Species in British Waters: a review and directory*; copyright Joint Nature Conservation Council – JNCC – UK.)

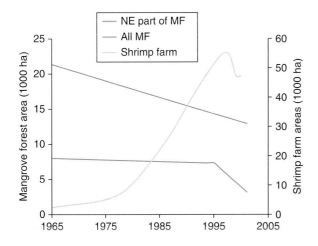

Figure 9.34

Relationship between mangrove forest (MF) areas and shrimp farms areas in the Mekong Delta, Vietnam. NE = northeastern. (From Thu & Populus 2007; reproduced with permission of Elsevier.)

500,000 ha of mangroves in existence at the turn of the twentieth century remain, whereas the area of aquaculture ponds which replace the mangroves have risen from 61,000 ha in the 1940s, to 230,000 in the mid 2000s. Similar destruction has been occurring in the Mekong Delta of Vietnam, where as Figure 9.34 shows, the rate of losses of mangroves to shrimp farms has accelerated significantly in the last decade or so (Thu & Populus 2006). Note that as we describe in the next chapter, restoring mangroves once they have been converted to aquaculture is extremely difficult.

Species invasions and introductions

Some of the most profound changes now occurring in coastal waters are caused by the human introduction of

Figure 9.36

Invasive seasquirt, *Styela clava*. (Photograph courtesy of Paul Naylor.)

non-native species. The constant movement of ships and goods creates huge potential for accidental introductions. However, there is also a steady, deliberate, introduction of species for commercial gain, sport, or simply because someone thinks they will enhance the local flora or fauna. The British fauna offers a well-studied example of the amount of movement of species. The probable relative frequency of the various methods of introduction of non-native fauna into British waters is illustrated in Figure 9.35. The most successful invaders have been annelid worms, crustaceans, molluscs, and seasquirts such as *Styela clava* (Figure 9.36). These introductions have their sources in all parts of the

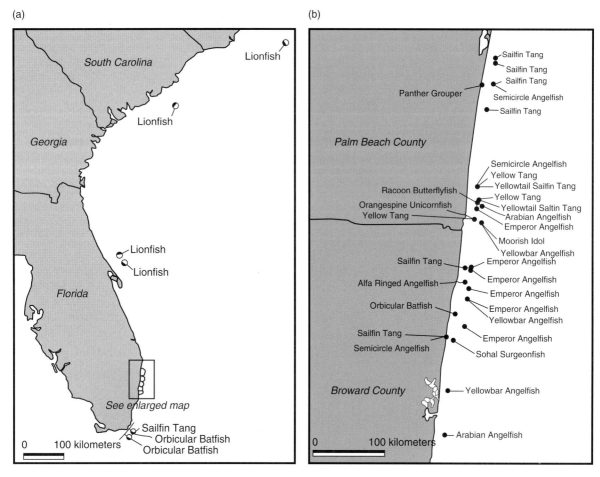

Figure 9.37

Locations of non-native marine fish sightings by sports divers along (a) the southeastern USA and (b) Palm Beach and Broward Counties in Florida. (From Semmens et al 2004; reproduced with permission of InterResearch.)

world, tropical and temperate, wherever maritime trade is practised. Fernandez (2008) describes similar problems on the west coast of America, where maritime trade under the auspices of the North American Free Trade Agreement (NAFTA), ships travelling the south–north route from Mexico via California to Canada are highly efficient shippers of invasive species, both in ballast water and via biofouling communities. Fernandez in fact describes the ships plying these routes as "mobile aquaria".

One increasingly serious source of invasive species involves the marine aquarium trade. Figure 9.37 shows the locations of non-native (exotic) fish species in parts of Florida (Semmens et al 2004), as seen by recreational SCUBA divers. No obvious correlations were found between the native homes of the species sighted and the incidence of shipping from these regions, but there was a clear relationship between the numbers of individuals of exotic fish sighted and the intensity of their import into the southern USA for tropical marine aquaria. Quite how these fish escape captivity and end up breeding successfully in the sea is less certain.

The introduction and impacts of major macro-algal invasive species into the Mediterranean have been described by Boudouresque (2002). The recent spread of an invasive strain of *Caulerpa taxifolia* (Figure 9.38), a species native to the Indian Ocean, shows the speed of spread that can occur. In 1984, this seaweed was accidentally released just below the Oceanographic Museum of Monaco and subsequently began to spread between ports along the Mediterranean coast where it smothers slower-growing native species. Aquarium strains at least of *C. taxifolia* have now invaded southern California and Australia, and it is likely to appear in Florida soon (Glardon et al 2008). Invasive species such as this one do of course provide habitats for other marine species which are native to a particular locality (Wright & Gibben 2008), but it is likely that such habitats may in the long term be suboptimal because of their exotic nature.

Successful invaders, both animal and algal, tend to have the following shared characteristics:

- Vegetative reproduction is usually the commonest if not the only method of reproduction, so only one viable propagule is sufficient to start a new colony.
- Vegetative reproduction is prolific.
- Habitat requirements are flexible.
- They tolerate the stresses of environmental fluctuations and extremes.
- There is a similarity between the native and recipient habitat.
- They are free from predators and diseases which occur in their native range.

Figure 9.38

The invasive green alga, *Caulerpa taxifolia*. (Photograph courtesy of A. Meinesz.)

Tourism and human disturbance

Tourism and general human disturbance from activities such as bait-digging is now beginning to have a notable impact in many parts of the world. Even simple activities such as walking can damage the intertidal. For example, some years ago, Neil (1990) showed the relationship between suspended sediment concentration and the number of walkers on a coral reef. He found that sediment levels produced in this way could exceed 1000 mg l^{-1}. To put the level into perspective the maximum level measured during a typhoon was about 200 mg l^{-1}. As well as stirring up sediments, reef walkers frequently come into contact with coral growths under the surface of the shallow water in which they are paddling or swimming. Leujak & Ormond (2008) examined the number of contacts, amount of damage, and remaining hard and soft coral cover in five sites near Sharm El Sheikh, Egypt, and showed that between 4% and 30% of corals were damaged by walkers (Figure 9.39). It seems that male snorkelers were the worst offenders, walking on the corals to rest or look around.

Most importantly, the effects of SCUBA divers can be considerable. Zakai & Chadwick-Furman (2002) demonstrated that the level of damage is related to the frequency of diving and differs between types of coral; *Millepora* (hydrozoan fire corals) and then *Acropora* (branching hard corals) were most affected. The results of such damage in terms of mean percent cover of all scleractinian corals is shown in Figure 9.40 (Hasler & Ott 2008). In Thailand, it was found that damage by fins was most common and severe, with experienced male photographers causing least damage overall (Worachananant et al 2008). Clearly, coral cover declines significantly at frequently dived sites compared with non-dived sites, but this effect varies with site and location on the reef. Both examples emphasize the

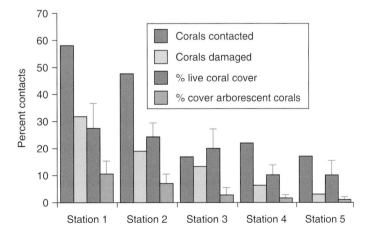

Figure 9.39

Effects of people walking on Red Sea reefs in terms of coral contacts, damage, and resulting live coral cover (±s.e.). Corals contacted = percent of observations with no damage; Corals damaged = percent of observations resulting in damage to hard or soft corals. (From Leujak & Ormond 2008; reproduced with permission of Elsevier.)

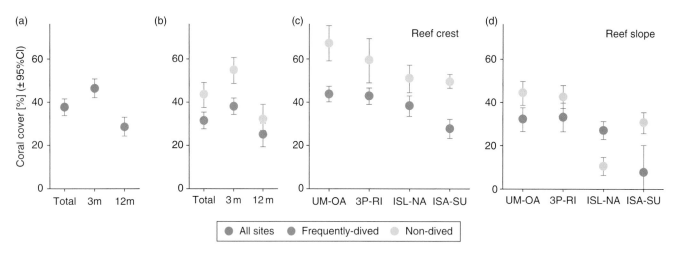

Figure 9.40

Mean percentage coral cover at eight sites in Dahab, northern Red Sea, in April 2006. (a) Pooled coral cover over all sites, (b) pooled coral cover at non-dived and frequently dived sites, (c) & (d) coral cover at four sites on reef crest and reef slope respectively. UM–OA = Um Sid–Oasis; 3P–RI = Three Pools–Rick's Reef; ISL–NA = Islands–Napoleon's Reef; ISA–SU = Islands outside–Suleiman Reef. (From Hasler and Ott 2008; reproduced with permission of Cambridge University Press.)

Figure 9.41

Bleached coral. (From: http://bluejakarta.wordpress.com)

fact that intensive SCUBA diving seriously threatens coral reefs. As we shall discuss in the tenth and final chapter, however, sport and tourist divers will usually only visit healthy reefs, which thus need divers to keep them healthy.

Climate change

An enormous quantity of literature has been generated over the last few years concerning the likely effects of climate change on marine ecosystems, and there has been much discussion of these potentially huge impacts. At the largest scale there have been claims that the melting of the polar ice caps will result in major changes in ocean currents and huge changes in ocean productivity. At a smaller, but still global scale, the melting of the ice caps will cause an increase in seawater level and cause a direct impact on coastal wetlands. One potential impact is called coast squeeze, in which the littoral zone is reduced because man builds coastal defences to stop the inundation of occupied land, and thus stops the development of new intertidal habitat. While changes at this scale are not yet proven and hopefully never will be, there is considerable evidence of recent change related to changes in water temperature. We introduced climate change and the world's oceans in Chapter 1, and we will only briefly reconsider three aspects here, i.e. increases in (a) seawater temperature, (b) seawater acidification, and (c) sea level, and the ways in which they may influence marine species.

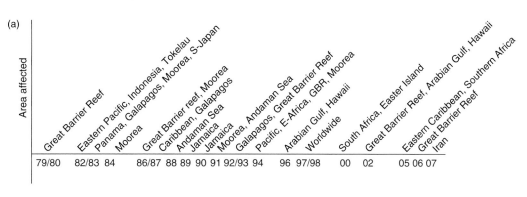

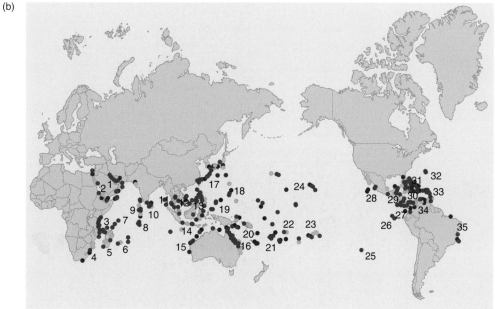

Figure 9.42

Coral bleaching worldwide. (a) Documented bleaching events. (b) Reef-base data of worldwide bleaching records (as of March 2008). Numbers refer to specific sites. Red circles = high; blue circles = medium; yellow circles = low; green circles = no bleaching. (From Baker et al 2008; reproduced with permission of Elsevier.)

We have mentioned coral bleaching (Figure 9.41) earlier in this book, and it is now a worldwide phenomenon (Figure 9.42) (Baker et al 2008). It arises due to the expulsion or death of the symbiotic zooxanthellae (*Symbiodinium*) from coral polyps, which seems in part at least to be related to temperature. Bleaching is most frequently reported when seawater temperature is elevated by 1–2°C for 5–10 weeks during the summer season (see Chapter 1) (Sampayo et al 2008). The long-term implications for coral reef survival are not clear at the moment, but it may be that many coral species are able to swap symbiont strains in favor of more temperature-tolerant ones in warmer water. Certainly, some reefs are able to recover from bleaching events even as temperatures continue to rise, but for those that are unable to do so, the effects of bleaching on coral communities may be serious, as argued by Munday et al (2008) (Figure 9.43). Here it can be seen that the effects of bleaching on coral and algal

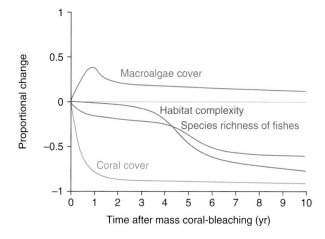

Figure 9.43

Conceptual diagram of changes in key attributes of coral reef habitats and fish communities following severe coral bleaching from which no recovery of corals takes place. (From Munday et al 2008; reproduced with permission of Wiley InterScience.)

cover, the habitat complexity, and hence the species richness of fish, are all inter-related. Habitat complexity provides three-dimensional mosaics for fish and invertebrates in which to live (see elsewhere in this book), and if dead coral skeletons are not replaced with living coral, their eventual erosion and collapse will significantly reduce complexity as well as structural integrity.

There are also numerous examples of changes in fish abundance related to temperature. Henderson (2007) reports a major shift in the structure of the fish community of the Bristol Channel, England, between the 1980s and the 2000s. Using a 25-year time series of monthly samples, he showed that the fish community of Bridgwater Bay in the outer Severn estuary is rapidly responding to changes in seawater temperature, salinity, and the North Atlantic Oscillation (NAO). The number of fish caught each year has followed an increasing trend, which could be related to increased temperature and decreased salinity. In contrast to this smooth change, there have been two discrete transitions in fish community structure around 1986 and 1993.

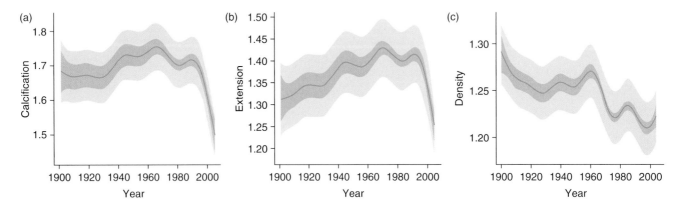

Figure 9.44

Variations of calcification (g cm^{-2} yr^{-1}), linear extension (cm yr^{-1}), and density (g cm^{-3}) in the hard coral *Porites* on the Great Barrier Reef, based on core samples. The dark blue shading indicates 95% confidence intervals for comparison between years, and light blue shading indicates 95% confidence interval for the predicted value for any given year. (From De'ath et al 2009; reproduced with permission of *Science* – AAAS.)

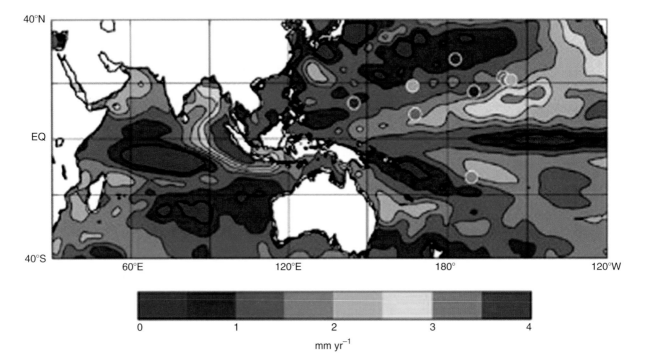

Figure 9.45

Reconstructed trends in sea level in the Indo-Pacific between 1950 and 2001. (From Church et al 2006; reproduced with permission of Elsevier.)

The first of these step changes, which altered the relative abundance of the dominant species, was linked to a change in the NAO. The second, which was caused by a change in the set of occasional visitor species, was linked to an increase in average seawater temperature. A marked increase in population variability for many fish in recent years is linked to increased seawater temperature. Temperature might not be the only factor which promotes coral bleaching; elevated CO_2 levels may also be related (Anthony et al 2008).

The average pH of seawater is around 8.2 or 8.3, varying with depth, latitude and season, but when CO_2 dissolves in seawater, carbonic acid (H_2CO_3) is formed, which then breaks down into bicarbonate (HCO_3^-) and hydrogen ions (H^+) which leads to a decrease in pH, so-called ocean acidification (Hofman & Schellnhuber 2009). A decrease of 0.1 in pH may not seem much, but in fact it represents a 10% change in acidity, so even deceptively small decreases in pH may have a significant effect on marine organisms that use various forms of calcium carbonate, such as aragonite, for their skeletal structures. As shown by De'ath et al (2009) (Figure 9.44), the calcification rate of *Porites* in Australia has declined by about 14% since 1900, due it is thought to the reduction in the ability of the coral to deposit calcium in water of a lower than normal pH. Many other types of marine organisms may be similarly affected, from phytoplankton and macro-algae to molluscs and echinoderms. If these organisms are unable to produce adequate skeletons, the consequences will be dire.

Finally, climate change is also expected to result in sea level rise. The sea has changed levels many times over geological time, and now it seems to be increasing steadily, as illustrated in Figure 9.45 for the Pacific and Indian Oceans (Church et al 2006). The global estimated average is around 2 mm per year, which if it remains constant for decades will inundate large areas of low-lying coastal habitats. Whether sessile species, such as hard corals, can grow sufficiently fast to compensate for their slow inundation remains to be seen. All of these climate change scenarios have their roots in part at least in man-made sources. Global warming, melting icecaps, and increased CO_2 emissions have all been blamed on anthropogenic activities such as the use of fossil fuels, and with little or no sign of appreciable declines in this, we must anticipate the impact on the world's oceans with some trepidation.

Chapter 10

Marine conservation

A Google search in October 2009 using the exact phrase "marine conservation" produced about 529,000 hits. A trawl through ISI Web of Knowledge for the same phrase at the same time revealed over 36,000 records. Clearly both the grey literature and the scientific community are working hard on the topic of marine conservation, and to distil all this knowledge and experience (and a certain amount of hearsay and propaganda) is a difficult task. In this chapter, we continue where the last chapter ended by discussing how to restore, rehabilitate, and manage marine ecosystems for their future survival and preservation. The first premise must of course be that the impacts described in Chapter 9 should if at all possible be minimized or avoided all together. However it is carried out, restoration and conservation of marine ecosystems is likely to be expensive and necessarily relatively small scale (Yeemin et al 2006), so avoiding damage is far better than attempting a cure.

Most research and development in the field of marine conversation and management has been carried out for fairly obvious reasons in and around coral reefs, and so we concentrate mainly on these, with a few temperate exceptions. Most if not all lessons learnt on reefs should be easily applicable to other marine habitats, arctic, tropical or temperate, shallow or deep.

Restoration and rehabilitation

In most circumstances, reefs do not recover naturally from anthropogenic impacts and disturbances (Rinkevich 2005). Several times in the book, we have shown illustrations of coral reefs degraded or destroyed by various impacts, and it is hopefully clear from these accounts that the first and foremost stage in the restoration and rehabilitation of reefs is to reintroduce healthy hard corals which will survive, grow, and finally reproduce into the foreseeable future. In order to do this, the causes of decline such as algal domination and overgrowth, pollution, temperature and light extremes, damage by tourism, fishing, dynamiting, poisoning, and so on, must be reduced or removed. Any leftover habitat damage must be repaired to provide clean and stable sites for coral re-colonization. After that, new coral recruitment can be encouraged by the creation of stable and complex structures that protrude above general substrate level to reduce abrasion and smothering (Fox et al 2005).

Stabilizing the substrate

One extreme form of damage as we discussed in Chapter 9 is dynamite (blast) fishing. In Southeast Asia, as Raymundo et al (2007) report, blast fishing creates rubble fields (Figure 10.1), which are too fragmented and mobile for new corals to settle on. These authors treated 17.5-m^2 rubble plots by laying out plastic nets with 2-cm mesh size onto the rubble and holding them in place with stakes hammered into the substrate. Holes were cut in the net to allow any surviving coral heads to be retained. Finally, 1-m-high pyramid-shaped rock piles were made on land using reef rock and cement and placed on the net underwater. Coral recruitment was then allowed to occur naturally from adjacent

Marine Ecology: Concepts and Applications, 1st edition. By M. Speight and P. Henderson. Published 2010 by Blackwell Publishing Ltd.

Figure 10.1

Rubble field left after blast fishing. (Photograph courtesy of Raphael Leiteritz.)

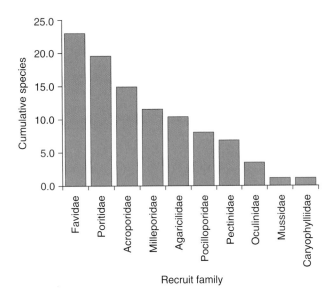

Figure 10.2

Composition of recruited coral community three years after plot establishment on blast rubble. (From Raymundo et al 2007; reproduced with permission of Elsevier.)

healthy reefs. Raymundo et al discovered that coral larvae settled both on the rock piles but also on the low-lying rubble underneath the restraining plastic net, and after 2 years, most experimental plots showed very good recruitment which varied with the coral family (Figure 10.2). So successful was this rehabilitation project that the fish communities which returned to the damaged reef were found to be very similar to nondamaged ones within a mere 3 years.

Of course, there has to be a supply of new coral recruits. As in the first example, these recruits may find their own way to the recently cleaned up sites, or more dependably

and rapidly, they can be produced elsewhere just like a gardener produces healthy seedlings in the nursery bed or greenhouse before planting out. The hope is that they will adapt their new locations and flourish. Once the living hard coral cover starts to proliferate, other crucial reef components including invertebrates and fish will hopefully colonize without further human intervention.

Nurseries, gardening, and transplanting

Just as in a plant nursery on land, it is possible to collect small fragments of coral (nubbins or ramets) by carefully removing "cuttings" from healthy coral colonies, placing them somewhere safe from wave action or predators, and allowing them to grow on before introducing them to degraded reef sites. Many different types of nursery structure exist, but Figure 10.3 illustrates two such frameworks from the Philippines, one suspended and the other on legs fixed to the seabed (Shaish et al 2008). Both systems seemed to be successful in providing for high coral fragment survival. Figure 10.4 shows the growth of three species and genera of corals from the Red Sea, indicating that coral fragments can indeed grow successfully and rapidly in culture (Shafir et al 2006). In this example, pruned coral fragments were super-glued to 2-cm plastic pins and attached to plastic frames suspended in midwater at 6 m depth off the coast of Eilat in the Israeli Red Sea. In general, the fragments showed a greater than 80% survival rate and an increase in volume of 147- to 163-fold after 306 days in the nursery. In all, several thousand coral fragments were farmed simultaneously, cheaply, and successfully.

Fragment size may influence the success of the gardening process for coral transplants. Smaller fragments take up less space of course and are easier to handle, but may be more prone to damage, disease and death than large ones. Forsman et al (2006) investigated the size-specific survival of two species of *Porites* in Hawaii in 3000-liter outdoor seawater tanks. Corals were collected from nearshore reefs and broken into fragments underwater which were then glued to unglazed ceramic tiles and the survival and growth (area expansion) measured subsequently. As Figure 10.5 shows, the original size of the fragment made little difference to their growth rate over 9 months. However, smaller fragments were more prone to damage by predators and grazers such as sea urchins. For branching corals at least, even the position on the donor colony from where the fragment is removed may influences survival. Soong & Chen (2003) found that fragments of *Acropora pulchra* grew faster when collected below the topmost points of a donor coral branch. In other words, collecting, caring for, and growing on of coral fragments is growth form and species dependent (just as with plants on land). In essence, coral gardening works and can provide copious quantities of young healthy corals for transplanting onto new or old reef sites.

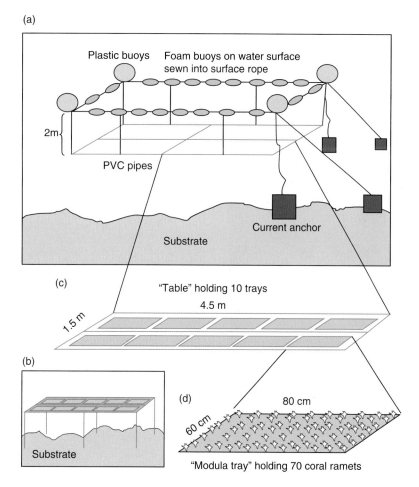

Figure 10.3

Diagram of two types of coral nursery structure used in the Philippines: (a) suspended on floats; (b) mounted on legs. (From Shaish et al 2008; reproduced with permission of Elsevier.)

Transplant spatial arrangements are equally important as when planting a forest of trees, and mathematical models have shown that even-spaced transplants in grids is likely to provide the fastest increase in coral cover, especially for r-selected coral species (Sleeman et al 2005). In general, it appears that corals with branching or foliose (leaf-like) forms have higher growth rates in transplantation situations, but lower survivorship, when compared to more massive species (Dizon & Yap 2006). These authors suggest that the latter types of coral are those that provide the framework of the reef, whilst the former fill in the gaps. Clearly, massive corals once transplanted will take appreciably longer to form a new reef system than the fast and delicate branching ones.

Up to now, we have discussed the restoration of hard (scleractinian) coral reefs; some work has also been done on the restoration of gorgonian (fan) corals. Gorgonians are common in tropical and subtropical seas, living close to the surface and down to considerable depths. Because of their delicate and finely branching nature, gorgonians are especially prone to being damaged, broken off or killed, by natural disturbances such as storms and hurricanes, but also

accidentally by divers and indeed intentionally by collectors. The red coral, *Paramuricea clavata*, lives on vertical and overhanging substrates in the northwestern Mediterranean. The species is technically known as sciaphilous, in that its lives in dark, shady habitats, as opposed to photophilous which describes organisms that do best in strong light conditions. Colonies of *P. clavata* tend only to grow two or may be three centimeters per year, which means that relatively large specimens can be decades old. Linares et al (2008) carried out experiments in the Medes Islands Marine Protected Area (located at the northern end of Spain's Costa Brava), where they attempted to transplant damaged fragments of *P. clavata* in order to boost population levels. The authors discovered that the method of transplantation that ensured the highest survival was to use PVC sticks embedded in epoxy stuck to the substrate as supports for the gorgonian fragments, but that overall survival rate was not high. Most transplants that survived did so in low light conditions under low competition pressure from algae, in other words, natural conditions. In general, this research showed that restoration of marine species and habitats can succeed, but success rates may be fairly low indicating the need for patience and determination.

As we have discussed in earlier chapters, coral reefs themselves cannot exhibit closed community structures and

assemblages without connectivity to associated ecosystems such as seagrass and mangroves (see next section). Hence the restoration of these latter habitats can be just as important to the reef as direct reef building. Mangroves are disappearing at a fast rate on a global scale, and though it is clear that preventing non-sustainable removal and harvesting is preferable, there are many instances of where mangrove restoration is the only alternative after damaging exploitation. Bosire et al (2008) review the functional status of restored mangroves, both as terrestrial (forest) habitats, but more importantly in the present context, as providers of marine habitats within the prop root communities. Clearly, it is not possible to have one without the other, and as Figure 10.6 illustrates, the steps towards mangrove restoration are complex and interrelated. Most importantly, site-specific considerations such as local traditions, socio economics, and faunal and floral ecology have to be borne in mind from the outset. Bosire et al are at pains to emphasize that local communities need to be involved with this type of restoration management; even school children can, for example, be encouraged to take care of young mangrove trees planted into the lagoons associated with their villages (Figure 10.7). The rest, it is hoped, can be left to time and natural ecology.

So far we have considered restoration in predominantly warm waters. However, cold temperate situations are just as deserving of restoration, both ecologically and commercially, as exemplified by oysters, *Crassostrea virginica*, in Chesapeake Bay. Long before mankind discovered the food (and hence economic) values of oysters and their cheaper relatives such as mussels, the ecological importance of

Figure 10.4

Examples of nursery-cultured coral colonies from Eilat in the Red Sea. (a) *Acropora eurystoma*, (b) *Pocillopora damicornis*, (c) *Stylophora pistillata*, 66, 144 and 200 days old. White bar = 2 cm. (Photograph by D. Gada; from Shafir et al 2006; reproduced with permission of Springer.)

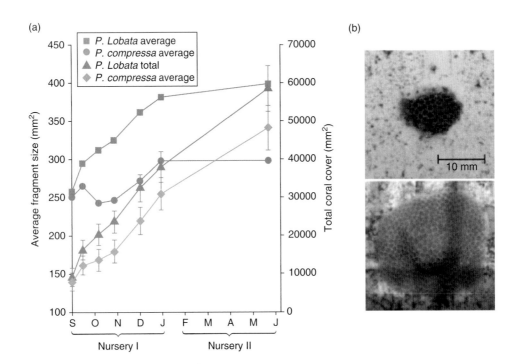

Figure 10.5

(a) Mean fragment size (area mm²) (±s.e.) for *Porites lobata* and *P. compressa* (y-axis 1) and total area covered by coral tissue (y-axis 2). (b) Example of *P. lobata* nubbin that quadrupled in area over 4 months, top to bottom. (From Forsman et al 2006; reproduced with permission of Elsevier.)

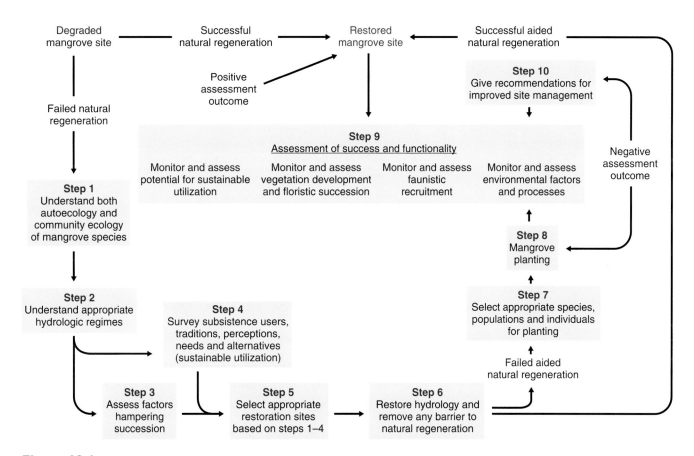

Figure 10.6

Ten steps to possible mangrove restoration depending on site condition. (From Bosire et al 2008; reproduced with permission of Elsevier.)

Figure 10.7

Planting out mangrove seedlings, Indonesia. (Photograph courtesy of Eddie Purwanto.)

three-dimensional structures (see Chapter 2) offered by communities of sessile bivalve molluscs was providing for marine biodiversity in shallow temperate seas. As Nestlerode et al (2007) report, oyster populations in Chesapeake Bay have declined dramatically over the years, due to over-exploitation, mismanagement, and parasite-caused diseases. Because oysters create their own substrates for the settlement of new recruits, their disappearance results in a system unsuitable for natural habitat rejuvenation. In order to restore healthy oyster beds, new substrates suitable for the settlement of oyster larvae (spat) must be created. In this set of field trials, mounds of shell fragments from both oysters, and the clam *Spisula solidissima*, were studied to see how they could accumulate new oyster populations. Oysters were found to be much more abundant on substrates consisting of oyster shell fragments, mainly due to the more suitable physical structure compared with clam-shell material. In addition, it was found that relatively large oysters are appearing on oyster shell substrates in the third year of the experiment, with all the associated ecological (e.g. age and size links to fecundity), and commercial, significance.

Connectivity

We have discussed the links between different marine habitats (mangroves, coral reefs, and seagrass, for example) and

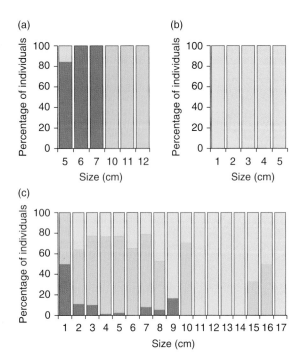

Figure 10.8

Size–frequency plots for fish in the British Virgin Islands in various coastal habitats. Brown = mangroves; green = seagrass; yellow = backreef; blue = forereef. (a) Horse-eye jack, *Caranx latus*. (b) Dusky damselfish, *Stegastes adustus*. (c) Yellowtail snapper, *Ocyurus chrysurus*. (From Morris 2004).

the movement of species of fish and invertebrates within and between such habitats in various chapters of this book, but it is important to remind ourselves how crucial the maintenance of this connectivity is to the overall health and sustainability of these communities (Mumby 2006; Dorenbosch et al 2007). Studies on fish communities in various reef-associated habitats certainly indicate that species assemblages vary between habitats, as shown by Unsworth et al (2007) on Indo-Pacific reefs in Sulawesi. Results indicated that fish assemblages were similar on reef flats and coral bombies, but differed from a second group found in various types of seagrass habitat. However, if juvenile reef fish have an obligate residence in seagrass or mangroves as young recruits or juveniles, no amount of reef restoration will bring them back if the other obligate habitats within the lifecycle are missing (Gratwicke et al 2006). Morris (2004) studied the distributions of various species of fish around the British Virgin Islands and found that horse-eye jack, *Caranx latus*, for example, is restricted ontogentically to mangroves and seagrass, the dusky damselfish, *Stegastes adustus*, only moves between backreef and forereef, whereas the yellowtail snapper, *Ocyurus chrysurus*, uses all four habitats sequentially as it grows (Figure 10.8). Clearly, removing mangroves or degrading seagrass beds is likely to have serious consequences for the last species in particular.

As well as restoring biodiversity, maintaining coastal connectivity can also enhance fisheries. Snappers in the example are certainly important food fish for local villages in the Caribbean, and work up and down the shoreline of tropical Queensland has shown a similar importance of connectivity between various types of coastal wetlands (mangroves, seagrass, saltmarsh, and mudflats) and commercial inshore fisheries (Meynecke et al 2008).

Artificial reefs

Humans have been putting artificial structures in the seas for many centuries, both accidentally as in the case of shipwrecks or intentionally as with harbors and oil platforms. Even if intentional, most of these structures have nothing directly to do with maintaining and enhancing marine biodiversity, though some are purposely designed to do just that. Whatever the source and primary purpose, all these items are collectively known as artificial reefs (ARs), and many of them have crucial roles to play in marine biodiversity restoration and conservation.

Seaman (2007) has produced a list of uses of artificial reefs, which includes aquaculture (also known as marine ranching), promotion of biodiversity, mitigation of environmental damage, enhancement of recreational SCUBA diving, eco-tourism development, artisanal and commercial fisheries production, protection of benthic habitats form trawling, and, last but not least, research. They may also protect and shelter juvenile fish, add to environmental complexity and stability, and provide new sites for larval settlement and recruitment (Relini et al 2007).

Marine structures with other original purposes

Almost anything solid and substantial placed in the sea cannot avoid becoming an artificial reef. It is the nature of marine species to seek out and settle on any and all suitable substrates (see Chapter 7). In this section we consider structures placed in the sea for purposes other than primary marine conservation, from marinas to breakwaters, oil and gas platforms and terminals, and shipwrecks.

Breakwaters, for example, provide excellent substrates for the build up of thriving new communities, as shown in Figure 10.9. This figure depicts the coral communities established on artificial cement breakwaters in Taiwan, which have been in place for around 30 years (Wen et al 2007). In this study, coral cover on the breakwaters ranged from 25% to 40% with between 107 and 103 species of scleractinian coral recorded. The walls and jetties of marinas not only provide similar sites for colonization as breakwaters, but they accumulate animals that like to live in safe proximity

(a) 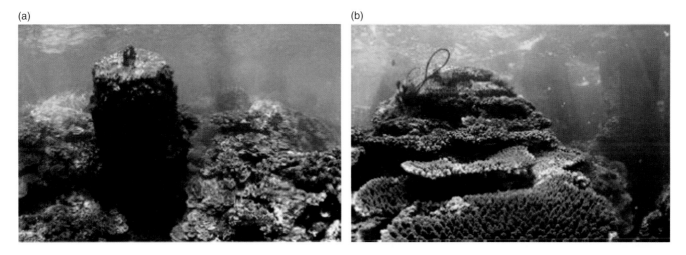 (b)

Figure 10.9

(a) Coral communities on circa 30-year-old breakwaters at Dou-Fu-Chia, Taiwan. (b) *Acropora* spp. at depths of 1–3 m on the breakwaters. (From Wen et al 2007; reproduced with permission of Springer.)

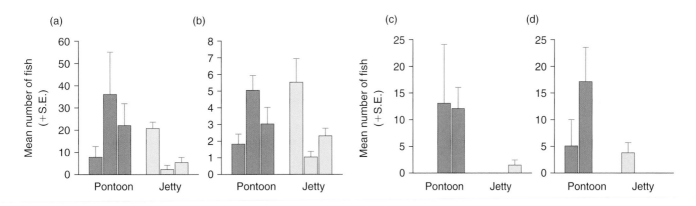

Figure 10.10

Mean (±s.e.) of (a) numbers of fish, (b) numbers of species, (c) luderick, *Girella tricuspidata*, and (d) eastern hulafish, *Trachinops taeniatus*, associated with different types of marinas in Sydney. Dark blue bars = marinas with floating pontoons; pale blue bars = marinas with fixed height jetties. (From Clynick 2008; reproduced with permission of Elsevier.)

of a solid, relatively massive structure even though they may not be directly attached to it. Various fish species show this type of behavior as exemplified in Figure 10.10. Clynick (2008) studied fish species that live in and around marinas in Sydney, Australia, and found that certain small species were much more common in marinas with floating pontoons than in otherwise equal marinas with jetties fixed to piles. The pontoons float and hence are always at the water surface whatever the state of the tide, whereas jetties can be some way out of water at low tide. Small fish requiring constant security, and safety and possibly food supply supplied by sessile organisms growing on the pontoons, favored the pontoons over the jetties.

Some of the largest fixed structures to be placed in our seas are piers, jetties, and oil or gas rigs and platforms (Figure 10.11). When located in mud or sand, they clearly enhance the three-dimensional complexity of marine habitats and

Figure 10.11

Pier legs underwater, Sulawesi. (Photograph Martin Speight.)

Figure 10.12

Thick encrustations of sponges and Christmas tree worms on pier legs, Caribbean. (Photograph Martin Speight.)

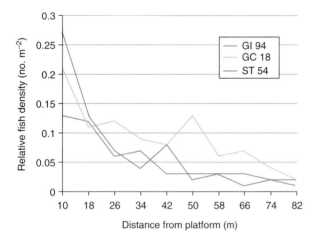

Figure 10.13

Mean relative horizontal fish densities up to a distance of 82 m from petroleum platforms in the Gulf of Mexico. ST 54 – depth 22 m; GI 94 – depth 60 m; GC 18 – depth 219 m. (From Stanley & Wilson 2000; reproduced with permission of Elsevier.)

hence encourage the colonization of both sessile (Figure 10.12.) and mobile species in abundance over the first few years after establishment (Fabi et al 2002). As might be expected, if such artificial reefs are placed in low complexity marine habitats, their area of influence in this context is likely to tail off the further away one goes from the rig or platform legs, as shown in Figure 10.13 (Stanley & Wilson 2000). Surveys of fish were carried out using hydroacoustics backed up by visual identifications at various distances away from the legs of petroleum platforms, and at different depths. Approximate exponential decays were found. Even the deepest fish populations, dominated at 219 m by the creole fish, *Paranthias furcifer*, showed some positive relationships with the artificial structures.

Oil and gas installations are sometimes such excellent artificial reefs that they surpass the diversity on adjacent natural coral reefs. Rilov & Benayahu (2000) studied oil jetties in Eilat, Israel. Divers carried out underwater water visual censuses (UVCs) at three natural reef sites and one oil terminal jetty which consisted of triplicate arrangements of 1-m-diameter pillars partially encircled by coiled barbed wire. Far from being a deterrent, the wire provided shelter sites for certain fish species. In addition, the strong vertical structures appeared to be particularly attractive to fish settlement and recruitment, when compared to the more gradual slopes of the natural reefs. Both fish abundance and species richness were significantly higher around the artificial structure. So successful are such structures that "rigs-to-reef" programs have been established in various parts of the world. Louisiana, for example, has the largest in the world (Kaiser 2006), which at the time of publication of the paper had created 83 artificial reefs using over 100 decommissioned oil and gas platforms in the Gulf of Mexico.

Commercial and indeed recreational fishermen view artificial reefs not as devices for enhancing species richness, but as means to increase catches. We have seen that these structures accumulate fish populations around them, and the term fish aggregation device sums up their benefits for fishing. These fish use artificial reefs for a variety of reasons, mainly for shelter and the "comfort" of something large and solid to have in the vicinity, and also as providers of extra food. In the latter case, for example, Leitão et al (2007) studied the food consumed by adult white seabream, *Diplodus argus*, by analyzing stomach contents, and found that the diet of other fish, crustaceans, bivalves, and gastropod molluscs were strongly associated with prey availability on artificial reefs, rather than from the surrounding natural substrate habitats. The shelter effects works not just for fish, but for other marine species such as spiny lobsters. In the Caribbean, *Panulirus argus* (Figure 6.12) is an important commercial species whose abundance and biomass can be markedly increased by the addition of artificial shelters (casitas) for juvenile lobsters, constructed out of PVC pipes and cement slabs (Briones-Fourzán et al 2007). Using these shelters, a sixfold increase in juvenile density and a sevenfold increase in biomass was achieved.

Some fishermen seem to be skeptical about the attractions of artificial reefs (Ramos et al 2007), but undoubtedly, their economic benefits in the context of commercial exploitation may be quite large. Whitmarsh et al (2008) analyzed the economic improvements derived from concrete artificial reefs for the fishing industry in the Algarve region of southern Portugal, using gill net capture data collected over more than a decade of exploitation. It seems clear that in two sites in two different locations, the value of fish catches was significantly higher on artificial reefs than from adjacent nonreef areas. There is a potential snag. If artificial reefs are truly fish aggregation devices, then it follows that increased populations around the structures must

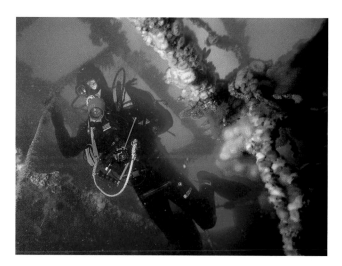

Figure 10.14

Diving the wreck of the *Hispania*, west coast of Scotland. (Photograph courtesy of Paul Naylor.)

be associated with reduced densities elsewhere, rather than an intrinsic increase in stock size (Sutton & Bushnell 2007). Localized over-exploitation on the reefs could easily result.

For many people, the most exciting type of structures forming artificial reefs are shipwrecks. Ships have sunk ever since sea traffic began, in times of peace and especially war (Figure 10.14). The numbers of wrecks on the bottom of the sea must number many thousands. In recent times, we have begun to deliberately sink ships for reefs following suitable treatment to make them safe for divers, overhead boats, and to prevent chemical pollution. Many countries including the USA, the UK, Canada, Australia, and New Zealand, have found a use for old warships for example which would otherwise be broken up for scrap, which is to scuttle them in selected spots for the benefit of marine organisms and SCUBA divers alike.

Wrecks quite obviously provide copious amounts of new, clean, hard substrates for the settlement of many sessile and sedentary species, whilst providing highly complex physical habitats as well as large (indeed, huge in many cases) objects for agoraphobic fish. In essence they should be the perfect artificial reef from all perspectives. They may also be the only source of hard substrates if they happen to sink (or be sunk) in areas of sandy or muddy seabed (Zintzen et al 2006); hitting a rock or coral reef is another matter of course, where a sunken ship merely adds to the local habitat complexity.

Once a ship goes down, accidentally or on purpose, organisms begin to colonize (see also Chapter 7). The length of time that a ship has been underwater will dictate the richness and diversity found on it, but in many cases, even a year is long enough for diverse communities to develop (Walker et al 2007). The ex-*HMAS Brisbane* is a 133-m guided missile destroyer that was purposely sunk in July 2005 in 27 m of water off the coast off the Sunshine Coast of southern Queensland. Just 1 year later, samples taken by Walker et al showed that a rich epifauna had colonized the wreck, dominated by barnacles, sponges, and bryozoans. Depth, surface orientation, and exposure dictated the community structures (Figure 10.15). This multivariate analysis which attempts to find patterns in data and represents them in terms of varying similarities, shows clearly that different parts of a large and highly complex three-dimensional structure such as a shipwreck support specific and different communities of organisms. Thus, for example, different assemblies of species can be identified inside and outside the wreck at the same depth (Figure 10.15d). As time progresses, shipwrecks accumulate more and more sessile and mobile faunas, and as might be expected, accumulation curves tend to plateau as the time that a wreck has spent on the seabed increases (Figure 10.16a) (Arena et al 2007). Note also that the community assemblages differ somewhat between shipwrecks and natural reefs even after years of association (Figure 10.16b). A hundred or more years is rather a long time for a wreck to stay intact in shallow waters, and Perkol-Finkel et al (2006) studied marine communities established on a 119-year-old wreck of a steel cargo vessel in the north of the Red Sea. Note that even old wrecks will only mimic adjacent natural coral reefs if these ships possess similar structural components and complexities, such as nooks, crannies, holes, solid substrates, and so on.

In economic terms, it seems clear that the most benefits from shipwrecks come from the dive tourism industry. Crabbe & McClanahan (2006) estimated that the economic benefit from enhanced dive tourism on the Kenya coast was between US$75,000 and US$174,000, with dive operators and their associated travel and catering industries the main beneficiaries. There is, as always, a final warning note. It may be that shipwrecks on coral reefs, far from being generally beneficial and indeed exciting to divers at least, may in fact be deleterious ecologically. In 1991, a fishing vessel went down on a pristine coral reef on Palmyra Atoll in the central Pacific Ocean (Work et al 2008). By 2005, a phase-shift had occurred, whereby the naturally diverse coral ecosystem had been overgrown by the aggressive corallimorph, *Rhodactis rhodostoma*. The authors suggest that a combination of physical disturbance and changes in nutrient and/or pollutant chemical balances favored the invader to the detriment of the natural habitat. Perhaps we should not always encourage the sinking of ships so readily and frequently.

Marine structures specially created for conservation

These days, a wide variety of devices are available which are specifically created to emulate natural marine habitats in terms of three-dimensional structure, relatively massive size, and solid substrates. These structures can be deployed

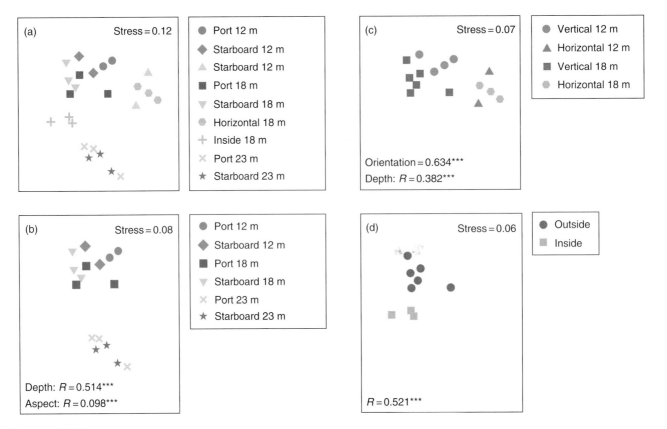

Figure 10.15

Non-metric multidimensional scaling (nMDS) comparing the epifaunal assemblages on the ex-*HMAS Brisbane*, 1 year after scuttling to create an artificial reef in Queensland waters. (a) All transects sampled showing depth, aspect, orientation, and location; (b) between three depths (12, 18, and 23 m); (c) between horizontal and vertical surfaces at 12 m and 18 m; (d) between locations inside and outside of the wreck at 18 m. (From Walker et al 2007; reproduced with permission of Wiley Interscience.)

either for research, or as components of marine conservation, restoration, and management. They may consist of many materials, from metal or plastic mesh and netting, fiberglass, plastic pipes (Figure 10.17a & b), to concrete blocks and spheres. It has been suggested that the precise structure of these artificial reefs is not critical when attempting to accumulate fish populations (Walker et al 2002), as long as an appropriate physical complexity is achieved. Nonetheless, most projects worldwide have employed one or other material arranged in predetermined ways. Several examples will illustrate this process and its effectiveness. Loh et al (2006) discuss the use of fiberglass artificial reef structures (called reef enhancement units – REUs) in Singapore. These units are a broad cone shape, approximately 70 cm bottom diameter and 50 cm high, with numerous regularly spaced holes, 3–5 cm diameter, over the surface. The fiberglass from which they are constructed is impregnated with sand and calcium carbonate to roughen the surfaces. These REUs can be anchored down using metal stakes hammered into the seabed. As can be seen from Figure 10.18, coral recruitment on these artificial reefs was variable in the Singapore study, but showed a clear progression over 18 months post-

establishment. Most importantly, scleractinian corals such as *Pocillopora damicornis* and various favids, acroporids, and poritids showed much better survival and growth on the REUs than on the surrounding coral rubble.

Fiberglass is of course light and easily handled, both above and below the surface, but most artificial reef projects employ concrete artificial reef structures. Italian marine biologists have been using various types of concrete structures as purpose-built artificial reefs for quite some years (Relini et al 2007), and altogether, 10 artificial reef sites have now been established in the Ligurian Sea. A common form of reef structure is a cube of concrete of various dimensions, with holes of different sizes penetrating the block, as used in the Loana Artificial Reef (LAR) project which began in 1986. Groups of concrete blocks were arranged into pyramids covering an area of seabed of around 350 ha in total, at depths between 17 and 25 m. The communities of fish and sessile invertebrates on the reefs were monitored over 15 or more years, and some of the results are shown in Figure 10.19. According to Relini et al, recorded data included over 150 species of algae, more than 200 species of benthic invertebrates, and 78 species of

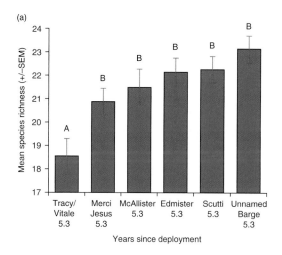

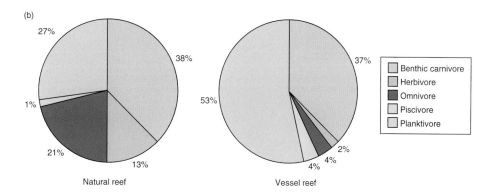

Figure 10.16

(a) Mean species richness (±s.e.) of fish on shipwrecks off southeast Florida according to age. Means with different letters are significantly different at $P < 0.05$ level. (b) Trophic composition as a percent of total fish abundance on shipwrecks versus natural reefs off southeast Florida. ((a,b) From Arena et al 2007; reproduced with permission of Springer.)

fish. As can be seen from the figures, fish assemblages detected by multidimensional scaling (MDS) show distinct clusters and dendrogram associations, with samples taken in the early 1990s, mid to late 1990s, and the 2000s, showing up as associated within their group and distinct from the other groups. A successional development of colonizers (see Chapter 7) is indicated, with a maturation of assemblages as time on the artificial reefs progresses.

Artificial reefs have also been used in mitigation projects where damage caused by one sort of development or habitat alteration is offset by enhancing the biodiversity and conservation of adjacent ecosystems. Dupont (2008) discussed the use of concrete reef structures in a mitigation project on the West Florida Shelf in the Gulf of Mexico, in order to balance out damage done by offshore oil pipeline construction. She compared the artificial reef units with limestone boulders and natural hard-bottomed habitats in terms of the abundance of commercial fish species (Figure 10.20). Five species were significantly more abundant on both types of artificial reef when compared with natural

areas, namely grey triggerfish (*Balistes capricus*), hogfish (*Lachnolaimus maximus*), grey snapper (*Lutjanus griseus*), gag grouper (*Mycetoperca microlepis*), and scamp grouper (*Mycetoperca phenax*). All of these species are basically carnivores, feeding on invertebrates and smaller fish associated with the artificial reef structures. Such structures are thus able to mimic natural reefs in terms of both epifauna and fish assemblages, and can indeed be used in mitigation and restoration projects as long as the pros and cons of individually cases are considered.

A final but widespread example of concrete structures as artificial reefs involves Reef Balls™ (copyright Reef Balls Foundation) (Figure 10.21a,b). According to the Foundation's website, Reef Balls are the most widely used type of purpose-built artificial reef structure in the world. They are made in molds from micro-silica concrete with a neutral pH and roughened surfaces. They are hollow with holes over the surface, and are available in a variety of sizes up to and including 2 m diameter weighing over 2000 kg. They can be used for a various purposes from planting mangrove trees,

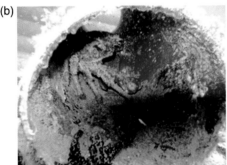

to reef restoration and regeneration, and at the largest end, breakwater construction (Reef Ball Foundation 2008 – Reef Rehabilitation document). Reef Balls undoubtedly do what they set out to do – to provide habitat diversity and clean hard substrates for reef organisms to colonize and settle. Jardeweski & de Almeida (2006) for example found the structures to be highly efficient at attracting fish species in a deployment project off the southern coast of Brazil, especially when close to existing natural reefs (Figure 10.22). The design of Reef Balls is highly customizable, and Sherman et al (2002) experimented in southeast Florida with filling the void inside each ball with concrete blocks, or attaching floating lines, both in attempts to alter the physical complexity of the artificial reefs. As Figure 10.23 shows, the Reef Balls containing concrete blocks showed higher fish abundance and species richness in almost all cases, when compared with empty balls or those with line attached. The

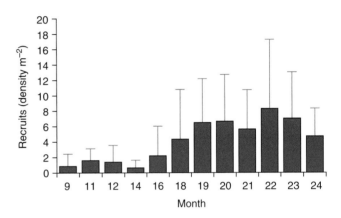

Figure 10.17

(a) Artificial reef made of plastic sewer pipes, Bay Islands, Honduras. Fish, sponges and corals are starting to colonize. (b) Spider crab, fish and sponges inhabiting sewer pipe artificial reef in the Bay Islands. ((a,b) Photographs Martin Speight.)

Figure 10.18

Density (± s.d.) of scleractinian coral recruits on the exterior surface of fibreglass reef enhancement units (REUs) at Raffles Lighthouse site, Singapore. (From Loh et al 2006; reproduced with permission of Wylie Interscience.)

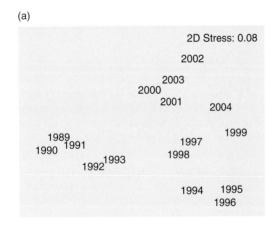

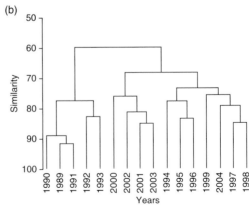

Figure 10.19

(a) Multi-dimensional scaling (MDS) of fish species abundance (based on Bray–Curtis Similarities) over 16 years of underwater census on the Loano Artificial Reef project.
(b) Dendrogram of similarity of fish species from 1989 to 2004 on the Loano Artificial Reef project. ((a,b) From Relini et al 2007; reproduced with permission of Springer.)

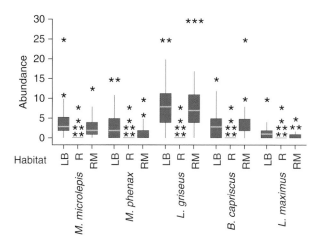

Figure 10.20

Sampling distribution plots (median, interquartile range, upper and lower limits, and outliers) of five commercially important fish species whose abundances were significantly higher (stars) at artificial reef sites (RM & LB) as compared to natural hard substrate habitats (R), on the West Florida Shelf. Asterisks denote data outliers. RM = reef modules; LB = limestone boulders. (From Dupont 2008; reproduced with permission of Taylor & Francis.)

Figure 10.21

Reef Ball™ structures. (Reproduced courtesy of Reef Ball Foundation www.reefball.org)

inner block material was likely to be providing extra refuges and habitats for reef fish. Reef Balls can be placed wherever required, even in deepwater using suitable boats, barges, cranes, or buoyancy devices. Reed et al (2005) employed them in long-term studies on deep-water coral reefs off the east coast of Florida. *Oculina varicosa* is a scleractinian hard coral that occurs in abundance between 60 m and 100 m off the southeastern USA. Communities in these depths are highly prone to damage by fishing and extraction, and various *Oculina* sites have now been designated as marine reserves. The figures show Reef Ball deployment in 100 m of water, with both mobile and sessile organisms taking up residence. Such structures would be very important in rehabilitating *Oculina* reefs.

We have now reached the stage where we need to try to amalgamate the various components of marine conservation and restoration into management packages, collectively known as Marine Protected Areas.

Marine protected areas (MPAs)

The term Marine Protected Area (MPA) is a coverall term for a wide variety of systems that attempt to reduce the pressures of human activities on habitats in the sea (Johnson et al 2008b). Over 4000 MPAs of one form or another exist in the world today (Vásárhelyi & Thomas 2008) with more formed almost every day. In particular, they try to reduce harvesting of species, whilst promoting long-term

sustainability of precious marine resources and their biodiversity (Mora et al 2006). There are many interpretations of the MPA, from purely voluntary to statutory and legally binding systems. Very few MPAs are actually marine nature reserves in the strictest sense of the term (Ballantine & Langlois 2008) where no exploitation is allowed at all, and these days they are just as much nature conservation "instruments" as fisheries-enhancing devices or tourist attractions (Stelzenmüller et al 2007). A key requirement for a successful MPA is, if possible, to keep everybody with an interest in the area (the stakeholders) happy. So, in any MPA structure it must be possible to meet the needs of fishermen, hoteliers, eco-tourism providers, and of course economists and conservationists; a tall order indeed. According to Frid et al (2008), "marine protected areas are generally designed and managed on the basis of the presence and extent of specific habitat types, or the habitats of important species." Unfortunately, in a large number of cases, MPAs are

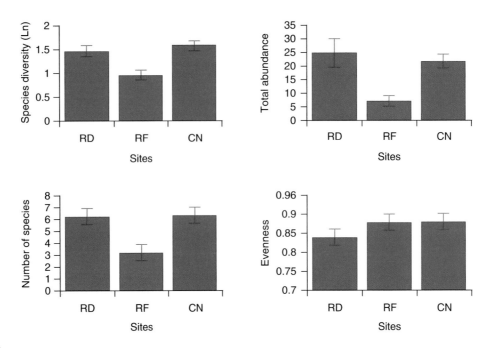

Figure 10.22

Community structure parameters (±s.e.) for three Reef Ball study sites in southern Brazil. RD = Reef Balls close to natural rocky shore; RF = Reef Balls 50+m from natural rocky shore; CN = natural rocky shore. (From Jardeweski & de Almeida 2006; reproduced with permission of Coastal Education & Research Foundation.)

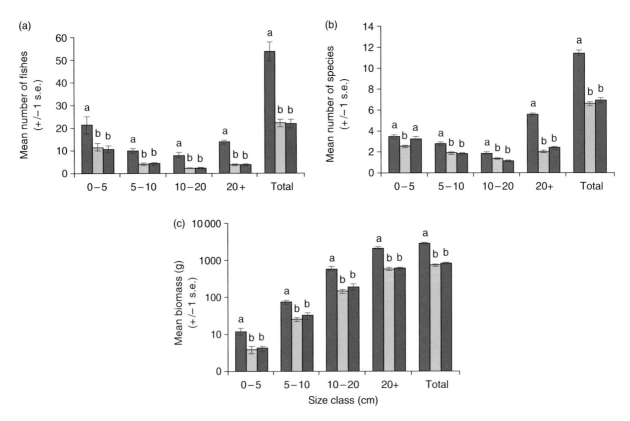

Figure 10.23

Mean (a) Number of individuals, (b) number of species, and (c) biomass (±s.e) of fish in different size classes on three types of Reef Ball™ artificial reef structures. Control (purple) = standard Reef Ball (1.3 × 1m) with inner empty space (void); Block (dark blue) = standard Reef Ball with inner space filled with concrete blocks; Streamer (light blue) = standard Reef Ball with 10-m floating line attached. (From Sherman et al 2002; reproduced with permission of Oxford University Press.)

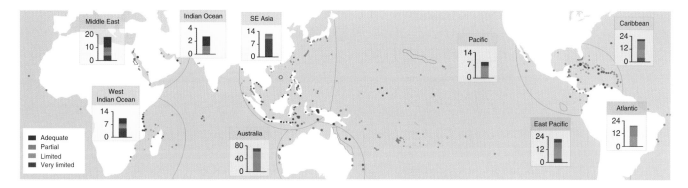

Figure 10.24

Location and condition of 980 Marine Nature Reserves (MPAs). (From Mora et al 2006; reproduced with permission of Science – AAAS.)

established on a wave of conservation euphoria and publicity, to be followed by a failure of interest or follow-up monitoring. Indeed, Figure 10.24. (Mora et al 2006) shows just how many MPAs out of approaching 1000 worldwide actually achieve their stated goals. The work of Mora et al evoked controversy when it was published and various other authors disagreed with one or more points raised (see *Science* vol. 314, 3, November 2006). Nonetheless, very few MPAs showed adequate success. The reasons for failure are possibly linked to lack of funding, lack of interest, or lack of ecological baselines on which to establish long-term monitoring and management tactics for changing pressures and expectations. In general it seems that we need more legislative coherence, more public engagement, and a better science base in order for MPAs to function (Johnson et al 2008b).

Effects and impacts of MPAs

Whether or not MPAs actually function in the various ways suggested is questionable, as Mumby & Steneck (2008) (Table 10.1) point out. The left-hand column in this table lists the aims and intentions of MPAs in coral ecosystems, whilst the right-hand column points out the status of the science behind these expectations. A lot of potentially beneficial effects appear not to be realized, at least in some of these coral reef examples. We have to assume, however, that there are grounds for optimism, and so before we go into the details of setting up and running marine nature reserves, it is important to see if all the effort is worthwhile, that they actually provide any or all of the services asked of them.

We can itemize these services under various categories, including: (a) enhancing the populations of animals and other organisms; (b) saving rare species; (c) maintaining food web structures and hence ecosystem stability; (d) providing income for regional and especially local people; and (e) acting as educational and research facilities for the promotion of marine conservation management. Take for example the work of Paddack & Estes (2000) who looked

Table 10.1 Status of knowledge about the effects of fully protected marine reserves in coral reef areas. (Various sources, from Mumby & Steneck 2008; reproduced with permission of Elsevier)

RESERVE IMPACT	STATUS OF SCIENCE
Increased fish and invertebrate biomass within borders	Confirmed and widely reported
Adult spillover to support adjacent fishery	Confirmed by a few studies but not others
Larval spillover to provide demographic support to nearby fished reefs	Expected but not demonstrated
Facilitation of trophic cascades that prevent urchin plagues (Indo-Pacific)	Confirmed by few studies so far
Facilitation of trophic cascades that increase fish grazing and reduce macroalgal cover (Caribbean)	Confirmed by few studies so far
Increased coral recruitment (Caribbean)	Confirmed by few studies so far
Increased recovery rate of coral populations	Expected but not demonstrated
Enhanced biodiversity	Mixed results (positive, negative and no impact reported)
Reduced direct impact of hurricanes or coral bleaching	Unlikely to occur
Reduced incidence of coral disease	Unknown

at the populations of rockfish, *Sebastes* species, in temperate water kelp forest MPAs off the coast of central California (Figure 10.25). Kelp rockfish as might be expected show close relationships with giant *Macrocystis* kelp; MPAs wherein fishing is stopped or limited (no- or partial-take zones) and which also assist the conservation of giant kelp

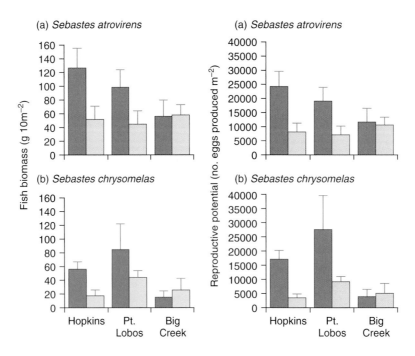

Figure 10.25

Average biomass and reproductive potential (±s.e.) per 10 m², for two species of rockfish in reserve (dark blue) and nonreserve (light blue) areas in three sites off the coast of central California. (From Paddack & Estes 2000; reproduced with permission of Ecological Society of America.)

forests show significantly bigger fish within reserves compared with outside them, and also (presumably because there are bigger or more abundant fish inside reserves) a much increased population fecundity. Clearly, the MPA system in this example promotes both standing crop population densities at one point in time, and also ensures a healthy recruitment into future generations of fish.

MPAs can influence a variety of ecosystem "metrics" from small animals to large animals, numbers of species, and species richness, as Barrett et al (2007) discovered during studies of fish populations inside and outside a marine reserve in Tasmania (Figure 10.26.). Numbers vary considerably, but in all cases, values of species richness and individual abundance are significantly higher within the reserve than outside, and indeed remain so over considerable periods of time. Commercial fish species, i.e. those that are routinely exploited by fishermen, would be expected to be particularly affected by reserve management that restricts harvesting. Tyler et al (2009) studied commercial fish populations off the coast of Tanzania (Figure 10.27) and found that the number of species increased with depth in fished areas, compared with the unfished areas where species richness declined with depth. Fishing has a significant impact on the structure of fish communities, and MPAs that enhance fish populations will also have an influence on other parts of marine ecosystems, since these fish must form part (top, middle, or bottom) of local or regional food webs. Triggerfish, *Balistapus undulatus* (Figure 10.28), is a multipurpose species where it occurs in the Indo-Pacific, since it

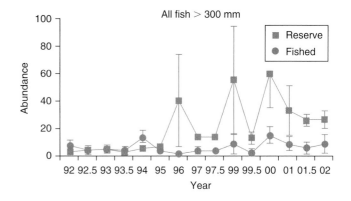

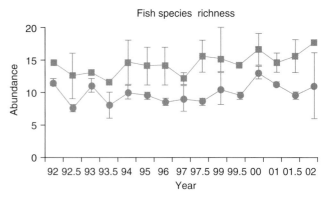

Figure 10.26

Comparisons between Tasmanian marine reserve and external reference sites of the mean abundance per site (numbers per 2000 m² ±s.e.) of fish parameters counted on surveys between 1992 and 2002. (From Barrett et al 2007; reproduced with permission of Elsevier.)

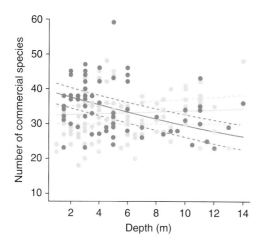

Figure 10.27

The effect of fishing and depth on commercial fish species richness on Pemba and Unguja (Zanzibar), East Africa. Light blue points indicate fished samples and dark blue points unfished samples; dashed lines are 95% confidence limits around each fitted line. (From Tyler et al 2009; reproduced with permission of Elsevier.)

Figure 10.28

Orange-lined triggerfish, *Balistapus undulatus*. (Photograph courtesy of Martin Leyendecker.)

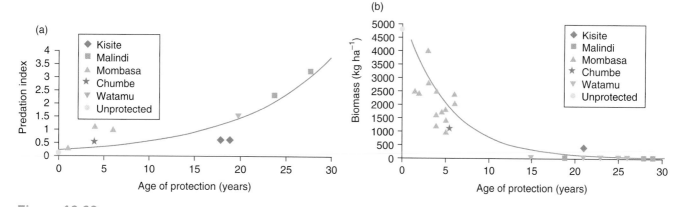

Figure 10.29

(a) Ages of East African MPAs and the density of triggerfish. (b) Ages of East African MPAs and the biomass of triggerfish prey, sea urchins. ((a,b) From McClanahan 2000; reproduced with permission of Elsevier.)

is a large, charismatic fish that attracts fee-paying dive tourists, is an excellent food fish for local fishermen, and is also a major predator of potentially damaging sea urchins. McClanahan et al (2000) report on the recovery of *B. undulatus* populations under the protection of five MPAs off the coasts of Kenya and Tanzania, with the accompanying effect on urchin populations (Figure 10.29a & b). As can be seen, the age of protection has a crucial impact on fish recovery and urchin demise – improvments do not happen overnight. The longest established MPA, Malindi, shows the largest numbers of fish and lowest numbers of urchins, whilst the newest reserve, Mombasa, differed little from unprotected sites in having very few fish and a lot of urchins. Notice however that repeated measures within one site over time show encouraging trends in fish recovery.

The importance of age of MPA, the duration of protection, has been shown in the more temperate climate of southern Europe (Figure 10.30) (Claudet et al 2008). These authors used 58 datasets from 19 European marine reserves. As we might now expect, the time since MPA establishment fundamentally influences fish species richness and population densities; older reserves contain more fish and more species of fish. At some stage of course, we must expect an asymptote on fish species richness and abundance; things cannot increase forever because of density dependence effects and the basic number of niches available for resource partitioning between species (see earlier in this book). MPA size is clearly also important (note the log scale on the size axis of the figure), a reflection of simple island biogeography. The larger the protected area is, the bigger, the more

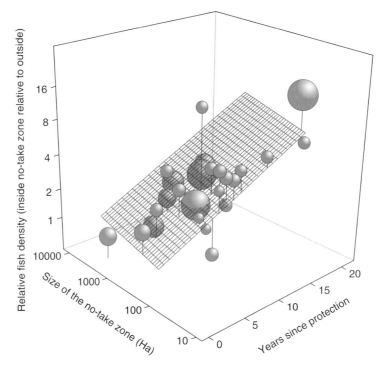

Figure 10.30

Effects of marine reserves on commercial fish densities as a function of years since protection and the size of the protected area (no-take zone) in southern Europe. Sizes of circles are proportional to weight of a particular study. (From Claudet et al 2008; reproduced with permission of Wiley-Blackwell.)

Figure 10.31

East African fish market. (Photograph courtesy of Josh Cinner.)

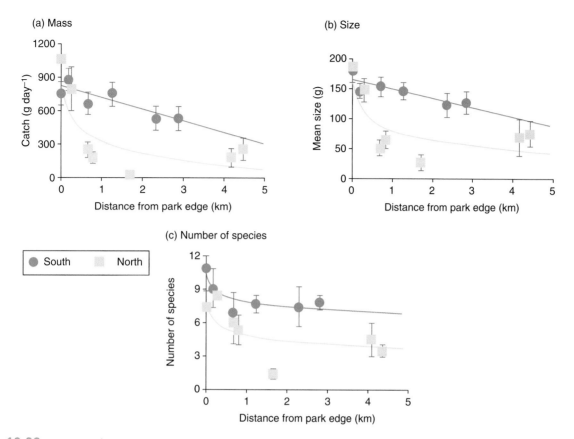

Figure 10.32

Edge effects and spillover in the Mombasa Marine Park: (a) total weight of fish catch, (b) mean size of fish, (c) number of species of fish, over a 14-day sampling period, as a function of distance from the park boundary on north and south borders (±s.e.). (From McClanahan & Mangi 2000; reproduced with permission of Ecological Society of America.)

abundant, and the more species-rich the fish. For some fish, especially those that mature late and move considerable distances, very large MPAs may be required. Blyth-Skyrme et al (2006) suggest that trophy fish in the western area of the English Channel, such as various rays, dab, flounder and plaice, may need protected areas over 500 km² before they can be adequately protected (see also Chapter 10). In general though, it seems that if you can wait long enough, MPAs should indeed be able to enhance fish populations significantly, and this is likely to be particular significant in the case of species targeted by fishing (Claudet et al 2006). All of the effects described so far can of course be equally applied in theory at least to populations and communities of invertebrates and other organisms, both sessile and mobile.

When local people are prevented from fishing in parts of an area, markets (Figure 10.31) are likely to decline and the new legislation can become very unpopular. However, once a MPA begins to hold high numbers of large organisms from lobsters to fish, movement and spillover are likely to take place. This will be particularly important for mobile species with high birth rates relative to fishing rates (Malvadkar & Hastings 2008). Several authors predict the gradients of fish abundance and biomass across reserve boundaries, from high levels within the reserves which decline into the nonreserve, potentially or actually fished areas (Forcada et al 2008; Harmelin-Vivien et al 2008). In 2000, McClanahan & Mangi used fish trapping to study the spillover of commercial fish species such as groupers, snappers, parrotfish, and surgeonfish from inside the Mombasa Marine Park off the Kenya coast into exploited (fished) areas adjacent to the park. Some of the results are shown in Figure 10.32. On the northern side of the Park, beach pull seine netting continued, a nonselective practice which can rapidly and efficiently deplete areas of fish large and small, whereas to the south, management reduced the fishing pressure by only allowing traditional line methods. As the figure shows clearly, the MPA increases catches, size and richness of fish, with clear effects of fishing intensity. Thus there are clear benefits to be had for exploitable

Table 10.2 Examples of techniques used for benthic habitat mapping for the establishment of Marine Protected Areas

ECOSYSTEM	COUNTRY/REGION	MAPPING TECHNIQUE	ITEMS ASSESSED	REFERENCE
Shallow semi-tropical	E. Australia, Moreton Bay	Towed video array	Abiotic surrogates of biodiversity (mud, current from ocean and rivers)	Stevens and Connolly (2004)
Rocks, sand, seagrass	Italy	Still digital photos and SCUBA visual surveys	Depth, aeomorphology, fish abundance	Greco et al (2004)
Coral, mangroves, seagrass	Turks and Caicos, S. Bahamas	High-res digital video and SCUBA visual surveys	Fish species, life stages, trophic levels	Mumby et al (2008)
Temperate rocky and sandy	S. Australia, Victoria	UW video, multibeam hydro-acoustics	Macroalgae and sessile invertebrates	Holmes et al (2008)
Coral reefs	SE. Florida, Broward County	Laser bathymetry, acoustic ground discrimination, subbottom profiling, aerial photography	Major reef components	Walker et al (2008)
Sand, mud and gravel	Ireland, Cork	Sediment grab sampling	Physical and chemical sediment properties, major biota	McBreen et al (2005)
Corals, seagrass	Pacific Panama	Satellite imaging, aerial imaging	Habitat classes (live coral, algae, rock, etc.)	Benfield et al (2007)
Corals	Egypt, S. Sinai	UW video, SCUBA point and line quadrats	Coral colonies, benthic substrate cover	Leujak and Ormond (2007)
Shallow arctic	E. Canada Scotian Shelf,	Hydro and geological surveys, sediment samples,	Depth, ice cover, sediment classes	Roff et al (2003)
Corals	Puerto Rico	Side scan sonar	Major marine taxa (corals, sponges, algae etc.)	Prada et al (2008)
Corals, seagrass	Solomon Islands	Aerial photos, local knowledge via questionnaires	Habitat types (seagrass, lagoons, reefs, sand etc.)	Aswani and Lauer (2006)
Cold-water corals (500 to 1100 m)	UK, Rockall Bank	Digital still photos and visual surveys from submersibles	Deep-water megafauna	Roberts et al (2008)
Corals	Florida, Biscayne National Park	Experimental Advanced Airborne Research Lidar (EAARL)	Reef rugosity	Kuffner et al (2007)

species populations by maintaining part of a region free of fishing pressure, thus enabling fish populations to proliferate to the overall benefit of all concerned. In addition, these no-take zones can attract the all-important dive-tourist to bolster the local income and enhance interest in biodiversity conservation.

MPAs can influence the impact on marine habitats caused by people enjoying themselves, as well as simply trying to stay alive. Sport fishing (angling) is a worldwide pastime, and in an example from Waikiki in Hawaii, Meyer (2007) showed that pole (rod) and line fishing were the dominant types, accounting for 70% of the total reef fish caught. Spear fishing was also popular. Some years ago, several small MPAs were set up mainly for snorkeling and SCUBA diving, within which recreational fishing was not allowed. Other adjacent areas allowed fishing in alternate years (so-called pulse fishing). The harvest of reef fish such as surgeonfish and tangs (and indeed octopuses) was very much reduced inside the protected areas. Interestingly, with fishing permitted only in alternate years, the yield from spear fishing at least during the "open" year was considerably higher than in the continually fished areas, perhaps an indication of stock recovery.

After all the ecology and conservation potential of MPAs have been considered, it makes sense to consider revenue – a management system which ticks all the right boxes in terms of biodiversity but at the same time makes money is attractive. We have already seen some aspects of the creation of MPAs on the East African coast. McClanahan et al (2005c) showed the important influence on revenue generation. Though the numbers of visitors varies from park to park and year to year, revenue has in general risen. The most successful park, Kisite, was generating over US$300,000 per annum in 2002, an enormously important source of income for local people. Crucially, all concerned from fishermen to hoteliers have to remain aware that any serious decline in the health of the MPAs and hence a reduction in (a) the productivity for locals and (b) the attractiveness to tourists of the marine ecosystems will "kill the goose that lays the golden egg." Everyone involved needs to be aware of the importance of retaining marine biodiversity, for people who want to eat it and visit it alike. We shall return to the socioeconomics of MPA management later in this section.

Surveys and monitoring for MPA status

So, how do we decide which tiny bits of the vast oceans are worthy and needy of conservation and protection? Clearly, random selection would not be appropriate, since we need to identify the most deserving sites or habitats for protection and conservation. Selection by species richness hotspots, rarity, representativeness, susceptibility to dam-

age, all need to be taken into account (Gladstone 2007a). Indeed, it is not even sufficient to "score" the species richness of a site ("biodiversity" – see Chapter 2), since low genetic diversity is something to avoid within protected areas (Bell & Okamura 2005). The snag is that most of these parameters are difficult to measure with any accuracy or dependability, and we need to find surrogates which are easier to assess but give a good picture of the variability of species richness, habitat types, and ecosystem functioning over a relatively large scale (Mumby et al 2008). Without doubt, one of the most important surrogates or predicators of species richness of fish and many invertebrates is rugosity, or habitat heterogeneity (Holmes et al 2008), a concept which we introduced in Chapter 2. Essentially the structurally more complex sites have more fish species. This type of measure should be sufficiently robust to use as a predictor of relative species richness, and if used at a seascape scale should provide surveys of diversity hotspots. Many other methods are available for producing surveys, or benthic habitat maps, of areas. Table 10.2 summarizes a whole variety of these from the literature that have been used to assess the conservation and protection values (ecological worth) of ecosystems all over the world. Most feed their data into geographical information systems (GIS) to produce colorful maps of sites to aid interpretion so that decisions, both at the establishment phase and also during the management of the area, may be made (Aswani & Lauer 2006). In this way, it becomes possible to produce detailed inventories of species, habitats and diversity on various spatial scales, and to investigate important parameters such as patch sizes, distributions, fragmentation, and so on (Rioja-Nieto & Sheppard 2008). An example of such habitat mapping comes from Puerto Rico (Figure 10.33) (Prada et al 2008). Here, a high resolution side-scan sonar device was installed in a small boat together with GPS and portable computer equipment, and an area of over 60 km^2 surveyed from the shore to between 11 km and 13 km offshore. In total, the survey took 48 days to complete. Ground-truth surveys were also carried out using SCUBA in order to identify in detail the major features detected by the sonar. According to the authors, the survey cost between US$250 and US$500 per km^2, making the system highly cost effective. The figure shows details of the amount and distribution of in total 21 habitat types, and provides the basis for pin-pointed studies and management decisions.

After all these high-tech systems have been employed, there remain one or two much more traditional ways of assessing the worth of a marine site for protection – we can ask the locals. Drew (2005) suggests that traditional people who have lived in an area and exploited its resources for many generations are likely to possess a wealth of knowledge about the habitats around them and the species living there. As mentioned already, any MPA which is likely to succeed in the long term has to take the needs of local people

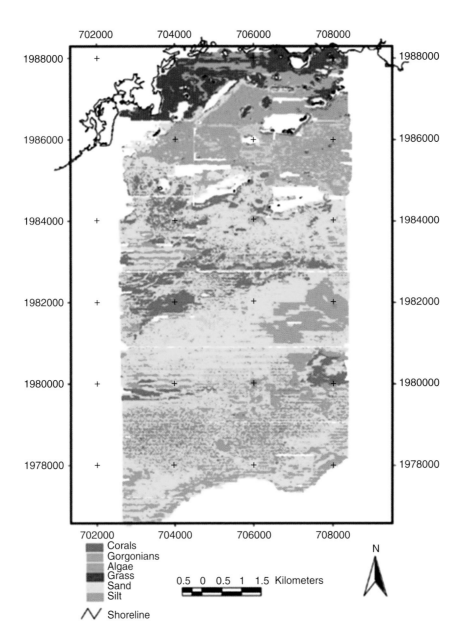

Figure 10.33
Community level habitat map of the shelf off La Parguera in southwest Puerto Rico, produced using visual interpretation of side scan sonar habitat mosaics. (From Prada et al 2008; reproduced with permission of Taylor & Francis.)

into account, and so interactions with them at the survey stage make a lot of sense.

However it is done, once the mapping and surveying has been carried out, and the scientific and conservation merits of areas have been prioritized, areas with equal merit need to be looked at from socioeconomic perspectives as well. Roberts et al (2003) have suggested a series of criteria that should be employed to select the locations of new MPAs (Table 10.3). These include economic, social, scientific and, above all perhaps, feasibility considerations. We shall illustrate many of these criteria in the final section in the

chapter which provides examples of MPA management systems from around the world.

Socioeconomics of MPA establishment and management

MPAs are costly things. Balmford et al (2004) estimated that if a global network of MPAs could be set up with a target of conserving between 20% and 30% of the world's seas, it would cost between US$5 billion and US$19 billion every

Table 10.3 Social and economic criteria used to select the locations of marine protected areas summarized from various sources. (From Roberts et al 2003; reproduced with permission of the Ecological Society of America)

VALUE	CRITERIA
Economic	Number of fishers dependent on the area Value for tourism Potential contribution of protection to enhancing or maintaining economic value
Social	Ease of access Maintenance of traditional fishing methods Presence of cultural artifacts/wrecks Heritage value Recreational value Educational value Esthetic appeal
Scientific	Amount of previous scientific work undertaken Regularity of survey or monitoring work done Presence of current research projects Educational value
Feasibility/practicality	Social/political acceptability Accessibility for education/tourism Compatibility with existing uses Ease of management Enforceability

year. Undoubtedly, therefore, income generation is crucial and fundamental to the sustainability of MPAs. Put simply, if no economic value can be placed on protection, then it is unlikely to survive for long. It has been suggested that lack of income is one of the most important reasons why MPAs fail (Gravestock et al 2008). These authors analyzed data from 79 MPAs in 36 countries and clustered the results into three broad categories according to the numbers of visitors, the focus of the area (conservation, recreation, fishing, and so on), location, and size (Table 10.4). The first (Cluster A) comprised MPAs in poorer countries with relatively few visitors, mainly concerned with fisheries. The second (Cluster B) contained MPAs with a wide range of functions, both recreational and commercial, whilst the third and smallest group (Cluster C) contained MPAs that were likely to be more involved with conservation than recreation or fishing. The major findings of this work concluded that MPAs have minimum income requirements to survive, which were closely linked to the size of the area and the numbers of visitors received. Smaller MPAs required more income-derived funding, and more visitors, not surprisingly, needed more money spent to keep them happy.

Thus, tourism is frequently the most important source of revenue, but visitors will only keep coming if the area

remains attractive. A good example of the income possibilities from tourism in a MPA comes from Mexico. The Puerto Morelos reef MPA is located on the Yucatan coast, approximately halfway between the tourist traps of Cancun and Cozumel. Rodriguez-Martinez (2008) provides estimates of the huge increases in tourists from 1998 to 2005, and the hotel rooms (with all the allied infrastructures) to entertain them (Table 10.5). The annual income resulting from this influx of visitors increased from US$8000 in 2001 to nearly US$300,000 in 2006. As we have mentioned previously, local people must be involved in MPA establishment, management, and hence sustainability. In the case of Puerto Morelos, Rodriguez-Martinez describes five stages in the establishment and management of the MPA, based on stakeholders (individuals with some sort of interest or potential benefit in the system) within the local community. The integration of locals, scientists, and politicians within the governance framework is crucial. Such stages can readily be applied to many other MPA scenarios all over the world, in both tropical and temperate regions, and the relative importance of various types of governance of MPAs has been summarized by Christie & White (2007) (Table 10.6). The complexities of these different categories tend to increase as they become more centralized and government-led. In some ways, it could be argued that the traditional ways are the best, or at least the simplest, but infrastructure to protect the most vulnerable and to be able to depend on the right long-term decisions only comes with recourse to socioeconomic and conservation decisions based on hard science. The rarest of MPA types, the privately owned ones, are perhaps the most successful, albeit on a small scale. The west coast of Chumbe Island (Figure 10.34) off the coast of Zanzibar in Tanzania, was gazetted as a no-take MPA in 1994. The area covers approximately 0.5 km² of coral reef, with the no-take zone extending from the shoreline to 300 m offshore. The east side of the small island, in contrast to the west side, allows open fishing at all times (Muthiga et al 2000). Income to run the MPA derives from eco-tourism, whereby extremely comfortable albeit expensive facilities provide exclusive marine wildlife experiences. Snorkeling is encouraged whilst SCUBA is not allowed. Local fishermen, whilst initially unhappy about having their fishing grounds reduced, seem to be mainly content with the no-take zone now, due to the extra general revenue generated by tourism, and also the perceived benefits to their fish stocks (Tyler 2005). If only all MPAs could all be like Chumbe we could all be considerably more optimistic for the future.

As would be anticipated, not everyone benefits equally from MPAs. Depending on the emphasis on the management goals, certain members of society may in fact be marginalized at the expense of others. For example, the needs and desires of tourists visiting a MPA may well differ from those of local resident fishermen (Oracion et al 2005). Whilst the former clearly want to see pristine

Table 10.4 Characterization of 79 MPAs from 36 countries. (From Gravestock et al 2008; reproduced with permission of Elsevier)

	CLUSTER A	CLUSTER B	CLUSTER C
Region (% breakdown)	95 South, 5 North	50 South, 50 North	77 North, 23 South
Area (% breakdown)	USA = 11 Caribbean = 6 Indo-Pacific = 83	USA = 27 Europe = 18 Caribbean = 32 Indo-Pacific = 18 Australia/NZ = 5	USA = 31 Europe = 23 Caribbean = 8 Indo-Pacific = 15 Australia/NZ = 23
Funds (% breakdown)	Government = 47 Foreign entities = 47 Visitors = 6	Government = 80 Foreign entities = 10 Visitors = 10	Government = 80 Visitors = 20
GNI per capita (USD)	4656	16,534	25,641
Size (mean ha)	21,012	160,429	67,908
Visitor numbers (counts)	49,940	259,674	145,062
Coral (Y/N)	94	50	54
Benefit to fisheries (Y/N)	89	23	15
No-take zone > 5% (Y/N)	64	41	58
Current income (USD mean)	219,834[*]	790,372	89,781[†]
Current income per hectare (USD mean)	46[*]	70	573[†]
Current income per visitor (USD mean)	294[*]	140	13[†]

[*] Excludes one outlying MPA with a current income per hectare 200 times larger than the average for the group.
[†] Excludes one outlying MPA with a current income per hectare 150 times larger than the average for the group.

underwater habitats with abundant big fish, the latter equally obviously need to catch the same fish to sell and eat. The trick of management is to keep both factions happy, perhaps via a system of co-management where fishermen, conservationists, and tourism specialists run the MPA together in a cooperative and mutually beneficial manner (Gelcich et al 2008).

Even then, it is not always clear what tourists perceive as "attractive," and how much potential interference they are prepared to put up with in terms of restrictions and rules to keep things that way (Petrosillo et al 2007). Diving tourists can be particularly demanding about what they want on a visit to a MPA (Fabinyi 2008). Spectacular underwater life and scenery are certainly required, but so are creature comforts such as nice hotels and nightlife after the dives, and in some circumstances, divers may actually conflict with conservation, as we have seen in Chapter 9.

In the final analysis, all these concepts and proposals are only as good as the ability to make them work in practice, which is where compliance and enforcement has to be considered. We shall see later in this chapter how individual MPA management systems tackle this problem, but one

example for now from Italy illustrates the potential benefits of enforcing rules within a MPA. Guidetti et al (2008) reviewed the effectiveness of fully no-take MPAs in promoting fish stocks, and found that in only 3 out of 15 cases were the no-fishing rules actually properly enforced. This high level of protection from illegal fishing was achieved via the combined efforts of reserve personnel, coast guards, and other marine police forces. Only high levels of enforcement produce the results expected, i.e. significantly enhanced fish stocks. There is no point in making rules and regulations if there is no one to enforce them.

Zoning

As we have seen throughout this book, marine ecosystems can be highly complex, variable and heterogeneous, both in terms of community types and species living within them. Thus, a single MPA may well contain many different habitats, not all of which are equally valuable for management and conservation and not equally exposed to impacts and pressures. Habitat mapping which we described in the

Table 10.5 Numbers of hotel rooms and tourists for the Puerto Morelos village and the Riviera Maya Tourist Corridor in Quintana Roo, Mexico, from 1998 to 2005. (From Rodriguez-Martinez 2008; reproduced with permission of Elsevier)

YEAR	PUERTO MORELOS		RIVIERA MAYA	
	HOTEL ROOMS	TOURISTS	HOTEL ROOMS	TOURISTS
1998	401	n.d.	10,095	595,050
1999	401	24,317*	12,653	801,521
2000	401	31,936*	15,297	1,215,727
2001	598	48,018*	18,731	1,504,052
2002	1134	101,641*	20,014	1,793,864
2003	1368	122,263*	22,624	2,021,989
2004	1452	150,089*	23,512	2,418,623
2005	1455	118,361*	26,980	2,194,765

n.d., no data.

Source: http://sedetur.qroo.gob.mx/estadistica; http://www.rivieramaya. com/esp/ es-EstadisticasProy.htm.

* No official data exist for the number of tourists that visit Puerto Morelos. They were estimated based on a proportion of the total number of tourists to the Riviera Maya.

previous section may therefore identify subsets of a MPA's total area (or indeed volume) which merit being handled differently in terms of protection and resource use. Take for example the coral reefs of the Las Perlas Archipelago MPA in the Gulf of Panama. Guzman et al (2008) have produced detailed habitat maps which pin-point the species richness and percentage cover of scleractinians and octocorals in relation to watersheds and rivers, human habitation, and spatial separation and isolation. They found that species richness does not necessarily spatially correlate directly with the cover of live coral (in fact the higher the cover, the lower the richness in this example). If it is considered important to preserve species richness, then certain parts of the MPA are clearly more important than others, but at the same time, likely pressures from river run-off and people pressure must also be taken into account. In this way, parts of the MPA with a combination of conservation worth and low external pressures can be chosen as protection or sanctuary zones, whereas others where people come first may become zones where sustainable exploitation may be permitted. It may even be that areas with high coral cover which is attractive to tourists can be designated as visitor zones, even thought the species richness may not be all that high.

This system of zoning MPAs into different areas with different pressures and purposes is a mainstay of marine conservation and protected area management the world over. Zoning "defines spatial objectives and accompanying restrictions in a format understandable to those who have a stake in protected area management and are on-going users of area resources" (Portman 2007), and it needs to integrate ecological, socioeconomic, and institutional/political concerns. A few examples will illustrate these points. St Barthelemy is an island complex in the northeastern Caribbean between Antigua and the Virgin Islands. The marine reserve on St Barts was launched in November 1999, and now consists of a series of zones covering over 1000 ha of protected islands, bays, and sensitive habitats (Figure 10.35) (Brosnan et al 2002). Three zones have been designated with different functions, purposes, and prohibitions. All three forbid fishing with traps, spear and net fishing, collecting corals or plants, lobster fishing, littering, water skiing, and jetskis. Anchoring is not allowed except with permission. Zones B and C do not allow SCUBA diving, though some snorkeling is permitted. Zone C is restricted to scientific observations only. Moving to the UK, Lundy Island is located at the mouth of the Bristol Channel, between the southwest of England and the tip of South Wales. It was the first statutory marine nature reserve in the UK, established in 1987. Figure 10.36 shows the various zones set up around Lundy (Landmark Trust 2008). Note the somewhat complex grid of prohibited and allowed activities, which indicates the intricate nature of recreational and especially commercial uses within the MPA. Lobster potting, a traditional local industry, is "not encouraged" in the Refuge Zone, and requires a permit between October 1 and June 30 in the Sanctuary Zone. The needs and aspirations of certain stakeholders have had to be given precedence over pure conservation and protection, a common and necessary compromise in MPA management in many parts of the world. Finally in this section we present an example of vertical, rather than horizontal zoning. The Huon Commonwealth Reserve in SE Australia contains rare examples of the seabed forming underwater peaks, seamounts that protrude upwards from the average depth in the region by between 1000 m and 2000 m to within 650 m of the surface. Their fauna (Figure 10.37) is extremely unusual and endemic, but in recent years has been severely damaged by deep-water fishing. In order to prevent this, and to protect the benthic habitats, two zones have been created following categories laid down by the IUCN (International Union for the Conservation of Nature). The first is a multiple use zone, a "Managed Resource Protected Area, mainly for the sustainable use of natural ecosystems" (IUCN Category VI), wherein fishing is still allowed within limits, and the second is a benthic sanctuary zone which is a "Strict Nature Reserve, managed mainly for science" (IUCN category 1A). Some fishing (though no trawling, seining, or dredging) is allowed in both zones, but only down to 500 m in the sanctuary zone since the protected regions are located deeper than this.

Table 10.6 Marine Protected Area governance systems, characteristics and examples. (From Christie & White 2007; reproduced with permission of Springer)

	TRADITIONAL: BASED ON PRE-COLONIAL MANAGEMENT SYSTEMS AND TRADITIONAL ECOLOGICAL KNOWLEDGE, TABOO SYSTEMS	BOTTOM-UP: LED PRIMARILY BY RESOURCE USERS, GENERALLY SMALL-SCALE, PARTICIPATORY	CO-MANAGEMENT: JOINT MANAGEMENT BY RESOURCE USERS AND GOVERNMENT	CENTRALIZED: LED BY GOVERNMENT AGENCY, CONSULTATIVE WITH RESOURCE USERS	PRIVATE: LED BY PRIVATE SECTOR
Africa			Mafia Island (Tanzania), Kenya		Chumbe Island, Tanzania
Asia		Apo Island (pre 1992), San Salvador Island (pre 1990), Philippines	Tubbataha National Marine Park, Apo Island and San Salvador Island (at present), Philippines		
Pacific Islands	Palau, American Samoa, Locally Managed Marine Areas (LMMA network) in Western Pacific				
Caribbean			Souffriere, St. Lucia		
United States				Florida Keys, United States	
Australia				Great Barrier Reef National Marine Park	

Management examples

Finally in this chapter, we present some examples of Marine Protected Areas from around the world, which bring together all the forgoing concepts and management tactics. Studies of the strategies employed in these examples should provide lessons and guidelines for other similar scenarios.

Figure 10.34

Chumbe Island, Zanzibar – a privately owned and managed MPA. (Photograph courtesy of Hal Thompson.)

Great Barrier Reef, Australia

The Great Barrier Reef (GBR) is the largest MPA in the world. It covers approximately $344,000\,km^2$ ($34,440,000\,ha$) off the coast of Queensland, and extends for around $2000\,km$. It was first set up in 1975, and became a World Heritage Site in 1981. It receives 1.6 million visitors per year, generating an income estimated to be around $1 billion p.a. (Great Barrier Reef Marine Park Authority 2009). The facilities for tourists are highly sophisticated (Figure 10.38), but the potential impacts of so many people is clearly huge. As well as tourism, all sorts of fishing, netting, collecting, boating, and shipping have to go on the reef, and hence zoning structures need to be equally complex and widespread, to allow all stakeholders access to the reef whilst at the same time protecting the delicate parts (Table 10.7). Prior to 2004, the percentage of no-take zones in the GBRMPA was around 4.6%, allowing fishing and other forms of exploitation of living creatures over a very large part of the reef. In 2004, the zoning plan was modified (Davis et al 2004) which increased the no-take zones to a full 33.4% of the MPA, in order to increase the level of protection very considerably. All users of the Park are able to consult online or hard-copy zoning maps according to the colored-coded zones (Table 10.7), and hence become familiar with what is and isn't allowed in specific areas. According to the GBRMP Authority, the General Use Zone "is to provide for conservation while providing for reasonable

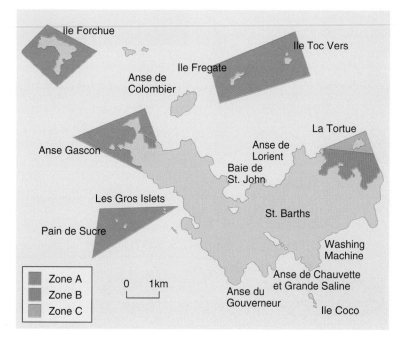

Figure 10.35

Map of St Barthelemy with the zones of the MPA marked. (From Brosnan et al 2002; Sustainable Ecosystems Institute.)

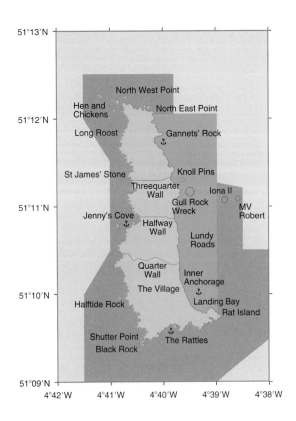

Figure 10.37

Deep-water fauna at 1115 m in the Huon Commonwealth Marine Reserve. (Reproduced courtesy of CSIRO Australia.)

Figure 10.38

Tourist facilities on the outer Great Barrier Reef. (Photograph Martin Speight.)

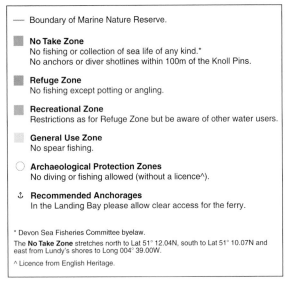

— Boundary of Marine Nature Reserve.

▪ **No Take Zone**
No fishing or collection of sea life of any kind.*
No anchors or diver shotlines within 100m of the Knoll Pins.

▪ **Refuge Zone**
No fishing except potting or angling.

▪ **Recreational Zone**
Restrictions as for Refuge Zone but be aware of other water users.

▪ **General Use Zone**
No spear fishing.

○ **Archaeological Protection Zones**
No diving or fishing allowed (without a licence^).

⚓ **Recommended Anchorages**
In the Landing Bay please allow clear access for the ferry.

* Devon Sea Fisheries Committee byelaw.
The **No Take Zone** stretches north to Lat 51° 12.04N, south to Lat 51° 10.07N and east from Lundy's shores to Long 004° 39.00W.

^ Licence from English Heritage.

Figure 10.36

Map of Lundy Island, UK, showing zoning. (From Peter Henderson.)

use". The Habitat Protection Zone "protects and manages sensitive habitats whilst still providing for reasonable use." The Conservation Park Zone "provides for conservation, but also allows reasonable use and enjoyment, with limited extractive use." The Buffer Zone "protects natural integrity and values, free from extractive use, with certain permitted activities including trolling (fishing) for pelagic species." The Scientific Research Zone "provides opportunities for research in relatively undisturbed areas." The Marine National Park Zone is rather similar to the Buffer Zone, "providing protection of natural integrity," with certain (somewhat different) permitted activities. Finally, the Preservation Zone is to "preserve natural integrity undisturbed by human activities." Figure 10.39 shows one zoning map for the Cairns area. Using GPS coordinates, users of the Park can easily find out where they can fish, extract, dive, and so on, so there can be no excuses for prohibited activities. Even with all this sophisticated management, there may still be room for improvement, especially for the very highly sensitive species such as marine turtles. Dryden et al (2008) suggested that some species, such as green, loggerhead and hawksbill turtles, required extremely high protection, especially from the side effects of commercial fishing, which even the new zoning schemes did not provide completely. Nevertheless, the GBRMPA is undoubtedly a hugely impressive and successful massive-scale

Table 10.7 Great Barrier Reef Marine Park zones (post-2004) (http://www.gbrmpa.gov.au/corp_site/management/zoning)

ZONE	% TOTAL AREA	ACTIVITIES ALLOWED	ACTIVITIES NOT ALLOWED
Preservation (pink)	>1%	Research (with permit)	Aquaculture, netting, boating, diving, harvesting aquarium fish/sea cucumbers/lobsters, spear-fishing, line fishing, tourism, traditional use, trawling, shipping etc.
Marine National Park (green)	33%	Boating, diving, research/shipping/tourism/traditional use (with permit)	Aquaculture, netting, trawling, harvesting aquarium fish/sea cucumbers/lobsters, spear-fishing, line fishing,
Scientific research (orange)	>1%	Boating, diving, research/shipping/tourism/traditional use (with permit)	Aquaculture, netting, trawling, harvesting aquarium fish/sea cucumbers/lobsters, spear-fishing, line fishing,
Buffer (olive green)	3%	Boating, diving, trolling, research/shipping/tourism/traditional use (with permit)	Aquaculture, netting, trapping, harvesting aquarium fish/sea cucumbers/lobsters, spear-fishing, line fishing, trawling
Conservation Park (yellow)		Aquaculture/harvesting aquarium fish/coral/research/tourism/shipping/traditional use (permit), limited spear-fishing with snorkel	Harvesting sea cucumber & lobster, netting, trawling
Habitat protection (dark blue)	28%	Netting, trapping, boating, diving, line fishing, trolling (for pelagics), aquaculture/harvesting aquarium fish/coral/research/tourism/shipping/traditional use (permit), limited spear-fishing with snorkel	Trawing
General use (light blue)	33%	As Habitat Protection plus trawling	None

Figure 10.39

Part of the Cairns section of the Great Barrier Reef Marine park zoning maps (see Table 10.7 for color codes). (Reproduced courtesy of Great Barrier Reef Marine park – GBRMP)

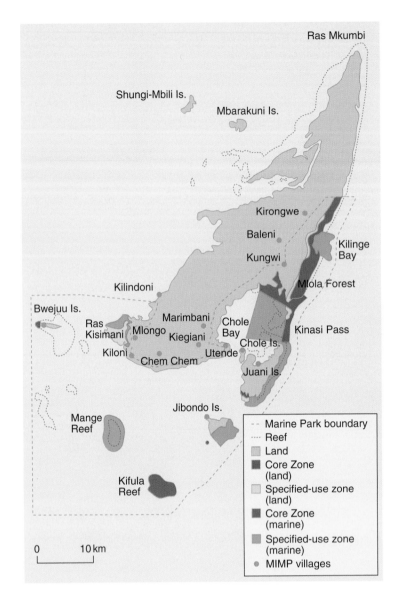

Figure 10.40

Map of Mafia Island, Tanzania, showing the zoning system in the Marine Park (MIMP). (Reproduced courtesy of gridnairobi.unep.org)

enterprise, which all others might like to follow, as long as the tactics and strategies employed here are appropriate for much smaller-scale projects.

Mafia Island, Tanzania

Mafia Island is located roughly 10 km off the coast of Tanzania. Land-based subsistence is based on the production of cashews and coconuts, the latter declining markedly these days. The coral reefs surrounding the island provide much better sources of food and revenue, including fish, crustacea (prawns and lobsters), and molluscs (squid and octopus). Coral and shell collection to sell as souvenirs on the mainland have also been significant in the past. In addition, tourism is now a very important source of revenue and thus a prime

reason for the maintenance of conservation efforts. The Mafia Island Marine park (MIMP), fully established in 1995, covers an area of about 822 km^2, with more than 75% below the high water mark (Francis et al 2002) (Figure 10.40). It is a multiple-use marine protected area, in that a variety of activities and aims are catered for, all of which contribute towards the sustainable livelihoods of more than 15,000 people who live in the park (www.mafiaisland.com 2009), as well as many more residing on the rest of the island. Because the Park has to cater for the many marine-based needs of the local community, fishing is allowed within it, but with limitations and a variety of prohibitions that limit the activities of fishermen, jet-skiers, turtle and dugong hunters, and the dumping of waste (Marine Parks & Reserves Authority, Tanzania 2009). Permit systems have been set up to restrict

Figure 10.41

Black spotted (dory) snapper, *Lutjanus fulviflamma*. (Photograph courtesy of Magdalena Fischhuber.)

the exploitation of octopus and lobsters, and the collection of sea-cucumbers using SCUBA has been banned. Boat surveillance is carried out by the community on a regular basis and confiscation of fishing gear and or fines are levied on people caught fishing illegally within the park. (United Republic of Tanzania (URT) 2000) Undoubtedly, such restrictions do the job of conservation of fish stocks. For example, the blackspot (or dory) snapper, *Lutjanus fulviflamma* (Figure 10.41), is one of the most important species caught in around the MIMP. Kamakuru et al (2004) showed that this species was four times more abundant, with six to ten times more biomass and nearly 40% larger individuals on average within the Park than when compared with adjacent fishing areas where no fishing restrictions were applied. Similarly, octopus harvesting within the Park yields bigger individual animals and more money for the community compared with outside (Guard & Mgaya 2002). Even turtle poaching has seen a substantial decrease in recent years, with members of the community employed to protect and monitor nesting beaches. (Muir 2005). Risks to the Park still exist. A serious coral bleaching event occurred in 1998 (as in many other parts of the world), resulting in large areas of reef around Mafia covered in dead coral and rubble (Garpe & Ohman 2007). Also,

whilst the management tactics for the MIMP legislate fishing by locals, fishing boats from mainland Tanzania and elsewhere raid the MIMP's plentiful resources. And finally there is tourism. As in many other MPAs around the world, fee-paying visitors are, at least in theory, a key component in the island's economy. Tourists seeking unspoilt reefs and forests are provided for by a variety of facilities such as hotels and dive operations, much of which are owned and run by off-islanders, so a large part of the revenue is not seen by the local community. In addition, loss of land or beach rights for local people, competition for resources such as water, and inappropriate behavior by tourists in terms of dress, etc. are cited as the down-side of tourism (Caplan 2009).

Darwin Mounds, UK

The Darwin Mounds were first encountered by deep-sea exploration in 1998. They are a series of sandy habitats located 185 km to the northwest of Scotland (Figure 10.42) (de Santo & Jones 2007a). The Mounds extend over an area of approximately 545 km² at depths of approximately 1000 m. They are dominated by colonies of the deep-water coral, *Lophelia pertusa*, which by growing on a sand base rather than a rocky substrate make the area particular special scientifically (Figure 10.43). The associated fauna is also extremely unusual, and therefore of great preservation merit. However, these and other similar sites have been found to be highly productive fishing grounds, and it was soon apparent after first discovery that deep-sea trawling was causing serious damage. It was feared that large areas of the Mounds could potentially be destroyed in this way (de Santo & Jones 2007a). When the Mounds were first discovered there was no legislation in place to protect them, since they were located in deep, offshore waters. In 1999 therefore, the Habitats Directive of the European Union (EU) enabled an extension from 10 nautical miles to a 200 nautical miles limit Exclusive Fisheries Zone

(EFZ), which meant that the UK was responsible for the conservation and protection of ecosystems within this greatly enlarged zone. This was followed in 2003 by a 6-month emergency ban on fishing on the Mounds. Permanent prohibitions of fishing on the Mounds came into force in 2004 (Table 10.8) (de Santo & Jones 2007b),

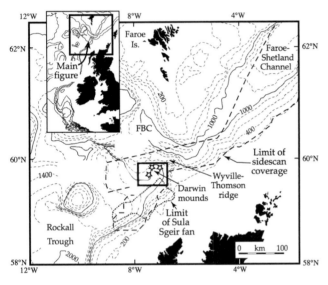

Figure 10.42

Location and extent of Darwin Mounds in the northern Rockall Trough to the northwest of Scotland. FBC = Faroe Bank Channel. (From Masson et al 2003, reproduced with courtesy of Elsevier.) 'Darwin Mounds' Rockall Trough

Table 10.8 Timeline of the designation of the Darwin Mounds MPA. 1 = Atlantic Frontier Environmental Network; 2 = Joint Nature Conservancy Council; 3 = European Commission; 4 = Common Fisheries Policy of the European Union. (From de Santo & Jones 2007b; reproduced with permission of Elsevier.)

1998	Discovery during AFEN (1) survey (revisited 2000, damage observed)
1999	Greenpeace Judgment
2001	Secretary of State Margaret Beckett announced discovery
2001	JNCC (2) Report #325 on implementing *Natura 2000* offshore
2002	Initial letter from UK to Commission (3) concerning Darwin Mounds
Jan. 2003	Revised CFP (4) Regulation 2371/2002 came into effect
Mar. 2003	Informal discussions with Commission and other Member States
Jun. 2003	Formal approach to Commission for action under Regulation 2371/2002
Jul. 2003	Formal request for closure
Aug. 2003	Emergency closure Regulation (for 6 months)
Sept. 2003	Proposal for permanent Regulation
Feb. 2004	Emergency closure extended a further 6 months
Mar. 2004	Closure made permanent

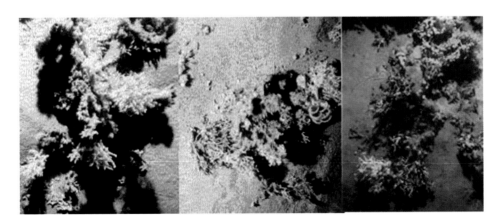

Figure 10.43

Darwin Mounds. (Photographs courtesy of Brain J. Bett, National Oceanography Centre, U.K.)

protecting an area of around 1380 km² from trawling. As a result of all this, the Darwin Mounds became the UK's first offshore Special Area of Conservation (SAC). Undoubtedly, the closure of the Darwin Mounds to bottom trawling was a political success (de Santo & Jones 2007a), but practical problems remain, in particular, enforcement. All fishing boats are required by law to carry

Vessel Monitoring Systems (VMS), which, via Geographical Positioning Systems (GPS), reveal the location of a boat all the time it is at sea. However, it is not illegal to actually be on the surface above the Mounds, nor is it illegal to trawl for pelagic species above the sea bottom, so unscrupulous trawlermen may be able to get away with fishing the prohibited sites undetected. One final point is that the protection of the Mounds only came about once damage was observed. It is another matter to force through protection legislation for pristine sites before any damage occurs (de Santo & Jones 2007a).

Florida Keys, USA

The Florida Keys is a string of islands that extend approximately 400 km southwest from the tip of Florida (Figure 10.44) (Keller & Causey 2005), ending up a mere 150 km away from the north coast of Cuba. It consists of a myriad of habitats including seagrass, sand banks, mangroves and above all, coral reefs; it is in fact the third longest barrier reef in the world, after the Great Barrier Reef in Australia and the Belize Barrier Reef in Meso-America (US Department of Commerce 2009). The Keys are home to an estimated 520 species of fish, 367 species of algae, 5 species of seagrass, 117 species of sponges, 89 species of polychaete

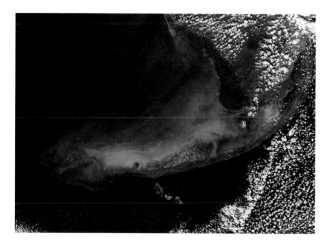

Figure 10.44

Satellite image of Florida Keys. (Photograph courtesy of Jeff Schmaltz, MODIS Rapid Response Team, NASA/GSFC.)

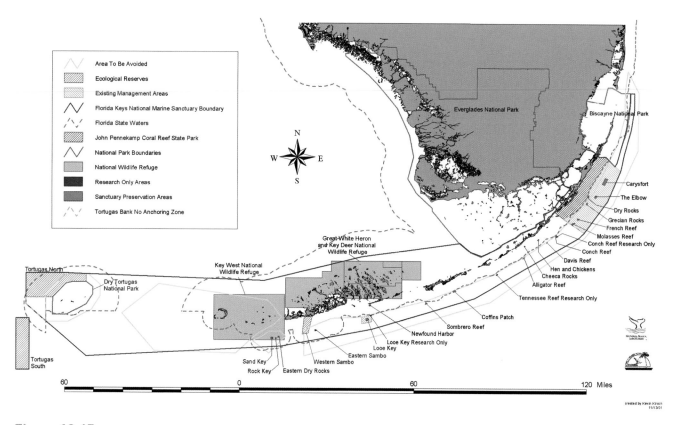

Figure 10.45

Map of the Florida Keys National Marine Sanctuary. (From Keller & Causey 2005; reproduced with permission of Elsevier.)

Table 10.9 Marine zones in the Florida Keys National Marine Sanctuary. (From US Department of Commerce 2009)

SANCTUARY PRESERVATION AREAS (SPAs)	Protect shallow & heavily used reefs where high conflicts occur between user groups & levels of visitor activity. They consist of discrete, relatively small areas. 18 SPAs; 2 km² in total
ECOLOGICAL RESERVES (ERs)	Encompass large, contiguous habitats which protect biodiversity by having minimal human disturbance. 2 ERs, Western Sambo & Tortugas; 548 km² in total
SPECIAL USE (RESEARCH ONLY) AREAS	Areas set aside for research, education, restoration & monitoring. Facilitate access or use of Sanctuary resources & reduce user conflicts. 4 Special Use Areas (Conch Reef, Tennessee Reef, Looe Key Patch Reef, & Eastern Sambo)
WILDLIFE MANAGEMENT AREAS (WMAs)	Seek to mimimize disturbance to highly sensitive wildlife (e.g. birds & turtles) & their habitats. Protection & preservation is maintained with limited user access. 27 WMAs 20 of which are co-managed by US Fish & Wildlife Service
EXISTING MANAGEMENT AREAS (EMAs)	Resource management areas established by NOAA et al before 1996 when the FKNMS plan was set up. 21 EMAs, e.g. Key Largo National Marine Sanctuary

worms, 109 species of echinoderms, 55 species of soft corals, and 65 species of hard corals. However, the reefs in the Keys have suffered all sorts of impacts for decades, including annual storms and hurricanes (Williams et al 2008), declining water quality due to current patterns and domestic waste water (Keller & Causey 2005), and huge increases in tourists over the last 20 years, with up to 4 million people per year visiting the Keys (US Department of Commerce 2009). The effects on the reef and other associated ecosystems have long been documented and highlighted (see Chapter 9), exemplified by substantial reductions in scleractinian coral genera such as *Acropora* (Somerfield et al 2008). Various MPAs (sanctuaries) were set up in the Keys from the 1960s onwards, with the John Pennekamp Coral Reef State Park being declared in 1960, hence becoming the world's first. In 1990, the US National Oceanic and Atmospheric Administration (NOAA) gathered all the Keys into one unit, the Florida Keys National Marine Sanctuary (FKNMS), which now covers around 9500 km² (Figure 10.45). The management plans for a MPA as large as this are by necessity complex and far reaching, encompassing everything from marking (buoying) of protected sites and enforcing the regulations appropriate to the local habitats, to scientific research and monitoring. As with any MPA, outreach and education are crucial and fundamental; people need to know what needs to be protected and why before they will take much interest in sustainability and keeping resources as pristine as possible. Nevertheless, with many tens of thousands of people living in the sanctuary, and millions more visiting annually, people pressure must be recognized and catered for. In order to achieve all these goals, there are five types of zone in the Sanctuary: Sanctuary Preservation Areas, Ecological Reserves, Special-use (Research-only) Areas, Wildlife Management Areas, and Existing Management Areas (Table 10.9), all of which have different rules and regulations, and all of which attempt to "limit consumptive activities whilst allowing (other) activities which do not threaten resource

Figure 10.46
Loggerhead Key in the Dry Tortugas National Park. (Photograph courtesy of Don Hickey, US Geological Survey.)

protection" (US Department of Commerce 2009). Take the Sanctuary Preservation Areas (SPAs) for example. Each one has its own zoning map available on the internet, which provides location details (how to find it), the major physical and biological features, and the restrictions and regulations which apply.

As with all such systems, proof of protection or preservation success takes a long time to obtain, but undoubtedly some aspects of the Florida Keys reef ecosystems are on the mend. At the very tip of the Keys, 110 km further into the Gulf of Mexico after the Keys road ends (Figure 10.45), lies the Dry Tortugas National Park, a no-take marine reserve covering approximately 520 km² in area (Delaney 2003) (Figure 10.46). Ault et al (2006) monitored the recovery of fish stocks in the reserve between 2000 and 2004, using intensive underwater surveys. They routinely measured the

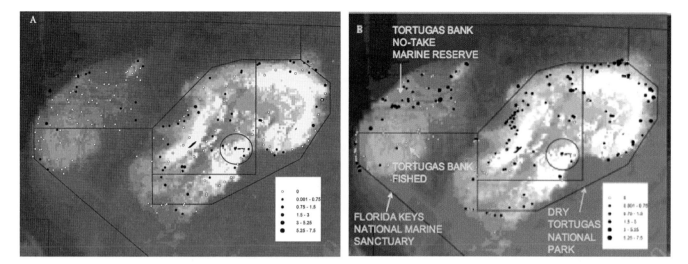

Figure 10.47

Spatial distribution of black grouper, *Mycteroperca bonaci* (mean number per primary sample unit) in 2 years of underwater surveys (A = 2000 and B = 2004) in fished and no-take areas in the Dry Tortugas. (From Ault et al 2006; reproduced with permission of *Bulletin of Marine Science*.)

abundances and locations of a total of 267 fish species and found that recoveries varied considerably according to species. Thus, for example, the numbers of certain labrids (wrasses) such as hogfish (*Lachnolaimus maximus*), and haemulids (grunts) such as white grunt *Haemulon plumieri* declined, some snappers (lutjanids) such as mutton snapper *Lutjanus analis* and damselfish (pomacentrids) such as purple reeffish *Chromis scotti*, increased. On the other hand, some parrotfish (scarids) such as stoplight parrotfish *Sparisoma viride* and angelfish (pomacentrids) such as blue angelfish *Holacanthus bermudensis* did not change significantly. Figure 10.47 shows the spatial distributions of black

grouper (*Mycteroperca bonaci*) over various parts of the no-take area and outside it. It is clear that densities of this commercially important species have increased over the years of protection, and as the authors point out, this sort of data suggests that fish stocks can be protected and enhanced in the Florida Keys through appropriate management and legislation.

In summary, protected areas are beginning to contribute to the protection and conservation of marine habitats and biodiversity. There will undoubtedly be far more development of Marine Protected Areas and refinement of management methods.

APPENDIX

Time-line of significant events in the Hudson ecosystem

YEAR	LEGISLATION, ETC.	INTRODUCED SPECIES	FISHERIES	OTHER ECOLOGICAL EVENTS
1609			Presence of "many salmon" noted by Henry Hudson in upper Hudson river	
1624–65			Dutch settlements started fisheries for shad and striped bass	
Mid-1600s			Increase in logging, agriculture, and industry	
1760				Piermont Marsh begins to be dominated by weedy taxa, as arboreal taxa decline
Early 1800s		Purple loosestrife introduced		
1800s		Mute swan introduced. Intentionally introduced fish species may include Atlantic salmon, brown trout, rainbow trout, goldfish, carp, smallmouth bass, largemouth bass, white and black crappie		
1819		Stonecat, white bass, and fantail darter reach Hudson via Champlain–Hudson Canal?		Champlain–Hudson Canal opened, joining Hudson to St Lawrence
1825		Emerald shiner, central mudminnow, redside dace, Brook silverside and greenside darter reach Hudson via Erie/New York Barge Canal?	Shad nursery and spawning grounds, which originally extended from Kingston to Glens Falls, restricted when State dam constructed at Troy	New York Barge Canal opened, joining Hudson to Lake Erie
1831/32		Carp introduced (?)		
1840s			Hudson oyster catches >1.5 m bushels annum^{-1}	Railroad bed construction begins, both banks of Hudson

239

YEAR	LEGISLATION, ETC.	INTRODUCED SPECIES	FISHERIES	OTHER ECOLOGICAL EVENTS
1842			Fyke and hoop nets replaced by more efficient stake gill net. Occasional Atlantic salmon recorded in Hudson Estuary by DeKay	
Mid-1800s			Sturgeon fishery develops fully	
1850			Reported that pollution by manufacturing wastes responsible for marked decline in oyster productivity of the waters in and adjacent to New York Harbor	
Pre-1896		Sea lamprey reaches Lake Ontario via Hudson		
1868			New York Fish Commission organized an artificial (shad) propagation operation with river fishermen	
1870			Reported that pollution by raw sewage having an adverse effect on oyster fisheries	
1875			Reduced catches promoted attempts to propagate sturgeon artificially	
1877			Disturbance of eggs and fry by ship traffic was first cited as a cause for the shad's decline	
1877–1901			American shad catches peak	
1880			Wild oyster catches fallen off to <1 m bushels annum^{-1}. Practice of oyster seeding and cultivation begun	
1883–95			NY State hatched and planted 33,322,500 shad fry in the Hudson; United States Fish Commission raised and planted 2000,000 eggs and 53,474,000 fry	
1884		Water chestnut introduced, Collins Lake, NY.		
1887	Law allowed State lands under water to be obtained under perpetual franchise. Added considerable impetus to the now common practice of oyster farming		American shad catch 2556,000 lb	

1888			Report of the US Fish Commission stated that the caviar industry was responsible for the decline in the sturgeon fishery	
1890			Reported landings of Atlantic sturgeon (across all Atlantic states) peak at 7.5 m	
1893			Cultivated oyster fishery increases greatly	
1896			Review of 1895 census of shad fishery in Hudson finds industry of even greater value ($184,898) than anticipated. Total of 3,471 nets in NY and NJ landed 1,155,610 shad weighing 4044,635 lb. Total East Coast shad landings 50,000,000 lb	
1896–1901			Hudson shad fishery very high and stable; 2–3000,000 lb annum^{-1}	
1898			Highest reported landings of Atlantic sturgeon in Hudson – 508,000 lb	
Early years of 1900s			Decreased shad runs occur	
1904–17			Hudson shad fishery drops steeply to 1916 low of 40,000 lb annum^{-1}	
1908	New State law required nets to be taken up at sunset Thursday and left out of water until Monday at sunrise in the hope that decreased fishing intensity would permit greater shad egg production and, subsequently, greater runs			
1911			Cultivated oyster catches peak	
1912	In the areas of the Hudson River, Raritan Bay and Staten Island Sound there were approximately 9000 acres (of oyster beds) under lease or franchise to 87 individuals or companies			
1913–22			Hudson sturgeon catches steady; 20–30,000 lb annum^{-1}	
1913–45			Catches of white perch high and reasonably stable: 15–60,000 lb annum^{-1}	

YEAR	LEGISLATION, ETC.	INTRODUCED SPECIES	FISHERIES	OTHER ECOLOGICAL EVENTS
1913–64			Atlantic tomcod catches go through three 20-year cycles of big increase – plateau – tail off to very low levels	
1918			Outstanding catch of American eel (182, 052 lb), compared to 1917 (5007 lb) and 1919 (14,657 lb). Atlantic tomcod catches peak at 7.526 lb	
1918–30			Hudson shad catches increase to stable level of 120,000–375,000 lb annum^{-1}	
1920s				Channel dredged to 27 feet between Hudson and Albany
1925			Sturgeon catches drop considerably, to 1–20% of previous levels	
1925–56			American smelt catches generally 50–850 lb annum^{-1}; spikes in 1929 and 1956, then nothing	
1926	State law prohibited the taking of either sturgeon between September 2 and May 30 and set size limits of 42 inches on Atlantic sturgeon and 20 inches on shortnosed sturgeon. (No evidence of its efficacy)			
1930s		Water chestnut infestation at pest proportions in upper Hudson		
1931–48			American eel catches roughly stable, varying from 12,000 to 60,000 lb annum^{-1}	
1933–47			Catches of herring high and relatively steady	
1935			Herring catch peak at 274,405 lb; white perch catch peaks – 60,552 lb. 71 regular fishermen on Hudson, 6278 yards of haul seines	
1936–49			American shad catches peak	
1936–59			Total Hudson shad fishery catches peak (2–3 m lb) and in 1950 drop back to 1–1.5 m lb	

Year				
1936		Survey reports only two plants of water chestnut downstream of Troy dam	First synoptic survey of fish (Greeley 1937) in Hudson undertaken	
1936–48			Total Hudson fish catch plateau at 1.2–2 m lb annum^{-1}	
1938			Fork length of 16 inches was enacted as the minimum legal size for striped bass taken; second huge peak of American eel catches (111,429 lb)	
1939–45			War years; catches of alewife, blueback herring, striped bass, American eel, shortnosed sturgeon, Atlantic sturgeon, white perch, Atlantic tomcod and American smelt peak	
1940s		Water chestnut invades Hudson		
1944			Last hatchery-raised shad fry planted in the Hudson. Hudson shad catch peaks at 3809,400 lb	
Mid 1940s		Water chestnut widespread throughout Hudson Estuary		
1945			Catch of American shad peaks at 2,091,300 lb, after 6 years of catches well >1 m lb; declines sharply post-1948; sturgeon catches recover a little, varying from 2000 to 10,000 lb annum^{-1}. American tomcod lowest at 35 lb. Total Hudson fish catch peaks at 2332,800 lb	
1946–64			White perch catches reduced, generally 2000–9000 lb annum^{-1}	
1947				GE begins using PCBs at Fort Edward
1948			After 1948, American eel catches drop away, recover slightly in early 1960s, then drop. Post-war peak sturgeon catch, 10,655 lb	
1950s				Dredged channel extended south to New York City
1952		McKeon begins purple loosestrife control experiments in Lower Hudson		GE begins using PCBs at Hudson Falls

YEAR	LEGISLATION, ETC.	INTRODUCED SPECIES	FISHERIES	OTHER ECOLOGICAL EVENTS
1952–60			Total Hudson fish catch drops sharply to 450–750,000 lb annum^{-1}	
1956			American smelt catch peaks at 2181 lb, disappears completely	
1959			Catches of striped bass peak this year (133,100 lb) and next (132,900 lb), then decline sharply	
1960s			River herring populations begin decline	Entire length of navigation channel deepened to 32 feet
1960–63			Total Hudson shad catches drop sharply	
1961			Hypothesis presented that the increasing sewage in estuaries may have been indirectly responsible for the frequent occurrence of dominant (striped bass) year-classes during the preceding several years. Total Hudson fish catch drops sharply to 150–350,000 lb annum^{-1}	
1963			Oyster catches reach low	
1964			Hudson fishery catches reach lowest ebb; American shad catches lowest since 1917 at 78,200 lb. Six regular fishermen, 385 yards of haul seines	
1965	Pure Waters Bond Act passed by State Legislature, to offset cost of building sewage treatment works		Subcommittee on Fisheries and Wildlife Conservation of the US Committee on Merchant Marine and Fisheries document fish mortalities at ConEd's Indian Point	
1967	Federal Water Pollution Control Act passed. Shortnose sturgeon listed as endangered throughout its range			
1968–84			Index of age 3 and older white perch indicates increasing abundance	
1969				Elevated levels of PCBs in Hudson biota first found

1970	Freshwater Wetland Protection Act and Wetlands Act – NJ			
1970s		*Botrylloides violaceus* introduced to Hudson	Dovel (1992) estimates shortnose sturgeon population at 13,000 fish	
1972			Angling record striped bass caught in NJ, weighs 78.5 lb	
1974–mid 1980s			Eggs, YSL and PYSL numbers of white perch increase	Discharge of suspended solids from water treatment in upper Hudson and Mohawk decreased by 56%
1975–78				Contaminated sediments with 160,000 lb of PCBs removed
1975	Freshwater Wetlands Act – NY; DEC implements monitoring for PCBs		Atlantic sturgeon listed under CITES Appendix II	
1976	Health advisory on eating PCB-contaminated fish first issued	Water chestnut eradication program ceases	Hudson Atlantic sturgeon year-class estimated between 14,500 and 36,000 fish	
1976	Troy Dam to Battery; all commercial fishing except for a few key species banned. Various restrictions lifted in 1980s and 1990s			
1976–95	Recreational fishing banned from Hudson Falls to Troy Dam			
1977	PCB discharges banned; Tidal Wetlands Act – NY; Clean Water Act		Dovel and Berggren (1983) estimate 25,000 Age 1 Atlantic sturgeon in Hudson	480,000 gallons of petroleum spilled, Hudson Highlands
Late 1970s				Improved sewage treatment removed problem of oxygen block at Albany Pool
1977–88				Reported decline in ppm of PCBs in Hudson Atlantic sturgeon
1978				PCB ppm peak in fish sampled at Stillwater/ Coveville, Catskill, Tappan Zee
1978 onwards			Indices of YOY and Age 1 white perch peak then decline; indices of American eel abundance decline threefold	

YEAR	LEGISLATION, ETC.	INTRODUCED SPECIES	FISHERIES	OTHER ECOLOGICAL EVENTS
1980	Settlement agreement, provision for design and construction of second striped bass hatchery		Production of young sturgeon begins to decline. Commercial Fisheries Monitoring Program started, to study returning shad catches	
1980–83			Striped bass abundance index from shad fishery by-catch = 1.25	
1980s		MSX oyster disease first occurs in Long Island Sound	Early 1980s. CPUE of Atlantic sturgeon, as by-catch of shad fishery, peaks and begins decline	1980s to early 1090s, upgrading of water treatment plants decreased discharges of raw sewage; summer dissolved oxygen (DO) levels increase
1982	Hudson River National Estuarine Research Reserve established			Mean PCB levels in Atlantic sturgeon of Hudson/ Raritan 2.35 ppm
1983			Recreational and commercial size limits for striped bass increased from 18 inches to 24 inches	
1984	200 miles from Hudson Falls to Battery designated Superfund site by EPA. FDA reduces upper limit of PCBs in fish and shellfish from 5 ppm to 2 ppm		Commercial in-river Shad catches across all Atlantic states peak (c. 3.2 m lb) and begin decline. 1984–86, adult numbers of white perch decrease sharply. Striped bass abundance index from shad fishery by-catch = 4.78	
1984–86			Hudson shad fishery catches increase sharply	
1984–88		Atlantic clam *Rangia cuneata* and American blenny *Hypsoblennius ionthas* introduced		
1985			Spawning stock survey of shad and striped bass started	
Mid 1980s		Water chestnut populations again at nuisance level		
Mid 1980s to 2000			Eggs, YSL and PYSL numbers of white perch stable	
1985–87			Striped bass abundance index from shad fishery by-catch = 8.36	

1985–92			Striped bass spawning stock Age 8+ index increases markedly, from 20% to >60%	
1986		Zebra mussel introduced into Lake St Clair, Detroit	Commercial shad gill-net fishery CPUE peaks (50.0) and begins decline. Relative abundance (CPUE) of Atlantic sturgeon in ConEd trawling survey peaks (0.120), begins decline. Moratorium on recreational (May 86–May 87) and commercial (May 86–Sep 90) striped bass fishing	
1987–2000			Shad fishery CPUE average relatively stable	
1987	NY–NJ Harbor Estuary Program designated; Hudson River Estuary Management Act passed	*Hemigrapsus sanguineus* – Asian Shore Crab, found in NJ	33 inch size limit imposed on recreational striped bass fishing	
1988–91			Striped bass abundance index from shad fishery by-catch = 13.6	
1988	Hudson River Estuary Management Advisory Committee appointed			
1989	Hudson River PCB Project Action Plan developed		American shad YOY abundance peaks in Hudson. Coastal commercial catches of shad (across all Atlantic states) peaks at c. 2 m lb, begins decline. Recreational limit on striped bass increased to 36 inches	
Early 1990s				New releases of PCBs from Fort Edward and Hudson Falls plants. Water treatment plants in lower Hudson upgraded to full secondary treatment
1990			Moratorium on striped bass commercial fishing lifted. Recreational limit on striped bass increased to 38 inches	200,000 gallons of kerosene spilled from grounded barge, Diamond Reef
1990–2000			American shad YOY abundance in general decline	
1991	Hudson Valley Greenway Council and Greenway Conservancy created; Atlantic striped bass Conservation Act; New ambient water quality standard for chlorine	May. Zebra mussel appeared in Hudson at Catskill		EPA estimates total PCB releases 1947–77 at 209,000–1330,000 lb. North River water treatment plant upgraded

YEAR	LEGISLATION, ETC.	INTRODUCED SPECIES	FISHERIES	OTHER ECOLOGICAL EVENTS
1992		End. Zebra mussels found everywhere between Troy and Haverstraw; population reaches 550 bn	1992–95 sphaeriid clams in decline; density, recruitment and condition of unionid clams decline steeply. Phytoplankton biomass falls by 80%, zooplankton biomass by 70%, macrobenthos biomass by 60%, water transparency rises by 45%, DO falls by 12%	Standing stock of phytoplankton in Hudson Estuary begins decline to 10–20% of previous levels (Oct.). New PCB releases at Hudson Falls discovered
1992–94			Striped bass abundance index from shad fishery by-catch = 30.42	
1992 on			Striped bass spawning stock Age 8+ index stable at 60–70%	
1992–96				Hudson-Raritan fish and invertebrate surveys
1993	Atlantic Coastal Marine Fisheries Cooperative Management Act; Dredged Material Management Forum convened		Relative abundance (CPUE) of shortnose sturgeon in ConEd trawling survey peaks (0.07), begins decline	PCB ppm reach peak from fish sampled at Thompson Island Pool
1994			Hudson Atlantic sturgeon year-class estimated at 9529 fish – of which only 4600 were of wild origin. 4929 farmed fish released in known nursery area	Further releases of PCBs discovered at Hudson Falls
1995–97			Population study of shortnose sturgeon carried out by NMFS and USACE. 1995 data estimates adult population at 55,265 fish	
1995		*Anguillicola crassus* seen in Texas and South Carolina	Striped bass population declared fully recovered	
1996	Clean Water/Clean Air Bond Act passed; Federal 2-year moratorium on taking and sale of Atlantic sturgeon. Hudson River Estuary Management Action Plan adopted by the New York State DEC.		Stock assessment of American shad, based on stock and landings, described as in decline, but not overfished. NY implements moratorium on Atlantic sturgeon harvest, NJ a zero quota. Blueback herring YOY abundance index highest at 50.16	
1997	Atlantic sturgeon recommended for addition to Federal list of Threatened and Endangered species	*A. crassus* seen in Chesapeake Bay and Hudson Estuary. 1997–2000, infestation increases exponentially		

1997–98		MSX decimates oyster fisheries		
1998	Coast-wide ban on possession of American sturgeon			
1999			Moratorium placed on new holders of blue crab fishing permits in NY State. Monitoring of blue crab catches started. Hudson River Estuary Management Advisory Committee recommend limited reopening of striped bass fishery between George Washington Bridge and the Bear Mountain Bridge	
2000			Shad fishery catches peak at CPUE of c. 35. Stock abundance of Atlantic striped bass (fish aged 1–12+) estimated at 53 m	
2001			Stock abundance of Atlantic striped bass estimated at 59.6 m	
2003?	40% cutback in ocean harvest of American shad scheduled		Tom Lake's current species list contains 210 spp., 149 genera, 73 families (some occurring only in tributaries)	
2005	Full closure of ocean harvest of American shad planned			

REFERENCES

Aceves-Medina, G., Saldierna-Martinez, R., Hinojosa-Medina, A., Jimenez-Rosenberg, S.P.A., Hernandez-Rivas, M.E. & Morales-Avila, R., 2008, Vertical structure of larval fish assemblages during diel cycles in summer and winter in the southern part of Bahia de La Paz, Mexico, *Estuarine Coastal and Shelf Science*, 76(4), pp. 889–901.

Adrianov, A.V., 2003, Marine biological diversity: patterns, processes and modern methodology, *Russian Journal of Nematology*, 11(2), pp. 119–26.

Aeby, G.S. & Santavy, D.L., 2006, Factors affecting susceptibility of the coral *Montastraea faveolata* to black-band disease, *Marine Ecology – Progress Series*, 318, pp. 103–10.

Aerts, L.A.M., 2000, Dynamics behind standoff interactions in three reef sponge species and the coral *Montastraea cavernosa*, *Marine Ecology-Pubblicazioni Della Stazione Zoologica Di Napoli I*, 21(3–4), pp. 191–204.

Ainsworth, T.D., Kvennefors, E.C., Blackall, L.L., Fine, M. & Hoegh-Guldberg, O., 2007, Disease and cell death in white syndrome of Acroporid corals on the Great Barrier Reef, *Marine Biology*, 151(1), pp. 19–29.

Alley, R. B., Clark, P. U., Huybrechts, P. & Joughin, I., 2005, Ice-sheet and sea-level changes. *Science*, 310(5747), pp. 456–60.

Almany, G.R., Berumen, M.L., Thorrold, S.R., Planes, S. & Jones, G.P., 2007, Local replenishment of coral reef fish populations in a marine reserve, *Science*, 316(5825), pp. 742–4.

Alvarez-Filip, L., Dulvy, N.K., Gill, J.A., Cote, I. M. & Watkinson, A.R., 2009, Flattening of Caribbean coral reefs: region-wide declines in architectural complexity. *Proceedings of the Royal Society B – Biological Sciences*, 276(1669), pp. 3019–25.

Angel, M.V., 1993, Biodiversity of the pelagic ocean, *Conservation Biology*, 7(4), pp. 760–72.

Angel, M. V. 1997. Pelagic biodiversity. In: Ormond, R.F.G., Gage, J.D. & Angel, M.V. (eds), *Marine Biodiversity*, pp. 35–68. Cambridge University Press, Cambridge, UK.

Angly, F.E., Felts, B., Breitbart, M., Salamon, P., Edwards, R.A., Carlson, C., Chan, A.M., Haynes, M., Kelley, S., Liu, H., Mahaffy, J.M., Mueller, J.E., Nulton, J., Olson, R., Parsons, R., Rayhawk, S., Suttle, C.A. & Rohwer, F., 2006, The marine viromes of four oceanic regions, *Plos Biology*, 4, pp. 2121–31.

Anthony, K.R.N. & Fabricius, K.E., 2000, Shifting roles of heterotrophy and autotrophy in coral energetics under varying turbidity, *Journal of Experimental Marine Biology and Ecology*, 252(2), pp. 221–53.

Anthony, K.R.N., Kline, D.I., Diaz-Pulido, G., Dove, S. & Hoegh-Guldberg, O. Ocean acidification causes bleaching and productivity loss in coral reef builders, *Proceedings of the National Academy of Sciences of the United States Of America*, 105(45), pp. 17442–6.

Arena, P.T., Jordan, L.K.B. & Spieler, R.E., 2007, Fish assemblages on sunken vessels and natural reefs in southeast Florida, USA, *Hydrobiologia*, 580, pp. 157–71.

Armstrong, C.W. & van den Hove, S., 2008, The formation of policy for protection of cold-water coral off the coast of Norway, *Marine Policy*, 32(1), pp. 66–73.

Aswani, S. & Lauer, M., 2006, Benthic mapping using local aerial photo interpretation and resident taxa inventories for designing marine protected areas, *Environmental Conservation*, 33(3), pp. 263–73.

Aubry, F.B., Berton, A., Bastianini, M., Socal, G. & Acri, F., 2004, Phytoplankton succession in a coastal area of the NW Adriatic, over a 10-year sampling period (1990–1999), *Continental Shelf Research*, 24(1), pp. 97–115.

Ault, J.S., Smith, S.G., Bohnsack, J.A., Luo, J.G., Harper, D.E. & McClellan, D.B., 2006, Building sustainable fisheries in Florida's coral reef ecosystem: Positive signs in the Dry Tortugas, *Bulletin of Marine Science*, 78(3), pp. 633–54.

Avila, E. & Carballo, J.L., 2006, Habitat selection by larvae of the symbiotic sponge *Haliclona caerulea* (Hechtel, 1965) (Demo-spongiae, Haplosclerida), *Symbiosis*, 41(1), pp. 21–9.

Ayre, D.J. & Hughes, T.P., 2000, Genotypic diversity and gene flow in brooding and spawning corals along the Great Barrier Reef, Australia, *Evolution*, 54(5), pp. 1590–605.

Azzellino, A., Gaspari, S., Airoldi, S. & Nani, B., 2008, Habitat use and preferences of cetaceans along the continental slope and the adjacent pelagic waters in the western Ligurian Sea, *Deep-Sea Research Part I – Oceanographic Research Papers*, 55(3), pp. 296–323.

Bailey, D.M., Ruhl, H.A. & Smith, K.L., 2006, Long-term change in benthopelagic fish abundance in the abyssal northeast Pacific Ocean, *Ecology*, 87(3), pp. 549–55.

Bailey, K.M., Abookire, A.A. & Duffy-Anderson, J.T., 2008, Ocean transport paths for the early life history stages of offshore-spawning flatfishes: a case study in the Gulf of Alaska, *Fish and Fisheries*, 9(1), pp. 44–66.

Baker, A.C., Glynn, P.W. & Riegl, B., 2008, Climate change and coral reef bleaching: An ecological assessment of long-term impacts, recovery trends and future outlook, *Estuarine Coastal and Shelf Science*, 80(4), pp. 435–71.

Balasubramanian, H. & Foster, R.S., 2007, Species and space: role of volume in organizing coral reef fish assemblages in SE Cuba, *Marine Ecology Progress Series*, 345, pp. 229–36.

Ballantine, W.J. & Langlois, T.J., 2008, Marine reserves: the need for systems, *Hydrobiologia*, 606, pp. 35–44.

Balmford, A., Gravestock, P., Hockley, N., McClean, C.J. & Roberts, C.M., 2004, The worldwide costs of marine protected areas, *Proceedings of the National Academy of Sciences of the United States of America*, 101(26), pp. 9694–7.

Bamber, R.N., 1990, Power-station thermal effluents and marine crustaceans, *Journal of Thermal Biology*, 15(1), pp. 91–6.

Bamber, R.N. & Seaby, R.M.H., 2004, The effects of power station entrainment passage on three species of marine planktonic crustacean, *Acartia tonsa* (Copepoda), *Crangon crangon* (Decapoda) and *Homarus gammarus* (Decapoda), *Marine Environmental Research*, 57(4), pp. 281–94.

Bamber, R.N. & Spencer, J.F., 1984, The benthos of a coastal power-station thermal discharge canal, *Journal of the Marine Biological Association of the United Kingdom*, 64(3), pp. 603–23.

Barnes, D.K.A. & Griffiths, H.J., 2008, Biodiversity and biogeography of southern temperate and polar bryozoans, *Global Ecology and Biogeography*, 17(1), pp. 84–99.

Barnett, P.R.O., 1971, Some changes in intertidal sand communities due to thermal pollution, *Proceedings of the Royal Society of London B*, 177, pp. 353–64.

Barrett, N.S., Edgar, G.J., Buxton, C.D. & Haddon, M., 2007, Changes in fish assemblages following 10 years of protection in Tasmanian marine protected areas, *Journal of Experimental Marine Biology and Ecology*, 345(2), pp. 141–57.

Bates, N.R. & Peters, A.J., 2007, The contribution of atmospheric acid deposition to ocean acidification in the subtropical North Atlantic Ocean, *Marine Chemistry*, 107(4), pp 547–558

Bay, L.K., Buechler, K., Gagliano, M. & Caley, M.J., 2006, Intraspecific variation in the pelagic larval duration of tropical reef fishes, *Journal of Fish Biology*, 68(4), pp. 1206–14.

Beaulieu, S.E., 2001, Colonization of habitat islands in the deep sea: recruitment to glass sponge stalks, *Deep-Sea Research Part I – Oceanographic Research Papers*, 48(4), pp. 1121–37.

Behrenfeld, M.J., Worthington, K., Sherrell, R.M., Chavez, F.P., Strutton, P., McPhaden, M. and Shea, D.M., 2006a, Controls on tropical Pacific Ocean productivity revealed through nutrient stress diagnostics. *Nature*, 442(7106), pp. 1025–8.

Behrenfeld, M.J., O'Malley, R.T., Siegel, D.A., McClain, C.R., Sarmiento, J.L., Feldman, G.C., Milligan, A.J., Falkowski, P.G., Letelier, R.M. & Boss, E.S., 2006b, Climate-driven trends in contemporary ocean productivity, *Nature*, 444(7120), pp. 752–5.

Bell, J.J. & Okamura, B., 2005, Low genetic diversity in a marine nature reserve: re-evaluating diversity criteria in reserve design, *Proceedings of the Royal Society B – Biological Sciences*, 272(1567), pp. 1067–74.

Bellwood, D.R. & Hughes, T.P., 2001, Regional-scale assembly rules and biodiversity of coral reefs, *Science*, 292(5521), pp. 1532–4.

Benfield, S.L., Guzman, H.M., Mair, J.M. & Young, J.A.T., 2007, Mapping the distribution of coral reefs and associated sublittoral habitats in Pacific Panama: a comparison of optical satellite sensors and classification methodologies. *International Journal of Remote Sensing* 28, pp. 5047–70.

Benton, M.J. & Emerson, B.C., 2007, How did life become so diverse? The dynamics of diversification according to the fossil record and molecular phylogenetics, *Palaeontology*, 50, pp. 23–40.

Berger, M.S. & Young, C.M., 2006, Physiological response of the cold-seep mussel *Bathymodiolus childressi* to acutely elevated temperature, *Marine Biology*, 149(6), pp. 1397–402.

Berkelmans, R. & van Oppen, M.J.H., 2006, The role of zooxanthellae in the thermal tolerance of corals: a 'nugget of hope' for coral reefs in an era of climate change, *Proceedings of the Royal Society B – Biological Sciences*, 273(1599), pp. 2305–12.

Berntsson, K.M., Jonsson, P.R., Larsson, A.I. & Holdt, S., 2004, Rejection of unsuitable substrata as a potential driver of aggregated settlement in the barnacle *Balanus improvisus*, *Marine Ecology Progress Series*, 275, pp. 199–210.

Beron-Vera, B., Crespo, E.A., Raga, J.A. & Fernandez, M., 2007, Parasite communities of common dolphins (*Delphinus delphis*) from Patagonia: The relation with host distribution and diet and comparison with sympatric hosts, *Journal of Parasitology*, 93, pp. 1056–60.

Bertasi, F., Colangelo, M.A., Abbiati, M. & Ceccherelli, V.U., 2007, Effects of an artificial protection structure on the sandy shore macrofaunal community: the special case of Lido di Dante (Northern Adriatic Sea), *Hydrobiologia*, 586, pp. 277–90.

Blachowiak-Samolyk, K., Kwasniewski, S., Dmoch, K., Hop, H. & Falk-Petersen, S., 2007, Trophic structure of zooplankton in the Fram Strait in spring and autumn 2003, *Deep-Sea Research Part Ii-Topical Studies in Oceanography*, 54(23–26), pp. 2716–28.

Blanchette, C.A., Broitman, B.R. & Gaines, S.D., 2006, Intertidal community structure and oceanographic patterns around Santa Cruz Island, CA, USA, *Marine Biology*, 149(3), pp. 689–701.

Blankenship, L.E., Yayanos, A.A., Cadien, D.B. & Levin, L.A., 2006, Vertical zonation patterns of scavenging amphipods from the Hadal zone of the Tonga and Kermadec Trenches, *Deep-Sea Research Part I – Oceanographic Research Papers*, 53(1), pp. 48–61.

Blaszkowski, C. & Moreira, G.S., 1986, Combined effects of temperature and salinity on the survival and duration of larval stages of pagurus-criniticornis (dana) (crustacea, paguridae), *Journal of Experimental Marine Biology and Ecology*, 103(1–3), pp. 77–86.

Blyth-Skyrme, R.E., Kaiser, M.J., Hiddink, J.G., Edwards-Jones, G. & Hart, P.J.B., 2006, Conservation benefits of temperate marine protected areas: Variation among fish species, *Conservation Biology*, 20(3), pp. 811–20.

Bolam, S.G., Schratzberger, M. & Whomersley, P., 2006, Macro- and meiofaunal recolonisation of dredged material used for habitat enhancement: Temporal patterns in community development, *Marine Pollution Bulletin*, 52(12), pp. 1746–55.

Bolstad, K.S. & O'Shea, S., 2004, Gut contents of a giant squid *Architeuthis dux* (Cephalopoda : Oegopsida) from New Zealand waters, *New Zealand Journal of Zoology*, 31(1), pp. 15–21.

Bonkosky, M., Hernandez-Delgado, E.A., Sandoz, B., Robledo, I.E., Norat-Ramirez, J. & Mattei, H., 2009, Detection of spatial fluctuations of non-point source fecal pollution in coral reef surrounding waters in southwestern Puerto Rico using PCR-based assays, *Marine Pollution Bulletin*, 58(1), pp. 45–54.

Borger, J.L., 2005, Scleractinian coral diseases in south Florida: incidence, species susceptibility, and mortality, *Diseases of Aquatic Organisms*, 67(3), pp. 249–58.

Borucinska, J.D. & Caira, J.N., 2006, Mode of attachment and lesions associated with trypanorhynch cestodes in the gastrointestinal tracts of two species of sharks collected from coastal waters of Borneo, *Journal of Fish Diseases*, 29(7), pp. 395–407.

Bosire, J.O., Dahdouh-Guebas, F., Walton, M., Crona, B.I., Lewis, R.R., Field, C., Kairo, J.G. & Koedam, N., 2008, Functionality of restored mangroves: A review, *Aquatic Botany*, 89(2), pp. 251–9.

Boudouresque, C.F., 2002, The spread of a non-native marine species, *Caulerpa taxifolia*. Impact on the Mediterranean biodiversity and possible economic consequences. Biodiversity and Tourism Symposium, Date: Sept. 20–23, 2000, Port Cros France. *Tourism, Biodiversity and Information*, pp. 75–87.

Bourlat, S.J., Nielsen, C., Economou, A.D. & Telford, M.J., 2008, Testing the new animal phylogeny: a phylum level molecular analysis of the animal kingdom. *Molecular Phylogenetics and Evolution* 49(1), pp. 23–31.

Boxaspen, K. 2006, A review of the biology and genetics of sea lice. ICES/NASCO Symposium on Interactions between Aquaculture and Wild Stocks of Atlantic Salmon and other Diadromous Fish Species, Oct. 18–21, 2005 Bergen, Norway. *ICES Journal of Marine Science*, 63(7), pp. 1304–16.

Boyd, P.W., Watson, A.J., Law, C.S., Abraham, E.R., Trull, T., Murdoch, R., Bakker, D.C.E., Bowie, A.R., Buesseler, K.O., Chang, H., Charette, M., Croot, P., Downing, K., Frew, R., Gall, M., Hadfield, M., Hall, J., Harvey, M., Jameson, G., LaRoche, J., Liddicoat, M., Ling, R., Maldonado, M.T., McKay, R.M., Nodder, S., Pickmere, S., Pridmore, R., Rintoul, S., Safi, K., Sutton, P., Strzepek, R., Tanneberger, K., Turner, S., Waite, A. & Zeldis, J., 2000, A mesoscale phytoplankton bloom in the polar Southern Ocean stimulated by iron fertilization, *Nature*, 407(6805), pp. 695–702.

Braby, C.E., Rouse, G.W., Johnson, S.B., Jones, W.J. & Vrijenhoek, R.C., 2007, Bathymetric and temporal variation among *Osedax* boneworms and associated megafauna on whale-falls in Monterey Bay, California, *Deep-Sea Research Part I – Oceanographic Research Papers*, 54, pp. 1773–91.

Bradbury, I.R., Laurel, B., Snelgrove, P.V.R., Bentzen, P. & Campana, S.E., 2008, Global patterns in marine dispersal estimates: the influence of geography, taxonomic category and life history, *Proceedings of the Royal Society B – Biological Sciences*, 275(1644), pp. 1803–9.

Bram, J.B., Page, H.M. & Dugan, J.E., 2005, Spatial and temporal variability in early successional patterns of an invertebrate assemblage at an offshore oil platform, *Journal of Experimental Marine Biology and Ecology*, 317(2), pp. 223–37.

Braun, C.L. & Smirnov, S.N., 1993, Why is water blue. *Journal of Chemical Education*, 70(8), pp. 612–15.

Brazelton, W.J., Schrenk, M.O., Kelley, D.S. & Baross, J.A., 2006, Methane- and sulfur-metabolizing microbial communities dominate the Lost City hydrothermal field ecosystem, *Applied and Environmental Microbiology*, 72(9), pp. 6257–70.

Briggs, J.C., 1991, Global species-diversity, *Journal of Natural History*, 25(6), pp. 1403–6.

Briggs, J.C., 1994, Species diversity: land and sea compared. *Systematic Biology*, 43(43), pp. 130–5.

Briggs, R.P., Armstrong, M.J., Dickey-Collas, M., Allen, M., McQuaid, N. & Whitmore, J., 2002, The application of fecundity estimates to determine the spawning stock biomass of Irish Sea *Nephrops norvegicus* (L.) using the annual larval production method, *ICES Journal of Marine Science*, 59(1), pp. 109–19.

Briones-Fourzán, P., Lozano-Alvarez, E., Negrete-Soto, F. & Barradas-Ortiz, C., 2007, Enhancement of juvenile Caribbean spiny lobsters: an evaluation of changes in multiple response variables with the addition of large artificial shelters, *Oecologia*, 151(3), pp. 401–16.

Brosnan, D., Buffam, I. & Brosnan, I., 2002, *Scientific Monitoring: reports and recommendations: The marine reserve of St Bartholemy, French West Indies*. Sustainable Ecosystems Institute, 28pp.

Brown, C.L. & Nunez, J.M., 1998, Disorders of development, *Fish Diseases and Disorders*, 2, pp. 1–17.

Bruno, J.F. & O'Connor, M.I., 2005, Cascading effects of predator diversity and omnivory in a marine food web, *Ecology Letters*, 8(10), pp. 1048–56.

Bruschetti, M., Luppi, T., Fanjul, E., Rosenthal, A. & Iribarne, O., 2008, Grazing effect of the invasive reef-forming polychaete *Ficopomatus enigmaticus* (Fauvel) on phytoplankton biomass in a SW Atlantic coastal lagoon, *Journal of Experimental Marine Biology and Ecology*, 354(2), pp. 212–9.

Brylinski, J.M., 1981, Report on the presence of *Acartia tonsa* Dana (Copepoda) in the harbour of Dunkirk (France) and its geographical distribution in Europe. *Journal of Plankton Research*, 3(2), pp. 255–60.

Bull, H.O., 1936, Studies on conditioned responses in fishes. VII. Temperature perception in teleosts, *Journal of the Marine Biological Association of the UK*, 21, pp. 1–27.

Buhl-Mortensen, L. & Mortensen, P.B., 2004, Symbiosis in deep-water corals, *Symbiosis*, 37(1–3), pp. 33–61.

Burkepile, D.E. & Hay, M.E., 2007, Predator release of the gastropod *Cyphoma gibbosum* increases predation on gorgonian corals, *Oecologia*, 154, pp. 167–73.

Busby, T.O. & Plante, C.J., 2007, Deposit feeding during tidal emersion by the suspension-feeding polychaete, *Mesochaetopterus taylori*, *Southeastern Naturalist*, 6(2), pp. 351–8.

Butt, A.A., Aldridge, K.E. & Sanders, C.V., 2004, Infections related to the ingestion of seafood. Part II: parasitic infections and food safety, *Lancet Infectious Diseases*, 4(5), pp. 294–300.

Byrne, M., 2006, Life history diversity and evolution in the Asterinidae, *Integrative and Comparative Biology*, 46(3), pp. 243–54.

Cabaço, S., Machas, R., Vieira, V. & Santos, R., 2008, Impacts of urban wastewater discharge on seagrass meadows (*Zostera noltii*), *Estuarine Coastal and Shelf Science*, 78(1), pp. 1–13.

Cabaitan, P.C., Gomez, E.D. & Alino, P.M., 2008, Effects of coral transplantation and giant clam restocking on the structure of fish communities on degraded patch reefs, *Journal of Experimental Marine Biology and Ecology*, 357(1), pp. 85–98.

Caira, J.N. & Euzet, L., 2001, Age of association between the nurse shark, *Ginglymostoma cirratum*, and tapeworms of the genus *Pedibothrium* (Tetraphyllidea: Onchobothriidae): implications from geography. *Biological Journal of the Linnean Society*, 72(4), pp. 609–14.

Caira, J.N. & Jensen, K., 2001, An investigation of the co-evolutionary relationships between onchobothriid tapeworms and their elasmobranch hosts, *International Journal for Parasitology*, 31(9), pp. 960–75.

Callaway, R., Alsvag, J., de Boois, I., Cotter, J., Ford, A., Hinz, H., Jennings, S., Kroncke, I., Lancaster, J., Piet, G., Prince, P. & Ehrich, S., 2002, Diversity and community structure of epibenthic invertebrates and fish in the North Sea, *ICES Journal of Marine Science*, 59(6), pp. 1199–214.

Caplan, P., 2009, http://mafia-island-tanzania.gold.ac.uk/

Carney, R.S., 2005, Zonation of deep biota on continental margins, *Oceanography and Marine Biology – an Annual Review*, 43, p. 211.

Carpenter, R.C. & Williams, S.L., 2007, Mass transfer limitation of photosynthesis of coral reef algal turfs, *Marine Biology*, 151(2), pp. 435–50.

Carreiro-Silva, M. & McClanahan, T.R., 2001, Echinoid bioerosion and herbivory on Kenyan coral reefs: the role of protection from fishing, *Journal of Experimental Marine Biology and Ecology*, 262(2), pp. 133–53.

Carroll, A., Harrison, P. & Adjeroud, M., 2006, Sexual reproduction of Acropora reef corals at Moorea, French Polynesia, *Coral Reefs*, 25(1), pp. 93–7.

Castellanos-Galindo, G.A. & Giraldo, A., 2008, Food resource use in a tropical eastern Pacific tidepool fish assemblage, *Marine Biology*, 153(6), pp. 1023–35.

Cerrano, C. & Bavestrello, G., 2008, Medium-term effects of die-off of rocky benthos in the Ligurian Sea. What can we learn from gorgonians? *Chemistry and Ecology*, 24, pp. 73–82.

Chaparro, O.R., Thompson, R.J. & Pereda, S.V., 2002, Feeding mechanisms in the gastropod *Crepidula fecunda*, *Marine Ecology – Progress Series*, 234, pp. 171–81.

Chen Zhong, Yang Hua-ping & Huang Chi-yue, 2007, Characteristics of cold seeps and structures of chemoautosynthesis-based communities in seep sediments, *Journal of Tropical Oceanography*, 26(6), pp. 73–82.

Cherry, D.S., Dickson, K.L., Cairns, J. Jr & Stauffer, J.R., 1977, Preferred, avoided, and lethal temperatures of fish during rising temperature conditions, *Journal of the Fisheries Research Board of Canada*, 34, 239–246.

Christie, P. & White, A.T., 2007, Best practices for improved governance of coral reef marine protected areas, *Coral Reefs*, 26(4), pp. 1047–56.

Church, J.A., Hunter, J.R., McInnes, K.L. & White, N.J., 2006a, Sea-level rise around the Australian coastline and the changing frequency of extreme sea-level events, *Australian Meteorological Magazine*, 55(4), pp. 253–60.

Church, J.A., White, N.J. & Hunter, J.R., 2006b, Sea-level rise at tropical Pacific and Indian Ocean islands, *Global and Planetary Change*, 53(3), pp. 155–68.

Clarke, A. & Crame, J.A., 1997, Diversity, latitude and time: Patterns in the shallow sea, *Marine Biodiversity*, pp. 122–47.

Clarke, D.G. & Wilber, D.H., 2000, *Assessment of Potential Impacts of Dredging Operations Due to Sediment Resuspension*. Dredging Operations & Environmental Research DOER-E9.

Claudet, J., Pelletier, D., Jouvenel, J.Y., Bachet, F. & Galzin, R., 2006, Assessing the effects of marine protected area (MPA) on a reef fish assemblage in a northwestern Mediterranean marine reserve: Identifying community-based indicators, *Biological Conservation*, 130(3), pp. 349–69.

Claudet, J., Osenberg, C.W., Benedetti-Cecchi, L., Domenici, P., Garcia-Charton, J.A., Perez-Ruzafa, A., Badalamenti, F., Bayle-Sempere, J., Brito, A., Bulleri, F., Culioli, J.M., Dimech, M., Falcon, J.M., Guala, I., Milazzo, M., Sanchez-Meca, J., Somerfield, P.J., Stobart, B.,

Vandeperre, F., Valle, C. & Planes, S., 2008, Marine reserves: size and age do matter, *Ecology Letters*, 11(5), pp. 481–9.

Clausen, I. & Riisgard, H.U., 1996, Growth, filtration and respiration in the mussel *Mytilus edulis*: No evidence for physiological regulation of the filter-pump to nutritional needs, *Marine Ecology Progress Series*, 141(1–3), pp. 37–45.

Claydon, J., 2005, Spawning aggregations of coral reef fishes: Characteristics, hypotheses, threats and management, *Oceanography and Marine Biology – an Annual Review, Vol 43*, 42, pp. 265–301.

Cloern, J.E., Jassby, A.D., Thompson, J.K. & Hieb, K.A., 2007, A cold phase of the East Pacific triggers new phytoplankton blooms in San Francisco Bay, *Proceedings of the National Academy of Sciences of the United States of America*, 104, pp. 18561–5.

Clynick, B.G., 2008, Characteristics of an urban fish assemblage: Distribution of fish associated with coastal marinas, *Marine Environmental Research*, 65(1), pp. 18–33.

Coale, K.H., Johnson, K.S., Fitzwater, S.E., Gordon, R.M., Tanner, S., Chavez, F.P., Ferioli, L., Sakamoto, C., Rogers, P., Millero, F., Steinberg, P., Nightingale, P., Cooper, D., Cochlan, W.P., Landry, M.R., Constantinou, J., Rollwagen, G., Trasvina, A. & Kudela, R., 1996, A massive phytoplankton bloom induced by an ecosystem-scale iron fertilization experiment in the equatorial Pacific Ocean, *Nature*, 383(6600), pp. 495–501.

Collins, M.A., 1995, "Dredging-induced Near-field Resuspended-sediment Concentrations and Source Strengths." Miscellaneous Paper D-95-2, US Army Engineer Waterways Experiment Station, Vicksburg, MS.

Colson, I. & Hughes, R.N., 2004, Rapid recovery of genetic diversity of dogwhelk (*Nucella lapillus* L.) populations after local extinction and recolonization contradicts predictions from life-history characteristics, *Molecular Ecology*, 13(8), pp. 2223–33.

Connell, J.H., 1961, Influence of interspecific competition and other factors on distribution of barnacle *Chthamalus stellatus*, *Ecology*, 42(4), pp. 710-&.

Connell, J.H., 1970, A predator-prey system in marine intertidal region.1. *Balanus-glandula* and several predatory species of *Thais*, *Ecological Monographs*, 40(1), p. 49.

Connell, J.H. & Slatyer, R.O., 1977, Mechanisms of succession in natural communities and their role in community stability and organization, *American Naturalist*, 111(982), pp. 1119–44.

Connolly, R.M., Hindell, J.S. & Gorman, D., 2005, Seagrass and epiphytic algae support nutrition of a fisheries species, *Sillago schomburgkii*, in adjacent intertidal habitats, *Marine Ecology Progress Series*, 286, pp. 69–79.

Constantino, R., Caspar, M.B., Tata-Regala, J., Carvalho, S., Curdia, J., Drago, T., Taborda, R. & Monteiro, C.C., 2009, Clam dredging effects and subsequent recovery of benthic communities at different depth ranges, *Marine Environmental Research*, 67(2), pp. 89–99.

Cooper, K.M., Frojan, C., Defew, E., Curtis, M., Fleddum, A., Brooks, L. & Paterson, D.M., 2008, Assessment of ecosystem function following marine aggregate dredging, *Journal of Experimental Marine Biology and Ecology*, 366(1–2), pp. 82–91.

Copley, J.T.P., Flint, H.C., Ferrero, T.J. & Van Dover, C.L., 2007, Diversity of melofauna and free-living nematodes in hydrothermal vent mussel beds on the northern and southern East Pacific Rise, *Journal of the Marine Biological Association of the United Kingdom*, 87(5), pp. 1141–52.

Cordes, E.E., Bergquist, D.C., Predmore, B.L., Jones, C., Deines, P., Telesnicki, G. & Fisher, C.R., 2006, Alternate unstable states: Convergent paths of succession in hydrocarbon-seep tubeworm-associated communities, *Journal of Experimental Marine Biology and Ecology*, 339(2), pp. 159–76.

Coughlan, J., 1970, Power stations and aquatic life. In: *Effects of Industry on the Environment. Proceedings of ECY Symposium. Field Studies Council.*

Courrat, A., Lobry, J., Nicolas, D., Laffargue, P., Amara, R., Lepage, M., Girardin, M. & Le Pape, O., 2009, Anthropogenic disturbance on nursery function of estuarine areas for marine species, *Estuarine Coastal and Shelf Science*, 81(2), pp. 179–90.

Coutant, C.C., 1977, Physiological considerations of future thermal additions for aquatic life. In: Marois, M. (ed.), *Proceedings of the World Conference Toward a Plan of Actions for Mankind*, vol. 3, *Biological Balance and Thermal Modication*, pp. 251–266. Pergamon Press, New York.

Cowen, R.K. & Sponaugle, S., 2009, Larval Dispersal and Marine Population Connectivity, *Annual Review of Marine Science*, 1, pp. 443–66.

Crabbe, M. & McClanahan, T.R., 2006, A biosocioeconomic evaluation of shipwrecks used for fishery and dive tourism enhancement in Kenya, *Western Indian Ocean Journal of Marine Science*, 5(1), pp. 35–53.

Crabbe, M.J.C., Martinez, E., Garcia, C., Chub, J., Castro, L. & Guy, J., 2008, Growth modelling indicates hurricanes and severe storms are linked to low coral recruitment in the Caribbean, *Marine Environmental Research*, 65(4), pp. 364–8.

Crain, C.M., Kroeker, K. & Halpern, B.S., 2008, Interactive and cumulative effects of multiple human stressors in marine systems, *Ecology Letters*, 11(12), pp. 1304–15.

Croll, D.A., Marinovic, B., Benson, S., Chavez, F.P., Black, N., Ternullo, R. & Tershy, B.R., 2005, From wind to whales: trophic links in a coastal upwelling system, *Marine Ecology Progress Series*, 289, pp. 117–30.

Cróquer, A. & Weil, E., 2009, Spatial variability in distribution and prevalence of Caribbean scleractinian coral and octocoral diseases. II. Genera-level analysis, *Diseases of Aquatic Organisms*, 83(3), pp. 209–22.

Cróquer, A., Weil, E., Zubillaga, A.L. & Pauls, S.M., 2005, Impact of a white plague-II outbreak on a coral reef in the archipelago Los Roques National Park, Venezuela, *Caribbean Journal of Science*, 41(4), pp. 815–23.

da Fonseca-Genevois, V., Somerfield, P.J., Neves, M.H.B., Coutinho, R. & Moens, T., 2006, Colonization and early succession on artificial hard substrata by meiofauna, *Marine Biology*, 148(5), pp. 1039–50.

Daly, B. & Konar, B., 2008, Effects of macroalgal structural complexity on nearshore larval and post-larval crab composition, *Marine Biology*, 153(6), pp. 1055–64.

Daniel, I., Oger, P. & Winter, R., 2006, Origins of life and biochemistry under high-pressure conditions, *Chemical Society Reviews*, 35(10), pp. 858–75.

Danovaro, R., Gambi, C., Dell'Anno, A., Corinaidesi, C., Fraschetti, S., Vanreusel, A., Vincx, M. & Gooday, A.J., 2008, Exponential decline of deep-sea ecosystem functioning linked to benthic biodiversity loss, *Current Biology*, 18(1), pp. 1–8.

Dalpadado, P., Ellertsen, B. & Johannessen, S., 2008, Inter-specific variations in distribution, abundance and reproduction strategies of krill and amphipods in the Marginal Ice Zone of the Barents Sea, *Deep-Sea Research Part II – Topical Studies In Oceanography*, 55(20–21), pp. 2257–2265.

David, V., Chardy, P. & Sautour, B., 2006, Fitting a predator-prey model to zooplankton time-series data in the Gironde estuary (France): Ecological significance of the parameters, *Estuarine Coastal and Shelf Science*, 67(4), pp. 605–17.

Davis, K.L.F., Russ, G.R., Williamson, D.H. & Evans, R.D., 2004, Surveillance and poaching on inshore reefs of the Great Barrier Reef Marine Park, *Coastal Management*, 32(4), pp. 373–87.

Davis, M.H. & Coughlan, J., 1978, Response of entrained plankton to low level chlorination at a coastal power station. In: Jolley, R.L., Hend, G. & Heyward Hamilton Jr. D. (eds), *Water Chlorination Environmental Impact and Health Effects*, vol. 2, *Proceedings of the 2nd Conference Gatlinburg, TN, USA, Oct 31–Nov 4, 1977*, xviii +

909pp. illus maps, pp. 369–76. Ann Arbor Science Publishers, Ann Arbor, MI.

de Castro, I.B., de Meirelles, C.A.O., Matthews-Cascon, H., Rocha-Barreira, C.A., Penchaszadeh, P., & Bigatti, G., 2008, Imposex in Endemic Volutid from Northeast Brazil (Mollusca: Gastropoda), *Brazilian Archives of Biology and Technology*, 51(5), pp. 1065–9.

De Iongh, H.H., Kiswara, W., Kustiawan, W. & Loth, P.E., 2007, A review of research on the interactions between dugongs (*Dugong dugon* Muller 1776) and intertidal seagrass beds in Indonesia, *Hydrobiologia*, 591, pp. 73–83.

De Santo, E.M. & Jones, P.J.S., 2007a, The Darwin Mounds: from undiscovered coral to the development of an offshore marine protected area regime, *Conservation and Adaptive Management of Seamount and Deep-Sea Coral Ecosystems*, pp. 147–56.

De Santo, E.M. & Jones, P.J.S., 2007b, Offshore marine conservation policies in the North East Atlantic: Emerging tensions and opportunities, *Marine Policy*, 31(3), pp. 336–47.

de Voogd, N.J. & Cleary, D.F.R., 2008, An analysis of sponge diversity and distribution at three taxonomic levels in the Thousand Islands/Jakarta Bay reef complex, West-Java, Indonesia, *Marine Ecology-Pubblicazioni Della Stazione Zoologica Di Napoli I*, 29(2), pp. 205–15.

De'ath, G., Lough, J.M. & Fabricius, K.E., 2009, Declining Coral Calcification on the Great Barrier Reef, *Science*, 323(5910), pp. 116–9.

Dean, T.A. & Jewett, S.C., 2001, Habitat-specific recovery of shallow subtidal communities following the Exxon Valdez oil spill, *Ecological Applications*, 11(5), pp. 1456–71.

Delaney, J.M., 2003, Community capacity building in the designation of the Tortugas Ecological Reserve, *Gulf and Caribbean Research*, 14(2), pp. 163–9.

delGiorgio, P.A., Cole, J.J. & Cimbleris, A., 1997, Respiration rates in bacteria exceed phytoplankton production in unproductive aquatic systems, *Nature*, 385(6612), pp. 148–51.

Denitto, F., Terlizzi, A. & Belmonte, G., 2007, Settlement and primary succession in a shallow submarine cave: spatial and temporal benthic assemblage distinctness, *Marine Ecology-Pubblicazioni Della Stazione Zoologica Di Napoli I*, 28, pp. 35–46.

Denton, G.R.W., Morrison, R.J., Bearden, B.G., Houk, P., Starmer, J.A. & Wood, H.R., 2009, Impact of a coastal dump in a tropical lagoon on trace metal concentrations in surrounding marine biota: A case study from Saipan, Commonwealth of the Northern Mariana Islands (CNMI), *Marine Pollution Bulletin*, 58(3), pp. 424–31.

DeVantier, L.M., De'ath, G., Turak, E., Done, T.J. & Fabricius, K.E., 2006, Species richness and community structure of reef-building corals on the nearshore Great Barrier Reef, *Coral Reefs*, 25(3), pp. 329–40.

Diaz, R. J. 2001. Overview of hypoxia around the world. *Journal of Environmental Quality* 30, 275–281.

Diaz, R.J. & Rosenberg, R., 2008, Spreading dead zones and consequences for marine ecosystems, *Science*, 321(5891), pp. 926–9.

Dickson, K.L., Hendrick.Ac, Crossman, J.S. & Cairns, J., 1974, Effects of intermittently chlorinated cooling-tower blowdown on fish and invertebrates, *Environmental Science & Technology*, 8(9), pp. 845–9.

Diez, S. & Bayona, J.M., 2009, Butyltin occurrence and risk assessment in the sediments of the Iberian Peninsula, *Journal of Environmental Management*, 90, pp. S25–30.

Digiano, F.A., Miller, C.T. & Yoon, J.Y., 1993, Predicting release of PCBs at point of dredging, *Journal of Environmental Engineering-Asce*, 119(1), pp. 72–89.

DiGiano, F.A., Miller, C.T. & Yoon, J., 1995, Dredging Elutriate Test (DRET) Development. Contract Report D-95-1, U.S. Army Engineer Waterways Experiment Station, Vicksburg, MS., NTIS No. AD A299 354.

Dittel, A.I., Perovich, G. & Epifanio, C.E., 2008, Biology of the vent crab *Bythograea thermydron*: A brief review, *Journal of Shellfish Research*, 27(1), pp. 63–77.

Dizon, R.M. & Yap, H.T., 2006, Effects of multiple perturbations on the survivorship of fragments of three coral species, *Marine Pollution Bulletin*, 52(8), pp. 928–34.

Doherty, P.J., Dufour, V., Galzin, R., Hixon, M.A., Meekan, M.G. & Planes, S., 2004, High mortality during settlement is a population bottleneck for a tropical surgeonfish, *Ecology*, 85(9), pp. 2422–8.

Dorenbosch, M., Verberk, W., Nagelkerken, I. & van der Veldel, G., 2007, Influence of habitat configuration on connectivity between fish assemblages of Caribbean seagrass beds, mangroves and coral reefs, *Marine Ecology Progress Series*, 334, pp. 103–16.

dos Santos, A. & Peliz, A., 2005, The occurrence of Norway lobster (*Nephrops norvegicus*) larvae off the Portuguese coast, *Journal of the Marine Biological Association of the United Kingdom*, 85(4), pp. 937–41.

Dovel, W.L. & Berggren, T.J., 1983, Atlantic sturgeon of the hudson estuary, new-york, *New York Fish and Game Journal*, 30(2), pp. 140–72.

Dovel, W.L., Pekovitch, A.W. & Berggren, T.J., 1992, Biology of the shortnose sturgeon (*Acipenser brevirostrum* lesueur, 1818) in the Hudson River estuary, new-york, *Estuarine Research in the 1980s*, pp. 187–216.

Dreanno, C., Kirby, R.R. & Clare, A.S., 2007, Involvement of the barnacle settlement-inducing protein complex (SIPC) in species recognition at settlement, *Journal of Experimental Marine Biology and Ecology*, 351(1–2), pp. 276–82.

Drew, J.A., 2005, Use of traditional ecological knowledge in marine conservation, *Conservation Biology*, 19(4), pp. 1286–93.

Dreyer, J.C., Knick, K.E., Flickinger, W.B. & Van Dover, C.L., 2005, Development of macrofaunal community structure in mussel beds on the northern East Pacific Rise, *Marine Ecology Progress Series*, 302, pp. 121–34.

Dryden, J., Grech, A., Moloney, J. & Hamann, M., 2008, Rezoning of the Great Barrier Reef World Heritage Area: does it afford greater protection for marine turtles? *Wildlife Research*, 35(5), pp. 477–85.

Duarte, Carlos M., Agusti, S., Aristegui, J., Gonzalez, N. & Anadon, R., 2001, Evidence for a heterotrophic subtropical northeast Atlantic, *Limnology and Oceanography*, 46(2), pp. 425–8.

Duarte, C.M. & Cebrian, J., 1996, The fate of marine autotrophic production, *Limnology and Oceanography*, 41(8), pp. 1758–66.

Dubilier, N., Bergin, C. & Lott, C., 2008, Symbiotic diversity in marine animals: the art of harnessing chemosynthesis, *Nature Reviews Microbiology*, 6(10), pp. 725–40.

Duckworth, A.R., Bruck, W.M., Janda, K.E, Pitts, T.P. & McCarthy, P.J., 2006, Retention efficiencies of the coral reef sponges *Aplysina lacunosa*, *Callyspongia vaginalis* and *Niphates digitalis* determined by Coulter counter and plate culture analysis. *Marine Biology Research*, 2(4), pp. 243–8.

Dulvy, N.K., Freckleton, R.P. & Polunin, N.V.C., 2004, Coral reef cascades and the indirect effects of predator removal by exploitation, *Ecology Letters*, 7(5), pp. 410–6.

Duncan, K.M., Martin, A.P., Bowen, B.W. & De Couet, H.G., 2006, Global phylogeography of the scalloped hammerhead shark (*Sphyrna lewini*), *Molecular Ecology*, 15(8), pp. 2239–51.

Duperron, S., Sibuet, M., MacGregor, B.J., Kuypers, M.M.M., Fisher, C.R. & Dubilier, N., 2007, Diversity, relative abundance and metabolic potential of bacterial endosymbionts in three *Bathymodiolus* mussel species from cold seeps in the Gulf of Mexico, *Environmental Microbiology*, 9(6), pp. 1423–38.

Duperron, S., Halary, S., Lorion, J., Sibuet, M. & Gaill, F., 2008, Unexpected co-occurrence of six bacterial symbionts in the gills of the cold seep mussel *Idas* sp. (Bivalvia : Mytilidae). *Environmental Microbiology*, 10(2), pp. 433–45.

Dupont, J.M., 2008, Artificial Reefs as Restoration Tools: A Case Study on the West Florida Shelf, *Coastal Management*, 36(5), pp. 495–507.

Dupuy, C., Le Gall, S., Hartmann, H.J. & Breret, M., 1999, Retention of ciliates and flagellates by the oyster *Crassostrea gigas* in French Atlantic coastal ponds: protists as a trophic link between bacterioplankton and benthic suspension-feeders, *Marine Ecology Progress Series*, 177, pp. 165–75.

Dworjanyn, S.A. & Pirozzi, I., 2008, Induction of settlement in the sea urchin *Tripneustes gratilla* by macroalgae, biofilms and conspecifics: A role for bacteria? *Aquaculture*, 274(2–4), pp. 268–74.

Ebert, T.A., 1996, The consequences of broadcasting, brooding, and asexual reproduction in echinoderm metapopulations, *Oceanologica Acta*, 19(3–4), pp. 217–26.

Elliott, M., Whitfield, A.K., Potter, I.C, Blaber, S.J.M., Cyrus, D.P., Nordlie, F.G. & Harrison, T.D., 2007, The guild approach to categorizing estuarine fish assemblages: a global review, *Fish and Fisheries*, 8(3), pp. 241–68.

Elmgren, R., Ejdung, G. & Ankar, S., 2001, Intraspecific food competition in the deposit-feeding benthic amphipod *Monoporeia affinis* – a laboratory study, *Marine Ecology Progress Series*, 210, pp. 185–93.

Elser, J.J., Bracken, M.E.S., Cleland, E.E., Gruner, D.S., Harpole, W.S., Hillebrand, H., Ngai, J.T., Seabloom, E.W., Shurin, J.B. & Smith, J.E., 2007, Global analysis of nitrogen and phosphorus limitation of primary producers in freshwater, marine and terrestrial ecosystems, *Ecology Letters*, 10, pp. 1135–42.

Engas, A., Lokkeborg, S., Ona, E. & Soldal, A.V., 1996, Effects of seismic shooting on local abundance and catch rates of cod (*Gadus morhua*) and haddock (*Melanogrammus aeglefinus*), *Canadian Journal of Fisheries and Aquatic Sciences*, 53(10), pp. 2238–49.

Epifanio, C.E., Perovich, G., Dittel, A.I. & Cary, S.C., 1999, Development and behavior of megalopa larvae and juveniles of the hydrothermal vent crab *Bythograea thermydron*, *Marine Ecology Progress Series*, 185, pp. 147–54.

Eppley, R.W., Renger, E.H. & Williams, P.M., 1976, Chlorine reactions with seawater constituents and inhibition of photosynthesis of natural marine phytoplankton, *Estuarine and Coastal Marine Science*, 4(2), pp. 147–61.

Escribano R., Daneri G., Farias L., Gallardo V.A., Gonzalez H.E., Gutierrez D., Lange C.B., Morales C.E., Pizarro O., Ulloa O., & Braun M., 2004, Biological and chemical consequences of the 1997–1998 El Nino in the Chilean coastal upwelling system: a synthesis. *Deep-Sea Research Part II-Topical Studies In Oceanography*, 51(20–21) pp2389–2411

Estes, J.A., Tinker, M.T., Williams, T.M. & Doak, D.F., 1998, Killer whale predation on sea otters linking oceanic and nearshore ecosystems, *Science*, 282(5388), pp. 473–6.

Etter, R.J. & Mullineaux, L.S., 2001, Deep-sea communities. In: Bertness, M.D., Gaines, S.D., Hay, M.E. (eds), *Marine Community Ecology*, pp. 367–393. Sinauer Associates, Sunderland, MA.

Fabi, G., Grati, F., Lucchetti, A. & Trovarelli, L., 2002, Evolution of the fish assemblage around a gas platform in the northern Adriatic Sea, *ICES Journal of Marine Science*, 59, pp. S309–15.

Fabinyi, M., 2008, Dive tourism, fishing and marine protected areas in the Calamianes Islands, Philippines, *Marine Policy*, 32(6), pp. 898–904.

Fabricius, K.E. & Dommisse, M., 2000, Depletion of suspended particulate matter over coastal reef communities dominated by zooxanthellate soft corals, *Marine Ecology Progress Series*, 196, pp. 157–67.

Fairclough, D.V., Clarke, K.R., Valesini, F.J. & Potter, I.C., 2008, Habitat partitioning by five congeneric and abundant *Choerodon* species (Labridae) in a large subtropical marine embayment, *Estuarine Coastal and Shelf Science*, 77(3), pp. 446–56.

Fernandez, L., 2008, NAFTA and member country strategies for maritime trade and marine invasive species, *Journal of Environmental Management*, 89(4), pp. 308–21.

Fettweis, M., Francken, F., Pison, V. & Van den Eynde, D., 2006, Suspended particulate matter dynamics and aggregate sizes in a high turbidity area, *Marine Geology*, 235(1–4), pp. 63–74.

Fiksen, O., Jorgensen, C., Kristiansen, T., Vikebo, F. & Huse, G., 2007, Linking behavioural ecology and oceanography: larval behaviour determines growth, mortality and dispersal, *Marine Ecology Progress Series*, 347, pp. 195–205.

Fine, M. & Loya, Y., 2003, Alternate coral-bryozoan competitive superiority during coral bleaching, *Marine Biology*, 142(5), pp. 989–96.

Finley, R.J. & Forrester, G.E., 2003, Impact of ectoparasites on the demography of a small reef fish, *Marine Ecology Progress Series*, 248, pp. 305–9.

Fleeger, J.W., Johnson, D.S., Galvan, K.A. & Deegan, L.A., 2008, Top-down and bottom-up control of infauna varies across the saltmarsh landscape, *Journal of Experimental Marine Biology and Ecology*, 357(1), pp. 20–34.

Flint, H.C., Copley, J.T.P., Ferrero, T.J. & Van Dover, C.L., 2006, Patterns of nematode diversity at hydrothermal vents on the East Pacific Rise, *Cahiers De Biologie Marine*, 47(4), pp. 365–70.

Fodrie, F.J., Herzka, S.Z., Lucas, A.J. & Francisco, V., 2007, Intraspecific density regulates positioning and feeding mode selection of the sand dollar *Dendraster excentricus*, *Journal of Experimental Marine Biology and Ecology*, 340(2), pp. 169–83.

Fonseca, G., Muthumbi, A.W. & Vanreusel, A., 2007, Species richness of the genus *Molgolaimus* (Nematoda) from local to ocean scale along continental slopes, *Marine Ecology-Pubblicazioni Della Stazione Zoologica Di Napoli I*, 28, pp. 446–59.

Fontaine, F.J., Wilcock, W.S.D. & Butterfield, D.A., 2007, Physical controls on the salinity of mid-ocean ridge hydrothermal vent fluids, *Earth and Planetary Science Letters*, 257(1–2), pp. 132–45.

Forcada, A., Bayle-Sempere, J.T., Valle, C. & Sanchez-Jerez, P., 2008, Habitat continuity effects on gradients of fish biomass across marine protected area boundaries, *Marine Environmental Research*, 66(5), pp. 536–47.

Forsman, Z.H., Rinkevich, B. & Hunter, C.L., 2006, Investigating fragment size for culturing reef-building corals (*Porites lobata* and *P-compressa*) in ex situ nurseries, *Aquaculture*, 261(1), pp. 89–97.

Fox, H.E., Mous, P.J., Pet, J.S., Muljadi, A.H. & Caldwell, R.L., 2005, Experimental assessment of coral reef rehabilitation following blast fishing, *Conservation Biology*, 19(1), pp. 98–107.

Fox, R.J. & Bellwood, D.R., 2007, Quantifying herbivory across a coral reef depth gradient, *Marine Ecology Progress Series*, 339, pp. 49–59.

Francis J., Agneta Nilsson, and Dixon Waruinge 2002 Marine Protected Areas in the Eastern African Region: How Successful Are They? Ambio 31(7) 503–511

Francour, P., Harmelin J-G., Pollard D. & Sartoretto S. 2001, A review of marine protected areas in the northwestern Mediterranean region: siting, usage, zonation and management. *Aquatic Conservation: Marine & Freshwater Systems* 11(3) pp. 155–188

Frank, K.T., Petrie, B. & Shackell, N.L., 2007, The ups and downs of trophic control in continental shelf ecosystems, *Trends in Ecology & Evolution*, 22(5), pp. 236–42.

Frederiksen, M., Edwards, M., Richardson, A.J., Halliday, N.C. & Wanless, S., 2006, From plankton to top predators: bottom-up control of a marine food web across four trophic levels, *Journal of Animal Ecology*, 75(6), pp. 1259–68.

Freidenberg, TL., Menge, BA., Halpin, PM, Webster M., & Sutton-Grier A (2007) Cross-scale variation in top-down and bottom-up control of algal abundance. *Journal of Experimental Marine Biology and Ecology*, 347(1–2), pp. 8–29.

Frid, C.L.J., Paramor, O.A.L., Brockington, S. & Bremner, J., 2008, Incorporating ecological functioning into the designation and management of marine protected areas, *Hydrobiologia*, 606, pp. 69–79.

Fuhrman, J.A., Steele, J.A., Hewson, I., Schwalbach, M.S., Brown, M.V., Green, J.L. & Brown, J.H., 2008, A latitudinal diversity gradient in planktonic marine bacteria, *Proceedings of the National Academy of Sciences of the United States of America*, 105(22), pp. 7774–8.

Fujiwara, Y., Kawato, M., Yamamoto, T., Yamanaka, T., Sato-Okoshi, W., Noda, C., Tsuchida, S., Komai, T., Cubelio, S.S., Sasakis, T., Jacobsen, K., Kubokawa, K., Fujikura, K., Maruyama, T., Furushima, Y., Okoshi, K., Miyake, H., Miyazaki, M., Nogi, Y., Yatabe, A. & Okutani, T., 2007, Three-year investigations into sperm whale-fall ecosystems in Japan, *Marine Ecology-Pubblicazioni Della Stazione Zoologica Di Napoli I*, 28(1), pp. 219–32.

Funch, P. & Kristensen, R.M., 1995, Cycliophora is a new phylum with affinities to Entoprocta and Ectoprocta, *Nature*, 378(6558), pp. 711–4.

Gage, J.D., 2004, Diversity in deep-sea benthic macrofauna: the importance of local ecology, the larger scale, history and the Antarctic, *Deep-Sea Research Part Ii-Topical Studies in Oceanography*, 51(14–16), pp. 1689–708.

Gardner TA, Cote IM, Gill JA, Grant A, Watkinson AR 2003 Long-term region-wide declines in Caribbean corals, *Science* 301(5635) pp. 958–960

Garpe, K.C. & Ohman, M.C., 2007, Non-random habitat use by coral reef fish recruits in Mafia Island Marine Park, Tanzania, *African Journal of Marine Science*, 29(2), pp. 187–99.

Garrett, C. & Kunze, E., 2007, Internal tide generation in the deep ocean, *Annual Review of Fluid Mechanics*, 39, pp. 57–87.

Garshelis, DL., Johnson, CB, 2001, Sea otter population dynamics and the Exxon Valdez oil spill: disentangling the confounding effects, *Journal of Applied Ecology*, 38(1), pp. 19–35.

Gaston, K.J. & Williams, P.H., 1996, Spatial patterns in taxonomic diversity, *Biodiversity: A biology of numbers and difference*, pp. 202–29.

Gattuso, J-P., Gentili, B., Duarte, C.M., Kleypas, J.A., Middelburg, J.J. & Antoine, D., 2006, *Biogeosciences*, 3(4), pp.489–513.

Gelcich, S., Kaiser, M.J., Castilla, J.C. & Edwards-Jones, G., 2008, Engagement in co-management of marine benthic resources influences environmental perceptions of artisanal fishers, *Environmental Conservation*, 35(1), pp. 36–45.

Gerlach, G., Atema, J., Kingsford, M.J., Black, K.P. & Miller-Sims, V., 2007, Smelling home can prevent dispersal of reef fish larvae, *Proceedings of the National Academy of Sciences of the United States of America*, 104(3), pp. 858–63.

Gilhooly, W.P., Carney, R.S. & Macko, S.A., 2007, Relationships between sulfide-oxidizing bacterial mats and their carbon sources in northern Gulf of Mexico cold seeps, *Organic Geochemistry*, 38(3), pp. 380–93.

Gladstone, W., 2007a, Requirements for marine protected areas to conserve the biodiversity of rocky reeffishes, *Aquatic Conservation-Marine and Freshwater Ecosystems*, 17(1), pp. 71–87.

Gladstone, W., 2007b, Temporal patterns of spawning and hatching in a spawning aggregation of the temperate reef fish *Chromis hypsilepis* (Pomacentridae), *Marine Biology*, 151(3), pp. 1143–52.

Glardon, C.G., Walters, L.J., Quintana-Ascencio, P.F., McCauley, L.A., Stam, W.T. & Olsen, J.L., 2008, Predicting risks of invasion of macroalgae in the genus *Caulerpa* in Florida, *Biological Invasions*, 10(7), pp. 1147–57.

Glenner, H. & Hebsgaard, M.B., 2006, Phylogeny and evolution of life history strategies of the Parasitic Barnacles (Crustacea, Cirripedia, Rhizocephala), *Molecular Phylogenetics and Evolution*, 41(3), pp. 528–38.

Godiksen, J.A., Hallfredsson, E.H. & Pedersen, T., 2006, Effects of alternative prey on predation intensity from herring *Clupea harengus* and sandeel *Ammodytes marinus* on capelin *Mallotus villosus* larvae in the Barents Sea, *Journal of Fish Biology*, 69(6), pp. 1807–23.

Golbuu, Y. & Richmond, R.H., 2007, Substratum preferences in planula larvae of two species of scleractinian corals, *Goniastrea retiformis* and *Stylaraea punctata*, *Marine Biology*, 152(3), pp. 639–44.

Goldman, J.C., Capuzzo, J.M. & Wong, G.T.F., 1978, Biological and chemical effects of chlorination at coastal power plants. In: Jolley, R.L., Hend, G. & Heyward Hamilton Jr, D. (eds), *Water Chlorination Environmental Impact and Health Effects*, vol. 2, *Proceedings of the 2nd Conference Gatlinburg, Tenn, USA, Oct 31-Nov 4, 1977*, xviii+909p illus maps, pp. 291–305. Ann Arbor Science Publishers, Ann Arbor, MI.

Goldson, A.J., Hughes, R.N. & Gliddon, C.J., 2001, Population genetic consequences of larval dispersal mode and hydrography: a case study with bryozoans, *Marine Biology*, 138(5), pp. 1037–42.

Goreau, T.J., 2008, Fighting algae in Kaneohe Bay, *Science*, 319(5860), p. 157.

Govenar, B. & Fisher, C.R., 2007, Experimental evidence of habitat provision by aggregations of *Riftia pachyptila* at hydrothermal vents on the East Pacific Rise, *Marine Ecology-Pubblicazioni Della Stazione Zoologica Di Napoli I*, 28(1), pp. 3–14.

Govenar, B., Freeman, M., Bergquist, D.C., Johnson, G.A. & Fisher, C.R., 2004, Composition of a one-year-old *Riftia pachyptila* community following a clearance experiment: Insight to succession patterns at deep-sea hydrothermal vents, *Biological Bulletin*, 207(3), pp. 177–82.

Graham, M.H., Vasquez, J.A. & Buschmann, A.H., 2007, Global ecology of the giant kelp *Macrocystis*: From ecotypes to ecosystems, *Oceanography and Marine Biology – an Annual Review, Vol 43*, 45, pp. 39–88.

Grassle, J.F., 1989, Species-diversity in deep-sea communities, *Trends in Ecology & Evolution*, 4(1), pp. 12–5.

Grassle, J.F. & Maciolek, N.J., 1992, Deep-sea species richness – regional and local diversity estimates from quantitative bottom samples, *American Naturalist*, 139(2), pp. 313–41.

Gratwicke, B. & Speight, M.R., 2005a, Effects of habitat complexity on Caribbean marine fish assemblages, *Marine Ecology Progress Series*, 292, pp. 301–10.

Gratwicke, B. & Speight, M.R., 2005b, The relationship between fish species richness, abundance and habitat complexity in a range of shallow tropical marine habitats, *Journal of Fish Biology*, 66(3), pp. 650–67.

Gratwicke, B., Petrovic, C. & Speight, M.R., 2006, Fish distribution and ontogenetic habitat preferences in non-estuarine lagoons and adjacent reefs, *Environmental Biology of Fishes*, 76(2–4), pp. 191–210.

Gravestock, P., Roberts, C.M. & Bailey, A., 2008, The income requirements of marine protected areas, *Ocean & Coastal Management*, 51(3), pp. 272–83.

Gray, J.S., 2001, Antarctic marine benthic biodiversity in a world-wide latitudinal context, *Polar Biology*, 24(9), pp. 633–41.

Great Barrier Reef Marine Park Authority (2009) www.gbrmpa.gov.au

Greco, S., Di Sciara, G.N. & Tunesi, L., 2004, 'Sistema Afrodite': an integrated programme for the inventorying and monitoring of the core zones of the Italian marine protected areas. *Aquatic Conservation – Marine and Freshwater Ecosystems* 14(Suppl. 1), pp. S119–22.

Griffin, J.N., De la Haye, K.L., Hawkins, S.J., Thompson, R.C. & Jenkins, S.R., 2008, Predator diversity and ecosystem functioning: Density modifies the effect of resource partitioning, *Ecology*, 89(2), pp. 298–305.

Grote, B., Ekau, W., Hagen, W., Huggett, J.A. & Verheye, H.M., 2007, Early life-history strategy of Cape hake in the Benguela upwelling region, *Fisheries Research*, 86(2–3), pp. 179–87.

Guan, L., Snelgrove, P.V.R. & Gamperl, A.K., 2008, Ontogenetic changes in the critical swimming speed of *Gadus morhua* (Atlantic cod) and Myoxocephalus scorpius (shorthorn sculpin) larvae and the role of temperature, *Journal of Experimental Marine Biology and Ecology*, 360(1), pp. 31–8.

Guard, M. & Mgaya, Y.D., 2002, The artisanal fishery for *Octopus cyanea* Gray in Tanzania, *Ambio*, 31(7–8), pp. 528–36.

Guidetti, P. & Dulcic, J., 2007, Relationships among predatory fish, sea urchins and barrens in Mediterranean rocky reefs across a latitudinal gradient, *Marine Environmental Research*, 63(2), pp. 168–84.

Guidetti, P., Milazzo, M., Bussotti, S., Molinari, A., Murenu, M., Pais, A., Spano, N., Balzano, R., Agardy, T., Boero, F., Carrada, G., Cattaneo-Vietti, R., Cau, A., Chemello, R., Greco, S., Manganaro, A., di Sciara, G.N., Russo, G.F. & Tunesi, L., 2008, Italian marine reserve effectiveness: Does enforcement matter? *Biological Conservation*, 141(3), pp. 699–709.

Guzman, H.M., Benfield, S., Breedy, O. & Mair, J.M., 2008, Broadening reef protection across the Marine Conservation Corridor of the Eastern Tropical Pacific: Distribution and diversity of reefs in Las Perlas Archipelago, Panama, *Environmental Conservation*, 35(1), pp. 46–54.

Halpern, B.S., Walbridge, S., Selkoe, K.A., Kappel, C.V., Micheli, F., D'Agrosa, C., Bruno, J.F., Casey, K.S., Ebert, C., Fox, H.E., Fujita, R., Heinemann, D., Lenihan, H.S., Madin, E.M.P., Perry, M.T., Selig, E.R., Spalding, M., Steneck, R. & Watson, R., 2008, A global map of human impact on marine ecosystems, *Science*, 319(5865), pp. 948–52.

Harborne, A.R., Mumby, P.J., Zychaluk, K., Hedley, J.D. & Blackwell, P.G., 2006, Modeling the beta diversity of coral reefs, *Ecology*, 87(11), pp. 2871–81.

Hare, C.E., DiTullio, G.R., Riseman, S.F., Crossley, A.C., Popels, L.C., Sedwick, P.N. & Hutchins, D.A., 2007, Effects of changing continuous iron input rates on a Southern Ocean algal assemblage, *Deep-Sea Research Part I – Oceanographic Research Papers*, 54(5), pp. 732–46.

Harii, S. & Kayanne, H., 2003, Larval dispersal, recruitment, and adult distribution of the brooding stony octocoral *Heliopora coerulea* on Ishigaki Island, southwest Japan, *Coral Reefs*, 22(2), pp. 188–96.

Harmelin-Vivien, M., Le Direach, L., Bayle-Sempere, J., Charbonnel, E., Garcia-Charton, J.A., Ody, D., Perez-Ruzafa, A., Renones, O., Sanchez-Jerez, P. & Valle, C., 2008, Gradients of abundance and biomass across reserve boundaries in six Mediterranean marine protected areas: Evidence of fish spillover? *Biological Conservation*, 141(7), pp. 1829–39.

Harrington, L., Fabricius, K., De'Ath, G. & Negri, A., 2004, Recognition and selection of settlement substrata determine post-settlement survival in corals, *Ecology*, 85(12), pp. 3428–37.

Hasler, H. & Ott, J.A., 2008, Diving down the reefs? Intensive diving tourism threatens the reefs of the northern Red Sea, *Marine Pollution Bulletin*, 56(10), pp. 1788–94.

Haupt, BJ., & Seidov, D (2007) Strengths and weaknesses of the global ocean conveyor: Inter-basin freshwater disparities as the major control, *Progress in Oceanography* 73 (3–4) 358–369.

Hayes, D.F., McLellan, T.N. & Truitt, C.L., 1988, Demonstrations of Innovative and Conventional Dredging Equipment at Calumet Harbor, Illinois, Miscellaneous Paper EL-88-1. US Army Engineer Waterways Experiment Station, Vicksburg, MS.

Heck, K.L. & Valentine, J.F., 2006, Plant-herbivore interactions in seagrass meadows, *Journal of Experimental Marine Biology and Ecology*, 330(1), pp. 420–36.

Helson, J.G. & Gardner, J.P.A., 2007, Variation in scope for growth: a test of food limitation among intertidal mussels, *Hydrobiologia*, 586, pp. 373–92.

Helson, J.G., Pledger, S. & Gardner, J.P.A., 2007, Does differential particulate food supply explain the presence of mussels in Wellington Harbour (New Zealand) and their absence on neighbouring Cook Strait shores? *Estuarine Coastal and Shelf Science*, 72(1-2), pp. 223–34.

Henderson, P.A., 2007, Discrete and continuous change in the fish community of the Bristol Channel in response to climate change, *Journal of the Marine Biological Association of the United Kingdom*, 87(2), pp. 589–98.

Henderson, P.A., 2003, *Practical Methods in Ecology*, pp. i–viii, 1–163.

Henry, L.A. & Roberts, J.M., 2007, Biodiversity and ecological composition of macrobenthos on cold-water coral mounds and adjacent off-mound habitat in the bathyal Porcupine Seabight, NE Atlantic, *Deep-Sea Research Part I – Oceanographic Research Papers*, 54(4), pp. 654–72.

Herbert, R.J.H. & Hawkins, S.J., 2006, Effect of rock type on the recruitment and early mortality of the barnacle *Chthamalus montagui*, *Journal of Experimental Marine Biology and Ecology*, 334(1), pp. 96–108.

Herbich, J.B. & de Vries, J.B., 1986, An Evaluation of the Effects of Operational Parameters on Sediment Resuspension During Cutterhead Dredging. Report CDS (Centre for Dredging Studies) 286, Texas A & M University.

Hereu, B., Zabala, M., Linares, C. & Sala, E., 2004, Temporal and spatial variability in settlement of the sea urchin *Paracentrotus lividus* in the NW Mediterranean, *Marine Biology*, 144(5), pp. 1011–8.

Herler, J. & Patzner, R.A., 2005, Spatial segregation of two common *Gobius* species (Teleostei : Gobiidae) in the northern Adriatic Sea, *Marine Ecology-Pubblicazioni Della Stazione Zoologica Di Napoli I*, 26(2), pp. 121–9.

Herrera, L. & Escribano, R., 2006, Factors structuring the phytoplankton community in the upwelling site off El Loa River in northern Chile, *Journal of Marine Systems*, 61(1–2), pp. 13–38.

Heyman, W.D., Kjerfve, B., Graham, R.T., Rhodes, K.L. & Garbutt, L., 2005, Spawning aggregations of *Lutjanus cyanopterus* (Cuvier) on the Belize Barrier Reef over a 6 year period, *Journal of Fish Biology*, 67(1), pp. 83–101.

Hill, V. & Cota, G., 2005, Spatial patterns of primary production on the shelf, slope and basin of the Western Arctic in 2002, *Deep-Sea Research Part Ii-Topical Studies in Oceanography*, 52(24–26), pp. 3344–54.

Hills, J.M. & Thomason, J.C., 2003, The 'ghost of settlement past' determines mortality and fecundity in the barnacle, *Semibalanus balanoides*, *Oikos*, 101(3), pp. 529–38.

Hodoki, Y. & Murakami, T., 2006, Effects of tidal flat reclamation on sediment quality and hypoxia in Isahaya Bay. *Aquatic Conservation – Marine and Freshwater Ecosystems*, 16(6), pp. 555–67.

Hofmann, M. & Schellnhuber, H.J., 2009, Oceanic acidification affects marine carbon pump and triggers extended marine oxygen holes, *Proceedings of the National Academy of Sciences of the United States of America*, 106(9), pp. 3017–22.

Holgate, S., Jevrejeva, S., Woodworth, P. & Brewer, S., 2007, Comment on "A semi-empirical approach to projecting future sea-level rise", *Science*, 317(5846), p. 2.

Holmes, G., Ortiz, J., Kaniewska, P. & Johnstone, R., 2008, Using three-dimensional surface area to compare the growth of two Pocilloporid coral species. *Marine Biology*, 155(4), pp. 421–7.

Hommes, S., Hulscher, S., Mulder, J.P.M., Otter, H.S. & Bressers, H.T.A., 2009, Role of perceptions and knowledge in the impact assessment for the extension of Mainport Rotterdam, *Marine Policy*, 33(1), pp. 146–55.

Horwood, J., O'Brien, C. & Darby, C., 2006, North sea cod recovery? *ICES Journal of Marine Science*, 63(6), pp. 961–8.

Houghton, J.D.R., Doyle, T.K., Wilson, M.W., Davenport, J. & Hays, G.C., 2006, Jellyfish aggregations and leatherback turtle foraging patterns in a temperate coastal environment, *Ecology*, 87(8), pp. 1967–72.

Howell, K.L., Billett, D.S.M. & Tyler, P.A., 2002, Depth-related distribution and abundance of seastars (Echinodermata : Asteroidea) in the Porcupine Seabight and Porcupine Abyssal Plain, NE Atlantic, *Deep-Sea Research Part I – Oceanographic Research Papers*, 49(10), pp. 1901–20.

Hubbard, D.K., 1989, Modern carbonate environments of St. Croix and the Caribbean: a general overview. In: Hubbard, D.K. (ed.), *12th Carribean Geological Conference*, pp. 85–94. Teague Bay, St. Croix: West Indies Laboratory.

Huggett, M.J., de Nys, R., Williamson, J.E., Heasman, M. & Steinberg, P.D., 2005, Settlement of larval blacklip abalone, *Haliotis rubra*, in response to green and red macroalgae, *Marine Biology*, 147(5), pp. 1155–63.

Huggett, M.J., Williamson, J.E., de Nys, R., Kjelleberg, S. & Steinberg, P.D., 2006, Larval settlement of the common Australian sea urchin *Heliocidaris erythrogramma* in response to bacteria from the surface of coralline algae, *Oecologia*, 149(4), pp. 604–19.

Hughes, R.G., 1978, Life-histories and abundance of epizoites of hydroid *Nemertesia antennina* (L), *Journal of the Marine Biological Association of the United Kingdom*, 58(2), pp. 313–32.

Hughes, T.P., 1994, Catastrophes, phase-shifts, and large-scale degradation of a caribbean coral-reef, *Science*, 265(5178), pp. 1547–51.

Hughes, T.P., Bellwood, D.R., Folke, C., Steneck, R.S. & Wilson, J., 2005, New paradigms for supporting the resilience of marine ecosystems, *Trends in Ecology & Evolution*, 20(7), pp. 380–6.

Hughes TP, Rodrigues MJ, Bellwood DR, Ceccarelli D, Hoegh-Guldberg O, McCook L, Moltschaniwskyj N, Pratchett MS, Steneck RS, Willis B 2007 Phase shifts, herbivory, and the resilience of coral reefs to climate change. *Current Biology* 17(4), pp. 360–365.

Huijbers, C.M., Mollee, E.M. & Nagelkerken, I., 2008, Post-larval French grunts (*Haemulon flavolineatum*) distinguish between seagrass, mangrove and coral reef water: Implications for recognition of potential nursery habitats, *Journal of Experimental Marine Biology and Ecology*, 357(2), pp. 134–9.

Huntley, J.W. & Kowalewski, M., 2007, Strong coupling of predation intensity and diversity in the Phanerozoic fossil record, *Proceedings of the National Academy of Sciences of the United States of America*, 104, pp. 15006–10.

Huston, M.A., 1985, Patterns of species-diversity on coral reefs, *Annual Review of Ecology and Systematics*, 16, pp. 149–77.

Hwang, S.J. & Song, J.I., 2007, Reproductive biology and larval development of the temperate soft coral *Dendronephthya gigantea* (Alcyonacea : Nephtheidae), *Marine Biology*, 152(2), pp. 273–84.

Hylleberg, J., 1975, Selective feeding by *Abarenicola pacifica* with notes on *Abarenicola vagabunda* and a concept of gardening in lugworms, *Ophelia*, 14(1–2), pp. 113–37.

Idjadi, J.A. & Karlson, R.H., 2007, Spatial arrangement of competitors influences coexistence of reef-building corals, *Ecology*, 88, pp. 2449–54.

Idjadi, J. & Karlson, R., 2009, Spatial aggregation promotes species coexistence among corals: Evidence from experiments and modeling. Annual Meeting of the Society for Integrative and Comparative Biology, Jan. 3–7, 2009 Boston MA. *Integrative and Comparative Biology*, 49, p. E83.

Iglesias-Rodriguez, M.D., Halloran, P.R., Rickaby, R.E.M., Hall, I.R., Colmenero-Hidalgo, E., Gittins, J.R., Green, D.R.H., Tyrrell, T., Gibbs, S.J., von Dassow, P., Rehm, E., Armbrust, E.V. & Boessenkool, K.P., 2008, Phytoplankton calcification in a high-CO2 world, *Science*, 320(5874), pp. 336–40.

Innocenti, G. & Galil, B.S., 2007, Modus vivendi: invasive host/parasite relations – *Charybdis longicollis* Leene, 1938 (Brachyura : Portunidae) and *Heterosaccus dollfusi* Boschma, 1960 (Rhizocephala: Sacculinidae), *Hydrobiologia*, 590, pp. 95–101.

Islam, M.S., Ueda, H. & Tanaka, M., 2005, Spatial distribution and trophic ecology of dominant copepods associated with turbidity maximum along the salinity gradient in a highly embayed estuarine system in Ariake Sea, Japan, *Journal of Experimental Marine Biology and Ecology*, 316(1), pp. 101–15.

Iso, S., Suizu, S. & Maejima, A., 1994, The lethal effect of hypertonic solutions and avoidance of marine organisms in relation to discharged brine from a desalination plant, *Desalination*, 97(1–3), pp. 389–99.

Jablonski, D., Roy, K. & Valentine, J.W., 2006, Out of the tropics: Evolutionary dynamics of the latitudinal diversity gradient, *Science*, 314(5796), pp. 102–6.

Jackson, J.B.C., Kirby, M.X., Berger, W.H., Bjorndal, K.A., Botsford, L.W., Bourque, B.J., Bradbury, R.H., Cooke, R., Erlandson, J., Estes, J.A., Hughes, T.P., Kidwell, S., Lange, C.B., Lenihan, H.S., Pandolfi, J.M., Peterson, C.H., Steneck, R.S., Tegner, M.J. & Warner, R.R., 2001, Historical overfishing and the recent collapse of coastal ecosystems, *Science*, 293(5530), pp. 629–38.

Jantzen, C., Wild, C., El-Zibdah, M., Roa-Quiaoit, H.A., Haacke, C. & Richter, C., 2008, Photosynthetic performance of giant clams, *Tridacna maxima* and *T-squamosa*, Red Sea, *Marine Biology*, 155(2), pp. 211–21.

Jardeweski, C.L.F. & de Almeida, T.C.M., 2006, Fish assemblage on artificial reefs in south brazilian coast, *Journal of Coastal Research*, 2, pp. 1210–4.

Jarnegren, J., Tobias, C.R., Macko, S.A. & Young, C.M., 2005, Egg predation fuels unique species association at deep-sea hydrocarbon seeps, *Biological Bulletin*, 209(2), pp. 87–93.

Jayaraj, K.A., Josia, J. & Kumar, P.D., 2008, Infaunal macrobenthic community of soft bottom sediment in a tropical shelf, *Journal of Coastal Research*, 24(3), pp. 708–18.

Jenkins, S.R., Murua, J. & Burrows, M.T., 2008, Temporal changes in the strength of density-dependent mortality and growth in intertidal barnacles, *Journal of Animal Ecology*, 77(3), pp. 573–84.

Jiang, Z.B., Zeng, J.N., Chen, Q.Z., Huang, Y.J., Liao, Y.B., Xu, X.Q. & Zheng, P., 2009, Potential impact of rising seawater temperature on copepods due to coastal power plants in subtropical areas, *Journal of Experimental Marine Biology and Ecology*, 368(2), pp. 196–201.

Jin, X., Gruber, N., Frenzel, H, Doney, S.C. & McWilliams, J.C., 2008, The impact on atmospheric CO_2 of iron fertilization induced changes in the ocean's biological pump. *Biogeosciences*, 5(2), pp. 385–406.

Jobling, M., 1981, Temperature tolerance and the final preferendum – rapid methods for the assessment of optimum growth temperatures. *Journal of Fish Biology*, 19, 439–55.

Johnson, D.W., 2006, Predation, habitat complexity, and variation in density-dependent mortality of temperate reef fishes, *Ecology*, 87(5), pp. 1179–88.

Johnson, J.K., 1972, Effect of turbidity on the rate of filtration and growth of the slipper limpet *Crepidula fornicata*, *Veliger*, 14(3), pp. 315–20.

Johnson, K.G., Jackson, J.B.C. & Budd, A.F., 2008a, Caribbean reef development was independent of coral diversity over 28 million years, *Science*, 319(5869), pp. 1521–3.

Johnson, M.P., Jessopp, M., Mulholland, O.R., McInerney, C., McAllen, R., Allcock, A.L. & Crowe, T.P., 2008b, What is the future for marine protected areas in Irish waters? *Biology and Environment-Proceedings of the Royal Irish Academy*, 108B(1), pp. 9–14.

Johnson, S.C. & Fast, M.D., 2004, Interactions between sea lice and their hosts, *Symp Soc Exp Biol* (55), pp. 131–59., discussion 243–5.

Johnson, S.C., Treasurer, J.W., Bravo, S., *et al.*, 2004, A review of the impact of parasitic copepods on marine aquaculture. 8th International Conference on Copepoda, Jul. 21–26, 2002 Natl. Taiwan Ocean Univ. Keelung Taiwan. *Zoological Studies*, 43(2), pp. 229–43.

Jokiel, P.L., Rodgers, K.S., Kuffner, I.B., Andersson, A.J., Cox, E.F., & Mackenzie, F.T., 2008, Ocean acidification and calcifying reef organisms: a mesocosm investigation *Coral Reefs*, 27(3), pp 473–483

Jompa, J. & McCook, L.J., 2002, Effects of competition and herbivory on interactions between a hard coral and a brown alga, *Journal of Experimental Marine Biology and Ecology*, 271(1), pp. 25–39.

Jones, G.P., Planes, S. & Thorrold, S.R., 2005, Coral reef fish larvae settle close to home, *Current Biology*, 15(14), pp. 1314–8.

Jordan-Garza, A.G., Rodriguez-Martinez, R.E., Maldonado, M.A. & Baker, D.M., 2008, High abundance of *Diadema antillarum* on a Mexican reef, *Coral Reefs*, 27(2), pp. 295-.

Jórgensen, H.B.H., Hansen, M.M., Bekkevold, D., Ruzzante, D.E. & Loeschcke, V., 2005, Marine landscapes and population genetic structure of herring (Clupea harengus L.) in the Baltic Sea, *Molecular Ecology*, 14(10), pp. 3219–34.

Jorundsdottir, K., Svavarsson, K. & Leung, K.M.Y., 2005, Imposex levels in the dogwhelk *Nucella lapillus* (L.) – continuing improvement at high latitudes, *Marine Pollution Bulletin*, 51(8–12), pp. 744–9.

Justine, J.L., 2007, Parasite biodiversity in a coral reef fish: twelve species of monogeneans on the gills of the grouper *Epinephelus maculatus* (Perciformes : Serranidae) off New Caledonia, with a description of eight new species of *Pseudorhabdosynochus* (Monogenea : Diplectanidae), *Systematic Parasitology*, 66(2), pp. 81–129.

Kaiser, M.J., 2006, The Louisiana artificial reef program, *Marine Policy*, 30(6), pp. 605–23.

Kamukuru, A.T., Mgaya, Y.D. & Ohman, M.C., 2004, Evaluating a marine protected area in a developing country: Mafia Island Marine Park, Tanzania. 3rd Western-Indian-Ocean-Marine-Science-Association Symposium (WIOMSA), Oct. 2003 Maputo Mozambique. *Ocean & Coastal Management*, 47(7-8), pp. 321–37.

Keller, B.D. & Causey, B.D., 2005, Linkages between the Florida Keys National Marine Sanctuary and the South Florida Ecosystem Restoration Initiative, *Ocean & Coastal Management*, 48(11–12), pp. 869–900.

Kemp, K.M., Jamieson, A.J., Bagley, P.M., McGrath, H., Bailey, D.M., Collins, M.A. & Priede, I.G., 2006, Consumption of large bathyal food fall, a six month study in the NE Atlantic, *Marine Ecology Progress Series*, 310, pp. 65–76.

Kimbro, D.L. & Grosholz, E.D., 2006, Disturbance influences oyster community richness and evenness, but not diversity, *Ecology*, 87(9), pp. 2378–88.

Kitting, C.L., Fry, B. & Morgan, M.D., Detection of inconspicuous epiphytic algae supporting food webs in seagrass meadows, *Oecologia*, 62(2), pp. 145–9.

Klum, P.P., Spinosa, J.S. & Fortes, M.D., 1992, The role of epiphytic periphyton and macroinvertebrate grazers in the trophic flux of a tropical seagrass community, *Aquatic Botany*, 43(4), pp. 327–49.

Knowlton, A.L. & Highsmith, R.C., 2000, Convergence in the time-space continuum: a predator-prey interaction, *Marine Ecology Progress Series*, 197, pp. 285–91.

Knowlton, N., 2001, Ecology – Coral reef biodiversity – Habitat size matters, *Science*, 292(5521), pp. 1493.

Knowlton, N. & Jackson, J.B.C., 2001, The ecology of coral reefs, (In) Bertness M.D., Gaines S., & Hay M.E., (eds), Marine Community Ecology, Sunderland(Mass) Sinauer, pp395–422.

Knowlton, N. & Jackson, J.B.C., 2008, Shifting baselines, local impacts, and global change on coral reefs, *PloS Biology*, 6(2), pp. 215–20.

Kocasoy, G., Mutlu, H.I. & Alagoz, B.A.Z., 2008, Prevention of marine environment pollution at the tourism regions by the application of a simple method for the domestic wastewater, *Desalination*, 226(1–3), pp. 21–37.

Kosobokova, K.N., Hirche, H.J. & Hopcroft, R.R., 2007, Reproductive biology of deep-water calanoid copepods from the Arctic Ocean, *Marine Biology*, 151(3), pp. 919–34.

Kowalik Z. (2004) Tide distribution & tapping into tidal energy, *Oceanologia* 46(3) 291–331

Krassoi, F.R., Brown, K.R., Bishop, M.J., Kelaher, B.P. & Summerhayes, S., 2008, Condition-specific competition allows coexistence of competitively superior exotic oysters with native oysters, *Journal of Animal Ecology*, 77(1), pp. 5–15.

Krkosek, M., Gottesfeld, A., Proctor, B., Rolston, D., Corr-Harris, C., & Lewis, M.A. 2007, Effects of host migration, diversity and aquaculture on sea/ice threats to Pacific salmon populations. *Proceedings of the Royal Society B – Biological Sciences*, 274, pp. 3141–49.

Kubodera, T. & Mori, K., 2005, First-ever observations of a live giant squid in the wild, *Proceedings of the Royal Society B – Biological Sciences*, 272(1581), pp. 2583–6.

Kubodera, T., Koyama, Y. & Mori, K., 2007, Observations of wild hunting behaviour and bioluminescence of a large deep-sea, eight-armed squid, *Taningia danae, Proceedings of the Royal Society B – Biological Sciences*, 274(1613), pp. 1029–34.

Kuffner, I.B., Brock, J.C., Grober-Dunsmore, R., Bonito, V.E., Hickey, T.D. & Wright, C.W., 2007, Relationships between reef fish communities and remotely sensed rugosity measurements in Biscayne National Park, Florida, USA, *Environmental Biology of Fishes*, 78(1), pp. 71–82.

Kuguru, B., Winters, G., Beer, S., Santos, S.R. & Chadwick, N.E., 2007, Adaptation strategies of the corallimorpharian *Rhodactis rhodostoma* to irradiance and temperature, *Marine Biology*, 151(4), pp. 1287–98.

Landman, N.H., Cochran, J.K., Cerrato, R., Mak, J., Roper, C.F.E. & Lu, C.C., 2004, Habitat and age of the giant squid (*Architeuthis sanctipauli*) inferred from isotopic analyses, *Marine Biology*, 144(4), pp. 685–91.

Larsson, A.I. & Jonsson, P.R., 2006, Barnacle larvae actively select flow environments supporting post-settlement growth and survival, *Ecology*, 87(8), pp. 1960–6.

Laurel, B.J., Hurst, T.P., Copeman, L.A. & Davis, M.W., 2008, The role of temperature on the growth and survival of early and late hatching Pacific cod larvae (*Gadus macrocephalus*), *Journal of Plankton Research*, 30(9), pp. 1051–60.

Lazzari, M.A., 2001, Dynamics of larval fish abundance in Penobscot Bay, Maine, *Fishery Bulletin*, 99(1), pp. 81–93.

Lecchini, D., Shima, J., Banaigs, B. & Galzin, R., 2005, Larval sensory abilities and mechanisms of habitat selection of a coral reef fish during settlement, *Oecologia*, 143(2), pp. 326–34.

Ledlie, M.H., Graham, N.A.J., Bythell, J.C., Wilson, S.K., Jennings, S., Polunin, N.V.C. & Hardcastle, J., 2007, Phase shifts and the role of herbivory in the resilience of coral reefs, *Coral Reefs*, 26(3), pp. 641–53.

Lee, J.T., Widdows, J., Jones, M.B. & Coleman, R.A., 2004, Settlement of megalopae and early juveniles of the velvet swimming crab *Necora puber* (Decapoda : Portunidae) in flow conditions, *Marine Ecology – Progress Series*, 272, pp. 191–202.

Leis, J.M., Wright, K.J. & Johnson, R.N., 2007, Behaviour that influences dispersal and connectivity in the small, young larvae of a reef fish, *Marine Biology*, 153, pp. 103–17.

Leitao, F., Santos, M.N. & Monteiro, C.C., 2007, Contribution of artificial reefs to the diet of the white sea bream (*Diplodus sargus*), *ICES Journal of Marine Science*, 64(3), pp. 473–8.

Leitao, F., Santos, M.N., Erzini, K. & Monteiro, C.C., 2008, The effect of predation on artificial reef juvenile demersal fish species, *Marine Biology*, 153(6), pp. 1233–44.

Leon, L.M. & Warnken, J., 2008, Copper and sewage inputs from recreational vessels at popular anchor sites in a semi-enclosed Bay (Qld, Australia): Estimates of potential annual loads, *Marine Pollution Bulletin*, 57(6–12), pp. 838–45.

Leon, T.M., 2001, Synopsis of the pycnogonids from Antarctic and Subantarctic waters, *Polar Biology*, 24(12), pp. 941–5.

Leoni, V., Vela, A., Pasqualini, V., Pergent-Martini, C. & Pergent, G., 2008, Effects of experimental reduction of light and nutrient enrichments (N and P) on seagrasses: a review, *Aquatic Conservation-Marine and Freshwater Ecosystems*, 18(2), pp. 202–20.

Lesoway, M.P. & Page, L.R., 2008, Growth and differentiation during delayed metamorphosis of feeding gastropod larvae: signatures of ancestry and innovation, *Marine Biology*, 153(4), pp. 723–34.

Lesser, M.P., Bythell, J.C., Gates, R.D., Johnstone, R.W. & Hoegh-Guldberg, O., 2007, Are infectious diseases really killing corals? Alternative interpretations of the experimental and ecological data, *Journal of Experimental Marine Biology and Ecology*, 346(1–2), pp. 36–44.

Lessios, H.A., Robertson, D.R. & Cubit, J.D., 1984, Spread of *Diadema* mass mortality through the caribbean, *Science*, 226(4672), pp. 335–7.

Lessios, H.A., Kessing, B.D. & Pearse, J.S., 2001, Population structure and speciation in tropical seas: Global phylogeography of the sea urchin *Diadema*, *Evolution*, 55(5), pp. 955–75.

Leujak, W. & Ormond, R.F.G., 2008, Reef walking on Red Sea reef flats – Quantifying impacts and identifying motives, *Ocean & Coastal Management*, 51(11), pp. 755–62.

Leujak, W. & Ormond, R.F.G., 2007, Comparative accuracy and efficiency of six coral community survey methods, *Journal of Experimental Marine Biology and Ecology*, 351, pp. 168–87.

Levin, L.A., 2006, Recent progress in understanding larval dispersal: new directions and digressions, *Integrative and Comparative Biology*, 46(3), pp. 282–97.

Levinton J.S., 2001, *Genetics, Paleontology, and Macroevolution*, 2nd edition, pp. i–xv, 1–617. Cambridge University Press, Cambridge, UK.

Li, J.H. & Kusky, T.M., 2007, World's largest known Precambrian fossil black smoker chimneys and associated microbial vent communities, North China: Implications for early life, *Gondwana Research*, 12(1–2), pp. 84–100.

Limbourn, A.J., Jones, G.P., Munday, P.L. & Srinivasan, M., 2007, Niche shifts and local competition between two coral reef fishes at their geographic boundary, *Marine and Freshwater Research*, 58(12), pp. 1120–9.

Linares, C., Coma, R. & Zabala, M., 2008, Restoration of threatened red gorgonian populations: An experimental and modelling approach, *Biological Conservation*, 141(2), pp. 427–37.

Linares, F., 2006, Effect of dissolved free amino acids (DFAA) on the biomass and production of microphytobenthic communities, *Journal of Experimental Marine Biology and Ecology*, 330(2), pp. 469–81.

Lindahl, U., Ohman, M.C. & Schelten, C.K., 2001, The 1997/1998 mass mortality of corals: Effects on fish communities on a Tanzanian coral reef, *Marine Pollution Bulletin*, 42(2), pp. 127–31.

Linse, K., Barnes, D.K.A. & Enderlein, P., 2006, Body size and growth of benthic invertebrates along an Antarctic latitudinal gradient, *Deep-Sea Research Part II – Topical Studies in Oceanography*, 53(8–10), pp. 921–31.

Linton, D.L. & Taghon, G.L., 2000, Feeding, growth, and fecundity of Abarenicola pacifica in relation to sediment organic concentration, *Journal of Experimental Marine Biology and Ecology*, 254(1), pp. 85–107.

Lipcius, R.N., Eggleston, D.B., Schreiber, S.J., Seitz, R.D., Shen, J., Sisson, M., Stockhausen, W.T. & Wang, H.V., 2008, Importance of metapopulation connectivity to restocking and restoration of marine species, *Reviews in Fisheries Science*, 16(1–3), pp. 101–10.

Lirman, D. & Manzello, D., 2009, Patterns of resistance and resilience of the stress-tolerant coral *Siderastrea radians* (Pallas) to suboptimal salinity and sediment burial, *Journal of Experimental Marine Biology and Ecology*, 369(1), pp. 72–7.

Lisbjerg, D. & Petersen, J.K., 2001, Feeding activity, retention efficiency, and effects of temperature and particle concentration on clearance rate in the marine bryozoan Electra crustulenta, *Marine Ecology Progress Series*, 215, pp. 133–41.

Little, C. & Kitching, J.A., 1996, *The Biology of Rocky Shores*. Oxford University Press, New York.

Liu, S. & Wang, W.X., 2002, Feeding and reproductive responses of marine copepods in South China Sea to toxic and nontoxic phytoplankton, *Marine Biology*, 140(3), pp. 595–603.

Lloyd, J., 1941, A survey of the fishing industry at Weston-super-Mare during the winter of 1940–41, *Proceedings of the Bristol Naturalists Society* 9, 316–27.

Loh, T.L., Tanzil, J.T.I. & Chou, L.M., 2006, Preliminary study of community development and scleractinian recruitment on fibreglass artificial reef units in the sedimented waters of Singapore, *Aquatic Conservation-Marine and Freshwater Ecosystems*, 16(1), pp. 61–76.

Longphuirt, S.N., Clavier, J., Grall, J., Chauvaud, L., Le Loc'h, F., Le Berre, I., Flye-Sainte-Marie, J., Richard, J. & Leynaert, A., 2007, Primary production and spatial distribution of subtidal microphytobenthos in a temperate coastal system, the Bay of Brest, France, *Estuarine Coastal and Shelf Science*, 74(3), pp. 367–80.

Longstaff, B.J. & Dennison, W.C., 1999, Seagrass survival during pulsed turbidity events: the effects of light deprivation on the seagrasses *Halodule pinifolia* and *Halophila ovalis*. 3rd International Seagrass Biology Workshop, Apr. 19–25, 1999 Philippines, Virgin Islands. *Aquatic Botany*, 65(1-4), pp. 105–21.

Lopez-Abellan, L. J., Balguerias, E.., Fernandez-Vergaz, V., 2002, Life history characteristics of the deep-sea crab Chaceon affinis population off Tenerife (Canary Islands). *Fisheries Research (Amsterdam)*, 58(2), pp. 231–9.

Lopez-Urrutia, A., San Martin, E., Harris, R.P. & Irigoien, X., 2006, Scaling the metabolic balance of the oceans, *Proceedings of the National Academy of Sciences of the United States of America*, 103(23), pp. 8739–44.

Lourido, A., Cacabelos, E. & Troncoso, J.S., 2008, Patterns of distribution of the polychaete fauna in subtidal soft sediments of the Ria de Aldan (north-western Spain), *Journal of the Marine Biological Association of the United Kingdom*, 88(2), pp. 263–75.

Lozano-Alvarez, E., Briones-Fourzan, P., Osorio-Arciniegas, A., Negrete-Soto, F. & Barradas-Ortiz, C., 2007, Coexistence of congeneric spiny lobsters on coral reefs: differential use of shelter resources and vulnerability to predators, *Coral Reefs*, 26(2), pp. 361–73.

Lund, D.C., Lynch-Stieglitz, J. & Curry, W.B., 2006, Gulf Stream density structure and transport during the past millennium. *Nature*, 444(7119), pp. 601–4.

Luo, X.J., Chen, S.J., Ni, H.G., Yu, M. & Mai, B.X., 2008, Tracing sewage pollution in the Pearl River Delta and its adjacent coastal area of South China Sea using linear alkylbenzenes (LABs), *Marine Pollution Bulletin*, 56(1), pp. 158–62.

Lutz, M.J., Caldeira, K., Dunbar, R.B. & Behrenfeld, M.J., 2007, Seasonal rhythms of net primary production and particulate organic carbon flux to depth describe the efficiency of biological pump in the global ocean, *Journal of Geophysical Research-Oceans*, 112(C10).

Macia, S., Robinson, M.P. & Nalevanko, A., 2007, Experimental dispersal of recovering *Diadema antillarum* increases grazing intensity and reduces macroalgal abundance on a coral reef, *Marine Ecology Progress Series*, 348, pp. 173–82.

Magurran, A.R., 1988, *Ecological Diversity and its Measurement*. Princeton University Press, New Jersey.

Malakoff, D., 2003, Marine research – Scientists counting on census to reveal marine biodiversity, *Science*, 302(5646), pp. 773–.

Maliao, R.J., Turingan, R.G. & Lin, J., 2008, Phase-shift in coral reef communities in the Florida Keys National Marine Sanctuary (FKNMS), USA, *Marine Biology*, 154(5), pp. 841–53.

Malme, C.I., Miles, P.R., Miller, G.W., Richardson, W.J., Roseneau, D.G., Thompson, D.H. & Greene Jr, C.R., 1989, Analysis and Ranking of the Acoustic Disturbance Potential of Petroleum Industry Activities and Other Sources of Noise in the Environment of Marine Mammals in Alaska. Bolt Beranek and Newman Systems and Technologies Corp., Cambridge, MA, USA. NTIS PB90-188673 Report No. 6945 to the US Department of the Interior, Minerals Management Service, Alaska OCS Office.

Malvadkar, U. & Hastings, A., 2008, Persistence of mobile species in marine protected areas, *Fisheries Research*, 91(1), pp. 69–78.

Margesini, R. & Nogi, Y., 2004, Psychropiezophilic microorganisms. *Cellular and Molecular Biology*, 50(4), pp. 429–36.

Mariani, S., Alcoverro, T., Uriz, MJ, et al. 2005a Early life histories in the bryozoan Schizobrachiella sanguinea: a case study. *Marine Biology*, 147(3), pp. 735–45

Mariani, S., Hutchinson, W.F., Hatfield, E.M.C., Ruzzante, D.E., Simmonds, E.J., Dahlgren, T.G., Andre, C., Brigham, J., Torstensen, E. & Carvalho, G.R., 2005b, North Sea herring population structure revealed by microsatellite analysis, *Marine Ecology Progress Series*, 303, pp. 245–57.

Mariani, S., Uriz, M.J., Turon, X. & Alcoverro, T., 2006, Dispersal strategies in sponge larvae: integrating the life history of larvae and the hydrologic component, *Oecologia*, 149(1), pp. 174–84.

Marine Parks and Reserves ity Tanzania (2009) www.marineparktz.com

Marrari, M., Daly, K.L. & Hu, C.M., 2008, Spatial and temporal variability of SeaWiFS chlorophyll a distributions west of the Antarctic Peninsula: Implications for krill production. *Deep-Sea Research Part II – Topical Studies In Oceanography*, 55(3-4), pp. 377–92.

Marsh, A.G., Mullineaux, L.S., Young, C.M. & Manahan, D.T., 2001, Larval dispersal potential of the tubeworm *Riftia pachyptila* at deep-sea hydrothermal vents, *Nature*, 411(6833), pp. 77–80.

Marshall, A.D., Kyne, P.M. & Bennett, M.B., 2008, Comparing the diet of two sympatric urolophid elasmobranchs (*Trygonoptera testacea* Muller & Henle and *Urolophus kapalensis* Yearsley & Last): evidence of ontogenetic shifts and possible resource partitioning, *Journal of Fish Biology*, 72(4), pp. 883–98.

Marshall, D.J. & Bolton, T.F., 2007, Effects of egg size on the development time of non-feeding larvae, *Biological Bulletin*, 212(1), pp. 6–11.

Martin, J.H., 1991, Iron still comes from above, *Nature*, 353(6340), pp. 123-.

Masson, D.G., Bett, B.J., Billett, D.S.M., Jacobs, C.L., Wheeler, A.J., & Wynn, R.B., 2003, The origin of deep-water, coral-topped mounds in the northern Rockall Trough, Northeast Atlantic, *Marine Geology*, 194(3–4), pp. 159–80.

Mattice, J.S. & Zittel, H.E., 1976, Site-specific evaluation of power-plant chlorination, *Journal Water Pollution Control Federation*, 48(10), pp. 2284–308.

Mattiucci, S., Nascetti, G, 2008 Advances and trends in the molecular systematics of anisakid nematodes, with implications for their evolutionary ecology and host-parasite co-evolutionary processes *Advances in Parasitology*, 66 pp. 47–148

May, R.C., 1974, Effects of temperature and salinity on yolk utilization in *Bairdiella icistia* (Jordan + Gilbert) (Pisces-Sciaenidae), *Journal of Experimental Marine Biology and Ecology*, 16(3), pp. 213–25.

May, R.M., 1993, Marine species richness – reply, *Nature*, 361(6413), pp. 598.

May, R.M., 1994, Conceptual aspects of the quantification of the extent of biological diversity. *Philosophical Transactions of the Royal Society of London Series B – Biological Sciences*, 345(1311), pp. 13–20.

Mayhew, P.J., Jenkins, G.B. & Benton, T.G., 2008, A long-term association between global temperature and biodiversity, origination and extinction in the fossil record, *Proceedings of the Royal Society B – Biological Sciences*, 275(1630), pp. 47–53.

McBreen, F., Wilson, J.G., Mackie, A.S.Y. & Aonghusa, C.N., 2008, Seabed mapping in the southern Irish Sea: Predicting benthic biological communities based on sediment characteristics. 41st European Marine Biology Symposium, Sep. 4–8, 2005, Univ. Coll. Cork, Dept. Zool., Ecol. & Plant Sci., Cork, Ireland. *Hydrobiologia*. 606, pp. 93–103.

McCauley, R.D., 1994, Seismic surveys. In: Neff J.M. & Young P.C. (eds), *Environmental Implications of Offshore Oil and Gas Development in Australia – the findings of an independent scientific review*. APEA, Sydney.

McClanahan, T.R., 2000, Recovery of a coral reef keystone predator, *Balistapus undulatus*, in East African marine parks, *Biological Conservation*, 94(2), pp. 191–8.

McClanahan, TR, 2008 Response of the coral reef benthos and herbivory to fishery closure management and the 1998 ENSO disturbance *Oecologia* 155(1) pp 169–177.

McClanahan, T.R. & Mangi, S., 2000, Spillover of exploitable fishes from a marine park and its effect on the adjacent fishery, *Ecological Applications*, 10(6), pp. 1792–805.

McClanahan, T., Davies, J. & Maina, J., 2005a, Factors influencing resource users and managers' perceptions towards marine protected area management in Kenya, *Environmental Conservation*, 32(1), pp. 42–9.

McClanahan, T.R., Mwaguni, S. & Muthiga, N.A., 2005b, Management of the Kenyan coast, *Ocean & Coastal Management*, 48(11–12), pp. 901–31.

McClanahan, T.R., Maina, J. & Davies, J., 2005c, Perceptions of resource users and managers towards fisheries management options in Kenyan coral reefs, *Fisheries Management and Ecology*, 12(2), pp. 105–12.

McClanahan, T.R., Carreiro-Silva, M. & DiLorenzo, M., 2007a, Effect of nitrogen, phosphorous, and their interaction on coral reef algal succession in Glover's Reef, Belize, *Marine Pollution Bulletin*, 54(12), pp. 1947–57.

McClanahan, T.R., Graham, N.A.J., Calnan, J.M. & MacNeil, M.A., 2007b, Toward pristine biomass: Reef fish recovery in coral reef marine protected areas in Kenya, *Ecological Applications*, 17(4), pp. 1055–67.

McClintock, J.B., Angus, R.A. & McClintock, F.E., 2007, Abundance, diversity and fidelity of macroinvertebrates sheltering beneath rocks during tidal emersion in an intertidal cobble field: Does the intermediate disturbance hypothesis hold for less exposed shores with smaller rocks? *Journal of Experimental Marine Biology and Ecology*, 352(2), pp. 351–60.

McCook, L.J., Jompa, J. & Diaz-Pulido, G., 2001, Competition between corals and algae on coral reefs: a review of evidence and mechanisms, *Coral Reefs*, 19(4), pp. 400–17.

McGehee, M.A., 2008, Changes in the Coral Reef Community of Southwest Puerto Rico 1995 to 2005, *Caribbean Journal of Science*, 44(3), pp. 345–54.

McLellan, T., Havis, R., Hayes, D. & Raymond, G., 1989, Field Studies of Sediment Resuspension Characteristics of Selected Dredges. Technical Report HL-89-9, US Army Engineer Waterways Experiment Station, Vicksburg, MS.

McWilliams, J.P., Cote, I.M., Gill, J.A., Sutherland, W.J. & Watkinson, A.R., 2005, Accelerating impacts of temperature-induced coral bleaching in the Caribbean, *Ecology*, 86(8), pp. 2055–60.

Mesias, J.M., Bisagni, J.J. & Brunner, A., 2007, A high-resolution satellite-derived sea surface temperature climatology for the western North Atlantic Ocean, *Continental Shelf Research*, 27(2), pp. 191–207.

Messieh, S.N., Wildish, D.J. & Peterson, R.H., 1981, Effect of suspended sediment on feeding by larval herring. *Can. Tech. Rep. Fish. Aquat. Sci.*, 1008:iv + 33 p.

Meyer, C.G., 2007, The impacts of spear and other recreational fishers on a small permanent Marine Protected Area and adjacent pulse fished area, *Fisheries Research*, 84(3), pp. 301–7.

Meynecke, J.O., Poole, G.C., Werry, J. & Lee, S.Y., 2008, Use of PIT tag and underwater video recording in assessing estuarine fish movement in a high intertidal mangrove and salt marsh creek, *Estuarine Coastal and Shelf Science*, 79(1), pp. 168–78.

Michel, P., Averty, B., Andral, B., Chiffoleau, J.F. & Galgani, F., 2001, Tributyltin along the coasts of Corsica (Western Mediterranean): A persistent problem, *Marine Pollution Bulletin*, 42(11), pp. 1128–32.

Miller, R.J., Adams, A.J., Ogden, N.B., Ogden, J.C. & Ebersole, J.P., 2003, *Diadema antillarum* 17 years after mass mortality: is recovery beginning on St. Croix? *Coral Reefs*, 22(2), pp. 181–7.

Miller, R.J., Adams, A.J., Ebersole, J.P. & Ruiz, E., 2007, Evidence for positive density-dependent effects in recovering *Diadema antillarum* populations, *Journal of Experimental Marine Biology and Ecology*, 349(2), pp. 215–22.

Mills, E.L., Strayer, D.L., Scheuerell, M.D. & Carlton, J.T., 1996, Exotic species in the Hudson River Basin: A history of invasions and introductions, *Estuaries*, 19(4), pp. 814–23.

Minello, T.J., Zimmerman, R.J. & Martinez, E.X., 1987, Fish predation on juvenile brown shrimp, *Penaeus aztecus* Ives – effects of turbidity and substratum on predation rates, *Fishery Bulletin*, 85(1), pp. 59–70.

Moksnes, P.O., 2002, The relative importance of habitat-specific settlement, predation and juvenile dispersal for distribution and abundance of young juvenile shore crabs *Carcinus maenas* L, *Journal of Experimental Marine Biology and Ecology*, 271(1), pp. 41–73.

Moksnes, P.O., 2004, Interference competition for space in nursery habitats: density-dependent effects on growth and dispersal in juvenile shore crabs Carcinus maenas, *Marine Ecology Progress Series*, 281, pp. 181–91.

Moller, L.F. & Riisgard, H.U., 2006, Filter feeding in the burrowing amphipod *Corophium volutator*. *Marine Ecology Progress Series*, 322, pp. 213–24.

Montgomery, J.C., Jeffs, A., Simpson, S.D., Meekan, M. & Tindle, C., 2006, Sound as an orientation cue for the pelagic larvae of reef fishes and decapod crustaceans, *Advances in Marine Biology, Vol 51*, 51, pp. 143–96.

Moore, C.M., Mills, M.M., Milne, A., Langlois, R., Achterberg, E.P., Lochte, K., Geider, R.J. & La Roche, J., 2006, Iron limits primary productivity during spring bloom development in the central North Atlantic, *Global Change Biology*, 12(4), pp. 626–34.

Mora, C. & Sale, P.F., 2002, Are populations of coral reef fish open or closed? *Trends in Ecology & Evolution*, 17(9), pp. 422–8.

Mora, C., Andrefouet, S., Costello, M.J., Kranenburg, C., Rollo, A., Veron, J., Gaston, K.J. & Myers, R.A., 2006, Coral reefs and the global network of marine protected areas, *Science*, 312(5781), pp. 1750–1.

Mora, C., Tittensor, D.P. & Myers, R.A., 2008, The completeness of taxonomic inventories for describing the global diversity and distribution of marine fishes. *Proceedings of the Royal Society B – Biological Sciences*, 275(1631), pp. 149–155.

Moreno, R.A., Rivadeneira, M.M., Hernandez, C.E., Sampertegui, S. & Rozbaczylo, N., 2008, Do Rapoport's rule, the mid-domain effect or the source-sink hypotheses predict bathymetric patterns of polychaete richness on the Pacific coast of South America? *Global Ecology and Biogeography*, 17(3), pp. 415–23.

Morgan, R.P., Rasin, V.J. & Noe, L.A., 1983, Sediment effects on eggs and larvae of striped bass and white perch, *Transactions of the American Fisheries Society*, 112(2), pp. 220–4.

Morris A.V., 2004, Unpublished DPhil thesis.

Mortensen, P.B., Buhl-Mortensen, L., Gebruk, A.V. & Krylova, E.M., 2008, Occurrence of deep-water corals on the Mid-Atlantic Ridge based on MAR-ECO data, *Deep-Sea Research Part Ii-Topical Studies in Oceanography*, 55(1–2), pp. 142–52.

Mouritsen, K.N. & Jensen, T., 2006, The effect of *Sacculina carcini* infections on the fouling, burying behaviour and condition of the shore crab, *Carcinus maenas*, *Marine Biology Research*, 2(4), pp. 270–5.

Mozetic, P., Malacic, V. & Turk, V., 2008, A case study of sewage discharge in the shallow coastal area of the Northern Adriatic Sea (Gulf of Trieste), *Marine Ecology-Pubblicazioni Della Stazione Zoologica Di Napoli I*, 29(4), pp. 483–94.

Muenhor, D., Satayavivad, J., Limpaseni, W., Parkpian, P., Delaune, R.D., Gambrell, R.P. & Jugsujinda, A., 2009, Mercury contamination and potential impacts from municipal waste incinerator on Samui Island, Thailand, *Journal of Environmental Science and Health Part a-Toxic/Hazardous Substances & Environmental Engineering*, 44(4), pp. 376–87.

Muir, C., 2005, Programme profile: Tanzania Turtle & Dugong Conservation Programme, *Marine Turtle Newsletter*(110), p. 9.

Mullineaux, L.S., Peterson, C.H., Micheli, F. & Mills, S.W., 2003, Successional mechanism varies along a gradient in hydrothermal fluid flux at deep-sea vents, *Ecological Monographs*, 73(4), pp. 523–42.

Mullineaux, L.S., Mills, S.W., Sweetman, A.K., Beaudreau, A.H., Metaxas, A. & Hunt, H.L., 2005, Vertical, lateral and temporal structure in larval distributions at hydrothermal vents, *Marine Ecology Progress Series*, 293, pp. 1–16.

Mumby, P.J., 2006, Connectivity of reef fish between mangroves and coral reefs: Algorithms for the design of marine reserves at seascape scales, *Biological Conservation*, 128(2), pp. 215–22.

Mumby, P.J. & Steneck, R.S., 2008, Coral reef management and conservation in light of rapidly evolving ecological paradigms, *Trends in Ecology & Evolution*, 23(10), pp. 555–63.

Mumby, P.J., Harborne, A.R., Williams, J., Kappel, C.V., Brumbaugh, D.R., Micheli, F., Holmes, K.E., Dahlgren, C.P., Paris, C.B. & Blackwell, P.G., 2007, Trophic cascade facilitates coral recruitment in a marine reserve, *Proceedings of the National Academy of Sciences of the United States of America*, 104(20), pp. 8362–7.

Mumby, P.J., Broad, K., Brumbaugh, D.R., Dahlgren, C.P., Harborne, A.R., Hastings, A., Holmes, K.E., Kappel, C.V., Micheli, F. & Sanchirico, J.N., 2008, Coral reef habitats as surrogates of species, ecological functions, and ecosystem services, *Conservation Biology*, 22(4), pp. 941–51.

Munday, P.L., Jones, G.P., Pratchett, M.S. & Williams, A.J., 2008, Climate change and the future for coral reef fishes, *Fish and Fisheries*, 9(3), pp. 261–85.

Munn, C.B., 2006, Viruses as pathogens of marine organisms from bacteria to whales, *Journal of the Marine Biology Association of the UK*, 86, 453–67.

Muthiga, N.A., Reidemiller, S., van der Elst, R., Mann-Lang, J., Horrill, C. & McClanahan, T.R., 2000, Management status and case studies. In: McClanahan, T.R., Sheppard, C.R.C, and Obura, D.O. (eds), *Coral Reefs of the Indian Ocean: their ecology and conservation*, pp. 471–505. Oxford University Press, New York.

Mydlarz, L.D., Jones, L.E. & Harvell, C.D., 2006, Innate immunity environmental drivers and disease ecology of marine and freshwater invertebrates, *Annual Review of Ecology Evolution and Systematics*, 37, pp. 251–88.

Myers, R.A. & Worm, B., 2003, Rapid worldwide depletion of predatory fish communities, *Nature*, 423(6937), pp. 280–3.

Myers, R.A., Baum, J.K., Shepherd, T.D., Powers, S.P. & Peterson, C.H., 2007, Cascading effects of the loss of apex predatory sharks from a coastal ocean, *Science*, 315(5820), pp. 1846–50.

Myhre, S. & Acevedo-Gutierrez, A., 2007, Recovery of sea urchin *Diadema antillarum* populations is correlated to increased coral and reduced macroalgal cover, *Marine Ecology Progress Series*, 329, pp. 205–10.

Nakagawa, S. & Takai, K., 2008, Deep-sea vent chemoautotrophs: diversity, biochemistry and ecological significance, *Fems Microbiology Ecology*, 65(1), pp. 1–14.

Nakamura, T. & Nakamori, T., 2007, A geochemical model for coral reef formation, *Coral Reefs*, 26(4), pp. 741–55.

Nakaoka, M., 2005, Plant-animal interactions in seagrass beds: ongoing and future challenges for understanding population and community dynamics, *Population Ecology*, 47(3), pp. 167–77.

Narváez D.A., Poulin E., Germán L., Hernádez E., Castilla J.C., & Navarrete S.A., 2004, Seasonal and spatial variation of nearshore hydrographic conditins in central Chile. *Continental Shelf Research* 24 pp. 279–292.

Nedwell, J., Turnpenny, A., Langworthy, J. & Edwards, B., 2003, Measurements of Underwater Noise During Piling at the Red Funnel Terminal, Southampton, and Observations of its Effect on Caged Fish. Fawley Aquatic Research Ltd, Fawley, UK.. Report Reference: 558 R 0207.

Neil, D., 1990, Potential for coral stress due to sediment resuspension and deposition by reef walkers, *Biological Conservation*, 52(3), pp. 221–27.

Nestlerode, J.A., Luckenbach, M.W. & O'Beirn, F.X., 2007, Settlement and survival of the oyster *Crassostrea virginica* on created oyster reef habitats in Chesapeake bay, *Restoration Ecology*, 15(2), pp. 273–83.

Nichols, O.C. & Hamilton, P.K., 2004, Occurrence of the parasitic sea lamprey, *Petromyzon marinus*, on western North Atlantic right whales, *Eubalaena glacialis*, *Environmental Biology of Fishes*, 71(4), pp. 413–7.

Nicoletti, L., Marzialetti, S., Paganelli, D. & Ardizzone, G.D., 2007, Long-term changes in a benthic assemblage associated with artificial reefs, *Hydrobiologia*, 580, pp. 233–40.

Nilsson, H.C. & Rosenberg, R., 2000, Succession in marine benthic habitats and fauna in response to oxygen deficiency: analysed by sediment profile-imaging and by grab samples, *Marine Ecology Progress Series*, 197, pp. 139–49.

Niquen, M., & Bouchon, M 2004 Impact of El Nino events on pelagic fisheries in Peruvian waters. *Deep-Sea Research Part II – Topical Studies In Oceanography*, 51(6–9), pp. 563–74.

Nogata, Y. & Matsumura, K., 2006, Larval development and settlement of a whale barnacle, *Biology Letters*, 2(1), pp. 92–3.

O'Connor, M.I., Bruno, J.F., Gaines, S.D., Halpern, B.S., Lester, S.E., Kinlan, B.P. & Weiss, J.M., 2007, Temperature control of larval dispersal and the implications for marine ecology, evolution, and conservation, *Proceedings of the National Academy of Sciences of the United States of America*, 104(4), pp. 1266–71.

O'Dor R.K. (2003) The Unknown Ocean: Baseline report of the census of Marine Life Programme. Consortium for Oceanographic Research and Education., Washington DC, 28pp.

O'Dor, R. & Gallardo, V.A., 2005, How to census marine life: ocean realm field projects, *Scientia Marina*, 69, pp. 181–99.

Ojeda, F.P. & Munoz, A.A., 1999, Feeding selectivity of the herbivorous fish Scartichthys viridis: effects on macroalgal community structure in a temperate rocky intertidal coastal zone, *Marine Ecology Progress Series*, 184, pp. 219–29.

Okamura, B., 1990, Particle-size, flow velocity, and suspension-feeding by the erect bryozoans *Bugula neritina* and *Bugula stolonifera*, *Marine Biology*, 105(1), pp. 33–8.

Olabarria, C., 2006, Faunal change and bathymetric diversity gradient in deep-sea prosobranchs from Northeastern Atlantic, *Biodiversity and Conservation*, 15, pp. 3685–702.

Onsrud, M.S.R., Kaartvedt, S., Rostad, A. & Klevjer, T.A., 2004, Vertical distribution and feeding patterns in fish foraging on the krill *Meganyctiphanes norvegica*, *ICES Journal of Marine Science*, 61(8), pp. 1278–90.

Oracion, E.G., Miller, M.L. & Christie, P., 2005, Marine protected areas for whom? Fisheries, tourism, and solidarity in a Philippine community, *Ocean & Coastal Management*, 48(3–6), pp. 393–410.

Ormond, R.F.G. & Roberts, C.M., 1999, Biodiversity of coral reef fish. In: Ormond, R.F.G., Gage, J.D. & Angel, M.V. (eds), *Marine Biodiversity: Patterns and Processes*, pp. 216–57. Cambridge University Press, UK.

Otis, M.J., 1994, New Bedford Harbor, Massachusetts dredging/disposal of PCB contaminated sediments. *Dredging 94—Proceedings of the Second International Conference on Dredging and Dredged Material Placement, American Society of Civil Engineers*, 1, 579–87.

Ottesen, O.H. & Bolla, S., 1998, Combined effects of temperature and salinity on development and survival of Atlantic halibut larvae, *Aquaculture International*, 6(2), pp. 103–20.

Ozaki, Y., Yusa, Y., Yamato, S. & Imaoka, T., 2008, Reproductive ecology of the pedunculate barnacle *Scalpellum stearnsii* (Cirripedia : Lepadomorpha : Scalpellidae), *Journal of the Marine Biological Association of the United Kingdom*, 88(1), pp. 77–83.

Paddack, M.J. & Estes, J.A., 2000, Kelp forest fish populations in marine reserves and adjacent exploited areas of central California, *Ecological Applications*, 10(3), pp. 855–70.

Paine RT (1974) Intertidal community structure: experimental studies on the relationship between a dominant competitor and its principal predator. *Oecologia* 15:93–120.

Palm, H.W. & Klimpel, S., 2007, Evolution of parasitic life in the ocean, *Trends in Parasitology*, 23(1), pp. 10–2.

Paltzat, D.L., Pearce, C.M., Barnes, P.A. & McKinley, R.S., 2008, Growth and production of California sea cucumbers (*Parastichopus californicus* Stimpson) co-cultured with suspended Pacific oysters (*Crassostrea gigas* Thunberg), *Aquaculture*, 275(1–4), pp. 124–37.

Park, S., Epifanio, C.E. & Iglay, R.B., 2005, Patterns of larval release by the Asian shore crab *Hemigrapsus sanguineus* (De Haan): Periodicity at diel and tidal frequencies, *Journal of Shellfish Research*, 24(2), pp. 591–5.

Paula, J., Dray, T. & Queiroga, H., 2001, Interaction of offshore and inshore processes controlling settlement of brachyuran megalopae in Saco mangrove creek, Inhaca Island (South Mozambique), *Marine Ecology Progress Series*, 215, pp. 251–60.

Pauley, G., 1997, Diversity and distribution of reef organisms. In: Birkeland, C. (ed.), *Life and Death of Coral Reefs*, pp. 298–353. Chapman & Hall, New York, Albany.

Pauly, D., Christensen, V., Dalsgaard, J., Froese, R. & Torres, F., 1998, Fishing down marine food webs, *Science*, 279(5352), pp. 860–3.

Pawlik, J.R., Steindler, L., Henkel, T.P., Beer, S. & Ilan, M., 2007, Chemical warfare on coral reefs: Sponge metabolites differentially affect coral symbiosis in situ, *Limnology and Oceanography*, 52(2), pp. 907–11.

Payne, J.R., Driskell, W.B., Short, J.W. & Larsen, M.L., 2008, Long term monitoring for oil in the Exxon Valdez spill region, *Marine Pollution Bulletin*, 56(12), pp. 2067–81.

Pearse, J.S. & Lockhart, S.J., 2004, Reproduction in cold water: paradigm changes in the 20th century and a role for cidaroid sea urchins, *Deep-Sea Research Part II-Topical Studies in Oceanography*, 51(14–16), pp. 1533–49.

Pearson, T.H. & Rosenberg, R. 1978, Macrobenthic succession in relation to organic enrichment of the marine environment. *Oceanography and Marine Biology: An Annual Review*, 16, pp. 229–311.

Pedros-Alio, C., 2006, Marine microbial diversity: can it be determined? *Trends in Microbiology*, 14(6), pp. 257–63.

Pequex, A.,Giles, R. & Marshall, W.S., 1988, NaCl transport in epithelia. In: Greger, R. (ed.), *Comparative and Environmental Physiology*, vol. 1, pp. 1–73. Springer, Berlin.

Perez, V., Fernandez, E., Maranon, E., Moran, X.A.G. & Zubkovc, M.V., 2006, Vertical distribution of phytoplankton biomass, production and growth in the Atlantic subtropical gyres, *Deep-Sea Research Part I – Oceanographic Research Papers*, 53(10), pp. 1616–34.

Perkol-Finkel, S. & Benayahu, Y., 2007, Differential recruitment of benthic communities on neighboring artificial and natural reefs, *Journal of Experimental Marine Biology and Ecology*, 340(1), pp. 25–39.

Perkol-Finkel, S., Shashar, N. & Benayahu, Y., 2006, Can artificial reefs mimic natural reef communities? The roles of structural features and age, *Marine Environmental Research*, 61(2), pp. 121–35.

Petersen, J.K., 2007, Ascidian suspension feeding, *Journal of Experimental Marine Biology and Ecology*, 342(1), pp. 127–37.

Petraitis, P.S., 2002, Effects of intraspecific competition and scavenging on growth of the periwinkle *Littorina littorea*. *Marine Ecology Progress Series*, 236, pp. 179–87.

Petraitis, P.S. & Dudgeon, S.R., 2005, Divergent succession and implications for alternative states on rocky intertidal shores, *Journal of Experimental Marine Biology and Ecology*, 326(1), pp. 14–26.

Petraitis, P.S. & Methratta, E.T., 2006, Using patterns of variability, to test for multiple community states on rocky intertidal shores, *Journal of Experimental Marine Biology and Ecology*, 338(2), pp. 222–32.

Petrosillo, I., Zurlini, G., Corliano, M.E., Zaccarelli, N. & Dadamo, M., 2007, Tourist perception of recreational environment and management in a marine protected area, *Landscape and Urban Planning*, 79(1), pp. 29–37.

Philippe, H., Derelle, R., Lopez, P., Pick, K., Borchiellini, C., Boury-Esnault, N., Vacelet, J., Renard, E., Houliston, E., Queinnec, E., Da Silva, C., Wincker, P., Le Guyader, H., Leys, S., Jackson, D., Schreiber, F., Erpenbeck, D., Morgenstern, B., Woerheide, G. & Manuel, M., 2009, Phylogenomics revives traditional views on deep animal relationships, *Current Biology*, 19(8), pp. 706–12.

Phillippi, A., Hamann, E. & Yund, P.O., 2004, Fertilization in an egg-brooding colonial ascidian does not vary with population density, *Biological Bulletin*, 206(3), pp. 152–60.

Phillips, N.E., 2005, Growth of filter-feeding benthic invertebrates from a region with variable upwelling intensity, *Marine Ecology Progress Series*, 295, pp. 79–89.

Picciano, M. & Ferrier-Pages, C., 2007, Ingestion of pico- and nanoplankton by the Mediterranean red coral *Corallium rubrum*, *Marine Biology*, 150, pp. 773–82.

Piniak, G.A., 2002, Effects of symbiotic status, flow speed, and prey type on prey capture by the facultatively symbiotic temperate coral Oculina arbuscula, *Marine Biology*, 141(3), pp. 449–55.

Poggiale, JC., Dauvin, JC 2001 Long-term dynamics of three benthic Ampelisca (Crustacea-Amphipoda) populations from the Bay of Morlaix (western English Channel) related to their disappearance after the 'Amoco Cadiz' oil spill *Marine Ecology Progress Series* 214, pp. 201–209.

Poore, G.C.B. & Wilson, G.D.F., 1993, Marine species richness, *Nature*, 361(6413), pp. 597–8.

Portman, M.E., 2007, Zoning design for ross-border marine protected areas: The Red Sea Marine Peace Park Case Study, *Ocean & Coastal Management*, 50, pp. 499–522.

Praca, E. & Gannier, A., 2008, Ecological niches of three teuthophageous odontocetes in the northwestern Mediterranean Sea, *Ocean Science*, 4(1), pp. 49–59.

Praca, E., Gannier, A., Das, K. & Laran, S., 2009, Modelling the habitat suitability of cetaceans: Example of the sperm whale in the northwestern Mediterranean Sea, *Deep-Sea Research Part I – Oceanographic Research Papers*, 56(4), pp. 648–57.

Prada, M.C., Appeldoorn, R.S. & Rivera, J.A., 2008, Improving coral reef habitat mapping of the Puerto Rico insular shelf using side scan sonar, *Marine Geodesy*, 31(1), pp. 49–73.

Prado, P., Alcoverro, T., Martinez-Crego, B., Verges, A., Perez, M. & Romero, J., 2007, Macrograzers strongly influence patterns of epiphytic assemblages in seagrass meadows, *Journal of Experimental Marine Biology and Ecology*, 350, pp. 130–43.

Pratchett, M.S., 2005a, Dietary overlap among coral-feeding butterflyfishes (Chaetodontidae) at Lizard Island, northern Great Barrier Reef, *Marine Biology*, 148(2), pp. 373–82.

Pratchett, M.S., 2005b, Dynamics of an outbreak population of Acanthaster planci at Lizard Island, northern Great Barrier Reef (1995–1999), *Coral Reefs*, 24(3), pp. 453–62.

Priede, I.G., Froese, R., Bailey, D.M., Bergstad, O.A., Collins, M.A., Dyb, J.E., Henriques, C., Jones, E.G. & King, N., 2006, The absence of sharks from abyssal regions of the world's oceans, *Proceedings of the Royal Society B – Biological Sciences*, 273(1592), pp. 1435–41.

Primavera, J.H., 2005, Mangroves, fishponds, and the quest for sustainability, *Science*, 310(5745), pp. 57–9.

Prowse, T.A.A., Sewell, M.A. & Byrne, M., 2008, Fuels for development: evolution of maternal provisioning in asterinid sea stars, *Marine Biology*, 153(3), pp. 337–49.

Ptacnik, R., Solimini, A.G., Andersen, T., Tamminen, T., Brettum, P., Lepisto, L., Willen, E. & Rekolainen, S., 2008, Diversity predicts stability and resource use efficiency in natural phytoplankton communities, *Proceedings of the National Academy of Sciences of the United States Of America*, 105(13), pp. 5134–8.

Purcell, J.E. & Sturdevant, M.V., 2001, Prey selection and dietary overlap among zooplanktivorous jellyfish and juvenile fishes in Prince William Sound, Alaska, *Marine Ecology Progress Series*, 210, pp. 67–83.

Qian, P.Y., Rittschof, D. & Sreedhar, B., 2000, Macrofouling in unidirectional flow: miniature pipes as experimental models for studying the interaction of flow and surface characteristics on the attachment of barnacle, bryozoan and polychaete larvae, *Marine Ecology Progress Series*, 207, pp. 109–21.

Queiroga, H., Cruz, T., dos Santos, A., Dubert, J., Gonzalez-Gordillo, J.I., Paula, J., Peliz, A. & Santos, A.M.P., 2007, Oceanographic and behavioural processes affecting invertebrate larval dispersal and supply in the western Iberia upwelling ecosystem, *Progress in Oceanography*, 74(2–3), pp. 174–91.

Qvarfordt, S., Kautsky, H. & Malm, T., 2006, Development of fouling communities on vertical structures in the Baltic Sea, *Estuarine Coastal and Shelf Science*, 67(4), pp. 618–28.

Rabalais, N.N., 2005, The potential for nutrient overenrichment to diminish marine biodiversity. In: Norse, E.A. & Crowder, L.B. (eds), *Marine Conservation Biology: The Science of Maintaining the Sea's Biodiversity*, pp. 109–122. Island Press. Washington, DC.

Raimondi, P.T., Lohse, D. & Blanchette, C., 2003, Unexpected dynamism in zonation and abundance revealed by long-term monitoring on rocky shores, *Ecological Society of America Annual Meeting Abstracts*, 88, p. 275.

Ramos, J., Santos, M.N., Whitmarsh, D. & Monteiro, C.C., 2007, Stakeholder perceptions regarding the environmental and socio-economic impacts of the Algarve artificial reefs, *Hydrobiologia*, 580, pp. 181–91.

Raymundo, L.J., Maypa, A.P., Gomez, E.D. & Cadiz, P., 2007, Can dynamite-blasted reefs recover? A novel, low-tech approach to stimulating natural recovery in fish and coral populations, *Marine Pollution Bulletin*, 54(7), pp. 1009–19.

Reaka-Kudla, M.L., 1996, Biodiversity of global coral reefs: How much is there and how much might we lose? *AAAS Annual Meeting and Science Innovation Exposition*, 162(0), p. A11.

Reaka-Kudla, M.L., 1997, The global biodiversity of coral reefs: A comparison with rainforests. In: Reaka-Kudla, M.L., Wilson, D.E. & Wilson, E.O. (eds), *Biodiversity II: Understanding and Protecting Our Natural Resources*, pp. 83–108. Joseph Henry/National Academy Press, Washington, DC.

Reed, J.K., Shepard, A.N., Koenig, C.C., Scanlon, K.M. & Gilmore, R.G., 2005, Mapping, habitat characterization, and fish surveys of the deep-water Oculina coral reef Marine Protected Area: a review of historical and current research, *Cold-Water Corals and Ecosystems*, pp. 443–65.

Reichelt-Brushett, A.J. & Harrison, P.L., 2000, The effect of copper on the settlement success of larvae from the scleractinian coral *Acropora tenuis*, *Marine Pollution Bulletin*, 41(7–12), pp. 385–91.

Reitzel, A.M., Sullivan, J.C., Brown, B.K., Chin, D.W., Cira, E.K., Edquist, S.K., Genco, B.M., Joseph, O.C., Kaufman, C.A., Kovitvongsa, K., Munoz, M.M., Negri, T.L., Taffel, J.R., Zuehlke, R.T. & Finnerty, J.R., 2007, Ecological and developmental dynamics of a host-parasite system involving a sea anemone and two ctenophores, *Journal of Parasitology*, 93(6), pp. 1392–402.

Relini, G., Relini, M., Palandri, G., Merello, S. & Beccornia, E., 2007, History, ecology and trends for artificial reefs of the Ligurian sea, Italy, *Hydrobiologia*, 580, pp. 193–217.

Rex, M.A., Etter, R.J. & Stuart, C.T., 1997, Large-scale patterns of species diversity in the deep-sea benthos, *Marine Biodiversity*, pp. 94–121.

Rex, M.A., Crame, J.A., Stuart, C.T. & Clarke, A., 2005a, Large-scale biogeographic patterns in marine mollusks: A confluence of history and productivity? *Ecology*, 86(9), pp. 2288–97.

Rex, M.A., McClain, C.R., Johnson, N.A., Etter, R.J., Allen, J.A., Bouchet, P. & Waren, A., 2005b, A source-sink hypothesis for abyssal biodiversity, *American Naturalist*, 165(2), pp. 163–78.

Ribes, M., Coma, R. & Gili, J.M., 1998, Seasonal variation of in situ feeding rates by the temperate ascidian *Halocynthia papillosa*. *Marine Ecology Progress Series*, 175, pp. 201–13.

Ricciardi, A. & Bourget, E., 1999, Global patterns of macroinvertebrate biomass in marine intertidal communities, *Marine Ecology – Progress Series*, 185, pp. 21–35.

Richards, Z.T., Beger, M., Pinca, S. & Wallace, C.C., 2008, Bikini Atoll coral biodiversity resilience five decades after nuclear testing, *Marine Pollution Bulletin*, 56(3), pp. 503–15.

Richardson, L.L. & Voss, J.D., 2005, Changes in a coral population on reefs of the northern Florida Keys following a coral disease epizootic, *Marine Ecology Progress Series*, 297, pp. 147–56.

Richardson, W.J., Greene, C.R., Malme, C.I. & Thomson, D.H., 1995, *Marine Mammals and Noise*. Academic Press, San Diego, CA.

Riisgard, H.U. & Schotge, P., 2007, Surface deposit feeding versus filter feeding in the amphipod *Corophium volutator*, *Marine Biology Research*, 3(6), pp. 421–7.

Rilov, G. & Benayahu, Y., 2000, Fish assemblage on natural versus vertical artificial reefs: the rehabilitation perspective, *Marine Biology*, 136(5), pp. 931–42.

Rinkevich, B., 2005, Conservation of coral reefs through active restoration measures: Recent approaches and last decade progress, *Environmental Science & Technology*, 39(12), pp. 4333–42.

Rioja-Nieto, R. & Sheppard, C., 2008, Effects of management strategies on the landscape ecology of a Marine Protected Area, *Ocean & Coastal Management*, 51(5), pp. 397–404.

Roberts, C.M., Andelman, S., Branch, G., Bustamante, R.H., Castilla, J.C., Dugan, J., Halpern, B.S., Lafferty, K.D., Leslie, H., Lubchenco, J., McArdle, D., Possingham, H.P., Ruckelshaus, M. & Warner, R.R., 2003, Ecological criteria for evaluating candidate sites for marine reserves, *Ecological Applications*, 13(1), pp. S199–214.

Roberts, J.M., Wheeler, A.J. & Freiwald, A., 2006, Reefs of the deep: The biology and geology of cold-water coral ecosystems, *Science*, 312(5773), pp. 543–7.

Roberts, J.M., Henry, L.A., Long, D. & Hartley J.P., 2008, Cold-water coral reef frameworks, megafaunal communities and evidence for coral carbonate mounds on the Hatton Bank, north east Atlantic. *FACIES*, 54(3), pp. 297–316.

Roberts, K., Granum, E., Leegood, R.C. & Raven, J.A., 2007, Carbon acquisition by diatoms, *Photosynthesis Research*, 93(1–3), pp. 79–88.

Robidart, J.C., Bench, S.R., Feldman, R.A., Novoradovsky, A., Podell, S.B., Gaasterland, T., Allen, E.E. & Felbeck, H., 2008, Metabolic versatility of the *Riftia pachyptila* endosymbiont revealed through metagenomics, *Environmental Microbiology*, 10(3), pp. 727–37.

Rocha, L.A. & Bowen, B.W., 2008, Speciation in coral-reef fishes, *Journal of Fish Biology*, 72(5), pp. 1101–21.

Rodriguez-Martinez, R.E., 2008, Community involvement in marine protected areas: The case of Puerto Morelos reef, Mexico, *Journal of Environmental Management*, 88(4), pp. 1151–60.

Rodriguez, S. & Cróquer, A., 2008, Dynamics of Black Band Disease in a *Diploria strigosa* population subjected to annual upwelling on the northeastern coast of Venezuela, *Coral Reefs*, 27(2), pp. 381–8.

Roff, J.C., Taylor, M.E. & Laughren, J., 2003, Geophysical approaches to the classification, delineation and monitoring of marine habitats and their communities. *Aquatic Conservation – Marine and Freshwater Ecosystems*, 13(1), pp. 77–90.

Rohde, K., 2002, Ecology and biogeography of marine parasites, *Advances in Marine Biology*, 43, pp. 1–86.

Ronce, O., 2007, How does it feel to be like a rolling stone? Ten questions about dispersal evolution, *Annual Review of Ecology Evolution and Systematics*, 38, pp. 231–53.

Rose, A.H. (ed.), 1967, *Thermobiology*. New York: Academic Press.

Rosser, N.L. & Gilmour, J.P., 2008, New insights into patterns of coral spawning on Western Australian reefs, *Coral Reefs*, 27(2), pp. 345–9.

Roughan, M., Garfield, N., Largier, J., Dever, E., Dorman, C., Peterson, D. & Dorman, J., 2006, Transport and retention in an upwelling region: The role of across-shelf structure, *Deep-Sea Research Part Ii-Topical Studies in Oceanography*, 53(25–26), pp. 2931–55.

Rutherford, S., D'Hondt, S. & Prell, W., 1999, Environmental controls on the geographic distribution of zooplankton diversity, *Nature*, 400(6746), pp. 749–53.

Rypien, K.L., 2008, African dust is an unlikely source of Aspergillus sydowii, the causative agent of sea fan disease, *Marine Ecology Progress Series*, 367, pp. 125–31.

Saier, B., 2001, Direct and indirect effects of seastars Asterias rubens on mussel beds (Mytilus edulis) in the Wadden Sea. *Journal of Sea Research*, 46(1), pp. 29–42.

Sala, E. & Knowlton, N., 2006, Global marine biodiversity trends, *Annual Review of Environment and Resources*, 31, pp. 93–122.

Samadi, S., Bottan, L., Macpherson, E., De Forges, B.R. & Boisselier, M.C., 2006, Seamount endemism questioned by the geographic distribution and population genetic structure of marine invertebrates, *Marine Biology*, 149(6), pp. 1463–75.

Sammarco, P.W., Winter, A. & Stewart, J.C., 2006, Coefficient of variation of sea surface temperature (SST) as an indicator of coral bleaching, *Marine Biology*, 149(6), pp. 1337–44.

Sampayo, E.M., Ridgway, T., Bongaerts, P. & Hoegh-Guldberg, O., 2008, Bleaching susceptibility and mortality of corals are determined by fine-scale differences in symbiont type, *Proceedings of the National Academy of Sciences of the United States of America*, 105(30), pp. 10444–9.

Sancho, G., Fisher, C.R., Mills, S., Micheli, F., Johnson, G.A., Lenihan, H.S., Peterson, C.H. & Mullineaux, L.S., 2005, Selective predation by the zoarcid fish Thermarces cerberus at hydrothermal vents, *Deep-Sea Research Part I – Oceanographic Research Papers*, 52(5), pp. 837–44.

Sanford, E., Holzman, S.B., Haney, R.A., Rand, D.M. & Bertness, M.D., 2006, Larval tolerance, gene flow, and the northern geographic range limit of fiddler crabs, *Ecology*, 87(11), pp. 2882–94.

Sarthou, G., Timmermans, K.R., Blain, S. & Treguer, P., 2005, Growth physiology and fate of diatoms in the ocean: a review, *Journal of Sea Research*, 53(1-2), pp. 25–42.

Sasal, P., Desdevises, Y., Durieux, E., Lenfant, P. & Romans, P., 2004, Parasites in marine protected areas: success and specificity of monogeneans, *Journal of Fish Biology*, 64(2), pp. 370–9.

Saunders, P.A., Deibel, D., Stevens, C.J., Rivkin, R.B., Lee, S.H. & Klein, B., 2003, Copepod herbivory rate in a large arctic polynya and its relationship to seasonal and spatial variation in copepod and phytoplankton biomass, *Marine Ecology Progress Series*, 261, pp. 183–99.

Scheffers, S.R, Haviser, J., Browne, T. & Scheffers, A., 2009, Tsunamis, hurricanes, the demise of coral reefs and shifts in prehistoric human populations in the Caribbean, *Quaternary International*, 195, pp. 69–87.

Schiel, D.R., Wood, S.A., Dunmore, R.A. & Taylor, D.I., 2006, Sediment on rocky intertidal reefs: Effects on early post-settlement stages of habitat-forming seaweeds, *Journal of Experimental Marine Biology and Ecology*, 331(2), pp. 158–72.

Schluter, P. & Uenzelmann-Neben, G., 2007, Seismostratigraphic analysis of the Transkei Basin: A history of deep sea current controlled sedimentation, *Marine Geology*, 240(1-4), pp. 99–111.

Schmidt, C., Le Bris, N. & Gaill, F., 2008, Interactions of deep-sea vent invertebrates with their environment: The case of *Rimicaris exoculata*, *Journal of Shellfish Research*, 27(1), pp. 79–90.

Schnetzer, A. & Steinberg, D.K., 2002, Natural diets of vertically migrating zooplankton in the Sargasso Sea, *Marine Biology*, 141(2), p. 403.

Schratzberger, M., Lampadariou, N., Somerfield, P., Vandepitte, L. & Vanden Berghe, E., 2009, The impact of seabed disturbance on nematode communities: linking field and laboratory observations, *Marine Biology*, 156(4), pp. 709–24.

Scott-Holland, T.B., Bennett, S.M. & Bennett, M.B., 2006, Distribution of an asymmetrical copepod, *Hatschekia plectropomi*, on the gills of *Plectropomus leopardus*, *Journal of Fish Biology*, 68(1), pp. 222–35.

Seaman, W., 2007, Artificial habitats and the restoration of degraded marine ecosystems and fisheries, *Hydrobiologia*, 580, pp. 143–55.

Seitz, R.D., Lipcius, R.N., Hines, A.H. & Eggleston, D.B., 2001, Density-dependent predation, habitat variation, and the persistence of marine bivalve prey, *Ecology*, 82(9), pp. 2435–51.

Sekar, R., Kaczmarsky, L.T. & Richardson, L.L., 2008, Microbial community composition of black band disease on the coral host *Siderastrea siderea* from three regions of the wider Caribbean, *Marine Ecology Progress Series*, 362, pp. 85–98.

Selleslagh, J. & Amara, R., 2007, Temporal variations in abundance and species composition of fish and epibenthic crustaceans of an intertidal zone: Environmental factor influence, *Cybium*, 31(2), pp. 155–62.

Semmens, B.X., Buhle, E.R., Salomon, A.K. & Pattengill-Semmens, C.V., 2004, A hotspot of non-native marine fishes: evidence for the aquarium trade as an invasion pathway, *Marine Ecology Progress Series*, 266, pp. 239–44.

Shaban, A., Hamze, M., El-Baz, F. & Ghoneim, E., 2009, Characterization of an oil spill along the Lebanese Coast by satellite images. *Environmental Forensics*, 10(1), pp. 51–9.

Shafir, S., Van Rijn, J. & Rinkevich, B., 2006, Steps in the construction of underwater coral nursery, an essential component in reef restoration acts, *Marine Biology*, 149(3), pp. 679–87.

Shaish, L., Levy, G., Gomez, E. & Rinkevich, B., 2008, Fixed and suspended coral nurseries in the Philippines: Establishing the first step in the "gardening concept" of reef restoration, *Journal of Experimental Marine Biology and Ecology*, 358(1), pp. 86–97.

Shank, T.M. & Halanych, K.M., 2007, Toward a mechanistic understanding of larval dispersal: insights from genomic fingerprinting of the deep-sea hydrothermal vent tubeworm *Riftia pachyptila*, *Marine Ecology-Pubblicazioni Della Stazione Zoologica Di Napoli I*, 28(1), pp. 25–35.

Shanks, A.L. & Brink, L., 2005, Upwelling, downwelling, and cross-shelf transport of bivalve larvae: test of a hypothesis, *Marine Ecology Progress Series*, 302, pp. 1–12.

Shephard, S., Goudey, C.A., Read, A. & Kaiser, M.J., 2009, Hydrodredge: Reducing the negative impacts of scallop dredging, *Fisheries Research*, 95(2–3), pp. 206–9.

Sherman, R.L., Gillian, D.S. & Spieler, R.E., 2002, Artificial reef design: void space, complexity, and attractants, *ICES Journal of Marine Science*, 59, pp. S196–200.

Shields, J.D. & Segonzac, M., 2007, New nemertean worms (Carcinonemertidae) on bythograeid crabs (Decapoda : Brachyura) from pacific hydrothermal vent sites, *Journal of Crustacean Biology*, 27(4), pp. 681–92.

Shields, J.D., Buchal, M.A. & Friedman, C.S., 1998, Microencapsulation as a potential control technique against sabellid worms in abalone culture, *Journal of Shellfish Research*, 17(1), pp. 79–83.

Shima, J.S., Osenberg, C.W. & St Mary, C.M., 2008, Quantifying site quality in a heterogeneous landscape: Recruitment of a reef fish, *Ecology*, 89(1), pp. 86–94.

Shimanaga, M., Nomaki, H. & Iijima, K., 2008, Spatial changes in the distributions of deep-sea "Cerviniidae" (Harpacticoida, Copepoda) and their associations with environmental factors in the bathyal zone around Sagami Bay, Japan, *Marine Biology*, 153(4), pp. 493–506.

Short, F., Carruthers, T., Dennison, W. & Waycott, M., 2007, Global seagrass distribution and diversity: A bioregional model, *Journal of Experimental Marine Biology and Ecology*, 350, pp. 3–20.

Shumway S.E, Dana M Frank, Lisa M Ewart, & J Evan Ward Effect of yellow loess on clearance rate in seven species of benthic, filter-feeding invertebrates, *Aquaculture Research*, 2003, 34, 1391–1402.

Silio-Calzada, A., Bricaud, A. & Gentili, B., 2008, Estimates of sea surface nitrate concentrations from sea surface temperature and chlorophyll concentration in upwelling areas: A case study for the Benguela system, *Remote Sensing of Environment*, 112(6), pp. 3173–80.

Silverman, H., Lynn, J.W. & Dietz, T.H., 1996, Particle capture by the gills of *Dreissena polymorpha*: Structure and function of latero-frontal cirri, *Biological Bulletin*, 191(1), pp. 42–54.

Simpson, S.D., Meekan, M., Montgomery, J., McCauley, R. & Jeffs, A., 2005, Homeward sound, *Science*, 308(5719), p. 221.

Sisson, C.G., 2005, Veligers from the nudibranch *Dendronotus frondosus* show shell growth and extended planktonic period in laboratory culture, *Hydrobiologia*, 541, pp. 205–13.

Skogen, M.D., Budgell, W.P. & Rey, F., 2007, Interannual variability in Nordic seas primary production, *ICES Journal of Marine Science*, 64, pp. 889–98.

Sleeman, J.C., Boggs, G.S., Radford, B.C. & Kendrick, G.A., 2005, Using agent-based models to aid reef restoration: Enhancing coral cover and topographic complexity through the spatial arrangement of coral transplants, *Restoration Ecology*, 13(4), pp. 685–94.

Slyusarev, G.S., 2008, Phylum Orthonectida: morphology, biology, and relationships to other multicellular animals, *Zhurnal Obshchei Biologii*, 69(6), pp. 403–27.

Smith, C.R. & Baco, A.R., 2003, Ecology of whale falls at the deep-sea floor, *Oceanography and Marine Biology – an Annual Review*, 43(41), pp. 311–54.

Smith, F. & Whitman, J.D., 1999, Species diversity in subtidal landscapes: Maintenance by physical processes and larval recruitment, *Ecology*, 80(1), pp. 51–69.

Sogin, M.L., Morrison, H.G., Huber, J.A., Mark Welch, D., Huse, S.M., Neal, P.R., Arrieta, J.M. & Herndl, G.J., 2006, Microbial diversity in the deep sea and the underexplored "rare biosphere", *Proceedings of the National Academy of Sciences of the United States of America*, 103, pp. 12115–20.

Somerfield, P.J., Jaap, W.C., Clarke, K.R., Callahan, M., Hackett, K., Porter, J., Lybolt, M., Tsokos, C. & Yanev, G., 2008, Changes in coral reef communities among the Florida Keys, 1996–2003, *Coral Reefs*, 27(4), pp. 951–65.

Sommer, U., 1997, Selectivity of *Idothea chelipes* (Crustacea : Isopoda) grazing on benthic microalgae, *Limnology and Oceanography*, 42(7), pp. 1622–8.

Sommer U (2000) Scarcity of medium-sized phytoplankton in the northern Red Sea explained by strong bottom-up and weak top-down control, *Marine Ecology Progress Series*, 197, 19–25.

Sommer, U., 2008, Trophic cascades in marine and freshwater plankton. *International Review of Hydrobiology*, 93(4–5), 506–16.

Sommer, U. & Lengfellner, K., 2008, Climate change and the timing, magnitude, and composition of the phytoplankton spring bloom, *Global Change Biology*, 14(6), pp. 1199–208.

Soong, K. & Chen, T.A., 2003, Coral transplantation: Regeneration and growth of Acropora fragments in a nursery, *Restoration Ecology*, 11(1), pp. 62–71.

Sprague, J.B. & Drury, D.E., 1969, Avoidance reactions of salmonid fish to representative pollutants. In: *Advances in Water Pollution Research, Proc. 4th Conf. Intern., Prague*, pp. 169–79.

Stachowicz, J.J., Bruno, J.F. & Duffy, J.E., 2007, Understanding the effects of marine biodiversity on communities and ecosystems, *Annual Review of Ecology Evolution and Systematics*, 38, pp. 739–66.

Stanley, D.R. & Wilson, C.A., 2000, Variation in the density and species composition of fishes associated with three petroleum platforms using dual beam hydroacoustics, *Fisheries Research*, 47(2–3), pp. 161–72.

Stanley, S.M., 2007, An analysis of the history of marine animal diversity, *Paleobiology*, 33, pp. 1–55.

Steen, H. & Scrosati, R., 2004, Intraspecific competition in *Fucus serratus* and *F evanescens* (Phaeophyceae : Fucales) germlings: effects of settlement density, nutrient concentration, and temperature, *Marine Biology*, 144(1), pp. 61–70.

Stelzenmüller, V., Maynou, F. & Martin, P., 2007, Spatial assessment of benefits of a coastal Mediterranean Marine Protected Area, *Biological Conservation*, 136(4), pp. 571–83.

Stevens, T. & Connolly, R.M., 2004, Testing the utility of abiotic surrogates for marine habitat mapping at scales relevant to management. *Biological Conservation*, 119(3), pp. 351–362.

Strathmann, M.F. & Strathmann, R.R., 2007, An extraordinarily long larval duration of 4.5 years from hatching to metamorphosis for teleplanic veligers *of Fusitriton oregonensis, Biological Bulletin*, 213, pp. 152–9.

Strathmann, R.R., 2007, Three functionally distinct kinds of pelagic development, *Bulletin of Marine Science*, 81(2), pp. 167–79.

Sugden, H., Lenz, M., Molis, M., Wahl, M. & Thomason, J.C., 2008, The interaction between nutrient availability and disturbance frequency on the diversity of benthic marine communities on the north-east coast of England, *Journal of Animal Ecology*, 77(1), pp. 24–31.

Sumida, P.Y.G., Tyler, P.A., Lampitt, R.S. & Gage, J.D., 2000, Reproduction, dispersal and settlement of the bathyal ophiuroid *Ophiocten gracilis* in the NE Atlantic Ocean, *Marine Biology*, 137(4), pp. 623–30.

Sussman, M., Willis, B.L., Victor, S. & Bourne, D.G., 2008, Coral pathogens identified for white syndrome (WS) epizootics in the Indo-Pacific, *PLoS ONE*, 3(6), pp. 1–14.

Sutton, S.G. & Bushnell, S.L., 2007, Socio-economic aspects of artificial reefs: Considerations for the great barrier reef marine park, *Ocean & Coastal Management*, 50, pp. 829–46.

Svavarsson, J., 2000, Imposex in the dogwhelk (*Nucella lapillus*) due to TBT contamination: Improvement at high latitudes. *Marine Pollution Bulletin*, 40(11), pp. 893–7.

Svensson, J.R., Lindegarth, M., Siccha, M., Lenz, M., Molis, M., Wahl, M. & Pavia, H., 2007, Maximum species richness at intermediate frequencies of disturbance: Consistency among levels of productivity, *Ecology*, 88(4), pp. 830–8.

Swanson, C., 1996, Early development of milkfish: Effects of salinity on embryonic and larval metabolism, yolk absorption and growth, *Journal of Fish Biology*, 48(3), pp. 405–21.

Tanimura, A., Kawaguchi, S., Oka, N., Nishikawa, J., Toczko, S., Takahashi, K.T., Terazaki, M., Odate, T., Fukuchi, M. & Hosie, G., 2008, Abundance and grazing impacts of krill, salps and copepods along the 140 degrees E meridian in the Southern Ocean during summer. *Antarctic Science*, 20(4), pp. 365–79.

Tation, M., Sahade, R., Mercuri, G., Fuentes, V.L., Antacli, J.C., Stellfeldt, A. & Esnal, G.B., 2008, Feeding ecology of benthic filter-feeders at Potter Cove, an Antarctic coastal ecosystem, *Polar Biology*, 31(4), pp. 509–17.

Taylor, J.D., 1997, Diversity and structure of tropical Indo-Pacific benthic communities: Relation to regimes of nutrient input, *Marine Biodiversity*, pp. 178–200.

Taylor, A.H., Harbour, D.S., Harris, R.P., Burkill, P.H. & Edwards, E.S., 1993, Seasonal succession in the pelagic ecosystem of the North-Atlantic and the utilization of nitrogen. *Journal of Plankton Research*, 15(8), pp. 875–91.

Temming, A., Floeter, J. & Ehrich, S., 2007, Predation hot spots: Large scale impact of local Aggregations, *Ecosystems*, 10, pp. 865–76.

Thacker, W.C., 2007, Estimating salinity to complement observed temperature: 1. Gulf of Mexico, *Journal of Marine Systems*, 65(1–4), pp. 224–48.

Thieltges, D.W. & Buschbaum, C., 2007, Vicious circle in the intertidal: Facilitation between barnacle epibionts, a shell boring polychaete and trematode parasites in the periwinkle *Littorina littorea*, *Journal of Experimental Marine Biology and Ecology*, 340(1), pp. 90–5.

Thornber, C., 2007, Associational resistance mediates predator-prey interactions in a marine subtidal system, *Marine Ecology-Pubblicazioni Della Stazione Zoologica Di Napoli I*, 28(4), pp. 480–6.

Thronson, A. & Quigg, A., 2008, Fifty-five years of fish kills in coastal Texas, *Estuaries and Coasts*, 31(4), pp. 802–13.

Thu, P.M. & Populus, J., 2007, Status and changes of mangrove forest in Mekong Delta: Case study in Tra Vinh, Vietnam, *Estuarine Coastal and Shelf Science*, 71(1–2), pp. 98–109.

Titlyanov, E.A., Yakovleva, I.M. & Titlyanova, T.V., 2007, Interaction between benthic algae (*Lyngbya bouillonii, Dictyota dichotoma*) and scleractinian coral *Porites lutea* in direct contact, *Journal of Experimental Marine Biology and Ecology*, 342(2), pp. 282–91.

Tjelmeland, S. & Lindstrøm, U., 2005, An ecosystem element added to the assessment of Norwegian spring-spawning herring: implementing predation by minke whales, *ICES Journal of Marine Science*, 62(2), pp. 285–94.

Toledo-Hernandez, C., Zuluaga-Montero, A., Bones-Gonzalez, A., Rodriguez, J.A., Sabat, A.M. & Bayman, P., 2008, Fungi in healthy and diseased sea fans (*Gorgonia ventalina*): is *Aspergillus sydowii* always the pathogen? *Coral Reefs*, 27(3), pp. 707–14.

Tolimieri, N., Haine, O., Jeffs, A., McCauley, R. & Montgomery, J., 2004, Directional orientation of pomacentrid larvae to ambient reef sound, *Coral Reefs*, 23(2), pp. 184–91.

Tomas, F., Turon, X. & Romero, J., 2005, Seasonal and small-scale spatial variability of herbivory pressure on the temperate seagrass *Posidonia oceanica*, *Marine Ecology Progress Series*, 301, pp. 95–107.

Trathan, P.N., Forcada, J., & Murphy, E.J., 2007, Environmental forcing and Southern Ocean marine predator populations: effects of climate change and variability *Philosophical Transactions Of The Royal Society B – Biological Sciences* 362 pp. 2351–2365

Tsuda, A., Takeda, S., Saito, H., Nishioka, J., Nojiri, Y., Kudo, I., Kiyosawa, H., Shiomoto, A., Imai, K., Ono, T., Shimamoto, A., Tsumune, D., Yoshimura, T., Aono, T., Hinuma, A., Kinugasa, M., Suzuki, K., Sohrin, Y., Noiri, Y., Tani, H., Deguchi, Y., Tsurushima, N., Ogawa, H., Fukami, K., Kuma, K. & Saino, T., 2003, A mesoscale iron enrichment in the western Subarctic Pacific induces a large centric diatom bloom, *Science*, 300(5621), pp. 958–61.

Turekian, K.K., 1968, Light availability in the coastal ocean: impact on the distribution of benthic photosynthetic organisms and their contribution to primary production. In: *Oceans*. Prentice-Hall, New York.

Turley, C.M., Roberts, J.M. & Guinotte, J.M., 2007, Corals in deepwater: will the unseen hand of ocean acidification destroy coldwater ecosystems? *Coral Reefs*, 26, pp. 445–8.

Turnpenny, A.W.H. & Bamber, R.N., 1983, The critical swimming speed of the sand smelt (*Atherina presbyter* Cuvier) in relation to capture at a power-station cooling water-intake, *Journal of Fish Biology*, 23(1), pp. 65–73.

Turnpenny, A.W.H. & Nedwell, J.R., 1994, *The Effects on Marine Fish, Diving Mammals and Birds of Underwater Sound Generated By Seismic Surveys*. FARL (Fawley Aquatic Research Ltd) Report Reference: FCR 089/94.

Tuya, F., Cisneros-Aguirre, J., Ortega-Borges, L. & Haroun, R.J., 2007, Bathymetric segregation of sea urchins on reefs of the Canarian Archipelago: Role of flow-induced forces, *Estuarine Coastal and Shelf Science*, 73(3–4), pp. 481–8.

Tyler, E.M., 2005, The effect of fully and partially protected marine reserves on coral reef fish populations in Zanzibar, Tanzania. Unpublished DPhil thesis, University of Oxford.

Tyler, E.H.M., Speight, M.R., Henderson, P. & Manica, A., 2009, Evidence for a depth refuge effect in artisanal coral reef fisheries, *Biological Conservation*, 142, 652–67.

Tyler, P.A., Pendlebury, S., Mills, S.W., Mullineaux, L., Eckelbarger, K.J., Baker, M. & Young, C.M., 2008, Reproduction of gastropods from vents on the East Pacific Rise and the Mid-Atlantic Ridge, *Journal of Shellfish Research*, 27(1), pp. 107–18.

Unsworth, R.K.F., Bell, J.J. & Smith, D.J., 2007, Tidal fish connectivity of reef and sea grass habitats in the Indo-Pacific, *Journal of the Marine Biological Association of the UK*, 87(5), pp. 1287–96.

US Department of Commerce (2009) www.commerce.gov

Uthicke, S., Schaffelke, B. & Byrne, M., 2009, A boom-bust phylum? Ecological and evolutionary consequences of density variations in echinoderms, *Ecological Monographs*, 79(1), pp. 3–24.

Uye, S. & Fleminger, A., 1976, Effects of various environmental factors on egg development of several species of *Acartia* in Southern California. *Marine Biology*, 38(3), pp. 253–62.

Vacelet, J. & Duport, E., 2004, Prey capture and digestion in the carnivorous sponge *Asbestopluma hypogea* (Porifera : Demospongiae), *Zoomorphology*, 123(4), pp. 179–90.

van der Molen, J., Rogers, S.I., Ellis, J.R., Fox, C.J. & McCloghrie, P., 2007, Dispersal patterns of the eggs and larvae of spring-spawning fish in the Irish Sea, UK, *Journal of Sea Research*, 58(4), pp. 313–30.

van der Veer, H.W., Feller, R.J., Weber, A. & Witte, J.I.J., 1998, Importance of predation by crustaceans upon bivalve spat in the intertidal zone of the Dutch Wadden Sea as revealed by immunological assays of gut contents, *Journal of Experimental Marine Biology and Ecology*, 231(1), pp. 139–57.

Vásárhelyi, C. & Thomas, V.G., 2008, Reflecting ecological criteria in laws supporting the Baja to Bering Sea marine protected areas network case study, *Environmental Science & Policy*, 11(5), pp. 394–407.

Venn, A.A., Loram, J.E. & Douglas, A.E., 2008, Photosynthetic symbioses in animals, *Journal of Experimental Botany*, 59(5), pp. 1069–80.

Vermeij, M.J.A., 2006, Early life-history dynamics of Caribbean coral species on artificial substratum: the importance of competition, growth and variation in life-history strategy, *Coral Reefs*, 25(1), pp. 59–71.

Vermeij, M.J.A., Sampayo, E., Broker, K. & Bak, R.P.M., 2003, Variation in planulae release of closely related coral species, *Marine Ecology Progress Series*, 247, pp. 75–84.

Verney, R., Deloffre, J., Brun-Cottan, J.C. & Lafite, R., 2007, The effect of wave-induced turbulence on intertidal mudflats: Impact of boat traffic and wind, *Continental Shelf Research*, 27(5), pp. 594–612.

Vizzini, S. & Mazzola, A., 2008, The fate of organic matter sources in coastal environments: a comparison of three Mediterranean lagoons, *Hydrobiologia*, 611, pp. 67–79.

Vogler, C., Benzie, J., Lessios, H., Barber, P. & Worheide, G., 2008, A threat to coral reefs multiplied? Four species of crown-of-thorns starfish, *Biology Letters*, 4(6), pp. 696–9.

Voight, J.R. & Sigwart, J.D., 2007, Scarred limpets at hydrothermal vents: evidence of predation by deep-sea whelks, *Marine Biology*, 152(1), pp. 129–33.

Volkenborn, N., Hedtkamp, S.I.C., van Beusekom, J.E.E. & Reise, K., 2007, Effects of bioturbation and bioirrigation by lugworms (*Arenicola marina*) on physical and chemical sediment properties and implications for intertidal habitat succession, *Estuarine Coastal and Shelf Science*, 74(1–2), pp. 331–43.

Vollmer, S.V. & Kline, D.I., 2008, Natural disease resistance in threatened staghorn corals, *PLoS One*, 3(11), article no. e3718.

Voss, J.D., Mills, D.K., Myers, J.L., Remily, E.R. & Richardson, L.L., 2007, Black band disease microbial community variation on corals in three regions of the wider Caribbean, *Microbial Ecology*, 54, pp. 730–9.

Wai, T.C. & Williams, G.A., 2005, The relative importance of herbivore-induced effects on productivity of crustose coralline algae: Sea urchin grazing and nitrogen excretion, *Journal of Experimental Marine Biology and Ecology*, 324(2), pp. 141–56.

Wakeford, M., Done, T.J. & Johnson, C.R., 2008, Decadal trends in a coral community and evidence of changed disturbance regime, *Coral Reefs*, 27(1), pp. 1–13.

Walker, B.K., Henderson, B. & Spieler, R.E., 2002, Fish assemblages associated with artificial reefs of concrete aggregates or quarry stone offshore Miami Beach, Florida, USA, *Aquatic Living Resources*, 15(2), pp. 95–105.

Walker, B.K., Riegl, B. & Dodge, R.E., 2008, Mapping coral reef habitats in southeast Florida using a combined technique approach, *Journal of Coastal Research*, 24(5), pp. 1138–50.

Walker, S.J., Schlacher, T.A. & Schlacher-Hoenlinger, M.A., 2007, Spatial heterogeneity of epibenthos on artificial reefs: fouling communities in the early stages of colonization on an East Australian shipwreck, *Marine Ecology-Pubblicazioni Della Stazione Zoologica Di Napoli I*, 28, pp. 435–45.

Wallace, C.C. & Muir, P.R., 2005, Biodiversity of the Indian Ocean from the perspective of staghorn corals (*Acropora* spp.), *Indian Journal of Marine Sciences*, 34(1), pp. 42–9.

Walter, R.P. & Haynes, J.M., 2006, Fish and coral community structure are related on shallow water patch reefs near San Salvador, Bahamas, *Bulletin of Marine Science*, 79(2), pp. 365–74.

Walton, W.C., MacKinnon, C., Rodriguez, L.F., Proctor, C. & Ruiz, G.A., 2002, Effect of an invasive crab upon a marine fishery: green crab, *Carcinus maenas*, predation upon a venerid clam, *Katelysia scalarina*, in Tasmania (Australia), *Journal of Experimental Marine Biology and Ecology*, 272(2), pp. 171–89.

Wang, J., Cota, G.F. & Comiso, J.C., 2005, Phytoplankton in the Beaufort and Chukchi Seas: Distribution, dynamics, and environmental forcing, *Deep-Sea Research Part Ii-Topical Studies in Oceanography*, 52(24–26), pp. 3355–68.

Wang, L.P. & Zheng, B.H., 2008, Toxic effects of fluoranthene and copper on marine diatom *Phaeodactylum tricornutum*, *Journal of Environmental Sciences-China*, 20(11), pp. 1363–72.

Watanabe, H., Kado, R., Kaida, M., Tsuchida, S. & Kojima, S., 2006, Dispersal of vent-barnacle (genus *Neoverruca*) in the Western Pacific, *Cahiers De Biologie Marine*, 47(4), pp. 353–7.

Waters, J.M., 2008, Marine biogeographical disjunction in temperate Australia: historical landbridge, contemporary currents, or both? *Diversity and Distributions*, 14(4), pp. 692–700.

Weir-Brush, J.R., Garrison, V.H., Smith, G.W. & Shinn, E.A., 2004, The relationship between gorgonian coral (Cnidaria: Gorgonacea) diseases and African dust storms, *Aerobiologia*, 20(2), pp. 119–26.

Wen, K. –C., Hsu, C. -M., Chen, K. -S., Liao, M. -H., Chen, C. -P., Chen, C. A. 2007 Unexpected coral diversity on the breakwaters: potential refuges for depleting coral reefs. *Coral Reefs*, 26(1), p. 127.

Wenngren, J. & Olafsson, E., 2002, Intraspecific competition for food within and between year classes in the deposit-feeding amphipod *Monoporeia affinis* – the cause of population fluctuations? *Marine Ecology Progress Series*, 240, pp. 205–13.

White, J.W., 2007, Spatially correlated recruitment of a marine predator and its prey shapes the large-scale pattern of density-dependent prey mortality, *Ecology Letters*, 10, pp. 1054–65.

Whitehead, L.F. & Douglas, A.E., 2003, Metabolite comparisons and the identity of nutrients translocated from symbiotic algae to an animal host, *Journal of Experimental Biology*, 206(18), pp. 3149–57.

Whitmarsh, D., Santos, M.N., Ramos, J. & Monteiro, C.C., 2008, Marine habitat modification through artificial reefs off the Algarve (southern Portugal): An economic analysis of the fisheries and the prospects for management, *Ocean & Coastal Management*, 51(6), pp. 463–8.

Williams D., McB., & Hatcher A.I., 1983, Structure of fish communities on outer slopes of inshore, mid-shelf and outer shelf reefs of the Great Barrier Reef. *Marine Ecology Progress Series* 10(3): pp. 239–250.

Williams, D.E., Miller, M.W. & Kramer, K.L., 2008, Recruitment failure in Florida Keys *Acropora palmata*, a threatened Caribbean coral, *Coral Reefs*, 27(3), pp. 697–705.

Williams, P.J.L., 1998, The balance of plankton respiration and photosynthesis in the open oceans, *Nature*, 394(6688), pp. 55–7.

Wilson, E.O., 1988, The current state of biological diversity. In: Wilson, E.O. (ed.), *Biodiversity*, pp. 3–18. National Academy Press, Washington, DC.

Wishner. K. F., Levin, L., Gowing, M., Mullineaux, L., 1990. The involvement of the oxygen minimum in benthic zonation on a deep seamount, *Nature*, 346, 57–59.

Worachananant, S., Carter, R.W., Hockings, M. & Reopanichkul, P., 2008, Managing the Impacts of SCUBA Divers on Thailand's Coral Reefs, *Journal of Sustainable Tourism*, 16(6), pp. 645–63.

Work, T.M., Aeby, G.S. & Maragos, J.E., 2008, Phase shift from a coral to a corallimorph-dominated reef associated with a shipwreck on Palmyra atoll, *PLoS ONE*, 3(8), pp. e2989, 1–5.

Worm, B., Lotze, H.K. & Myers, R.A., 2003, Predator diversity hotspots in the blue ocean, *Proceedings of the National Academy of Sciences of the United States of America*, 100(17), pp. 9884–8.

Worm, B., Sandow, M., Oschlies, A., Lotze, H.K. & Myers, R.A., 2005, Global patterns of predator diversity in the open oceans, *Science*, 309(5739), pp. 1365–9.

Wright, J.I. & Gribben, P.E., 2008, Predicting the impact of an invasive seaweed on the fitness of native fauna, *Journal of Applied Ecology*, 45(5), pp. 1540–9.

Yahel, G., Eerkes-Medrano, D.I. & Leys, S.P., 2006, Size independent selective filtration of ultraplankton by hexactinellid glass sponges, *Aquatic Microbial Ecology*, 45(2), pp. 181–194.

Yasuhara, M., Kato, M., Ikeya, N. & Seto, K., 2007, Modern benthic ostracodes from Lutzow-Holm, Bay, East Antarctica: paleoceanographic, paleobiogeographic, and evolutionary significance, *Micropaleontology*, 53(6), pp. 469–96.

Yasuhara, M., Cronin, T.M., deMenocal, P.B, *et al.*, 2008, Abrupt climate change and collapse of deep-sea ecosystems, *Proceedings of the National Academy of Sciences of the United States Of America*, 105(5), pp. 1556–60.

Yeemin, T., Sutthacheep, M. & Pettongma, R., 2006, Coral reef restoration projects in Thailand, *Ocean & Coastal Management*, 49(9–10), pp. 562–75.

Young, C.R., Fujio, S. & Vrijenhoek, R.C., 2008, Directional dispersal between mid-ocean ridges: deep-ocean circulation and gene flow in *Ridgeia piscesae*, *Molecular Ecology*, 17(7), pp. 1718–31.

Yukihira, H., Klumpp, D.W. & Lucas, J.S., 1999, Feeding adaptations of the pearl oysters *Pinctada margaritifera* and *P-maxima* to variations in natural particulates, *Marine Ecology Progress Series*, 182(182), pp. 161–73.

Zakai, D. & Chadwick-Furman, N.E., 2002, Impacts of intensive recreational diving on reef corals at Eilat, northern Red Sea, *Biological Conservation*, 105(2), pp. 179–87.

Zander, C.D., 2007, Parasite diversity of sticklebacks from the Baltic Sea, *Parasitology Research*, 100(2), pp. 287–97.

Zekely, J., Van Dover, C.L., Nemeschkal, H.L. & Bright, M., 2006, Hydrothermal vent meiobenthos associated with mytilid mussel aggregations from the Mid-Atlantic Ridge and the East Pacific Rise, *Deep-Sea Research Part I – Oceanographic Research Papers*, 53(8), pp. 1363–78.

Zhuang, S.H., 2006, Species richness, biomass and diversity of macroalgal assemblages in tidepools of different sizes, *Marine Ecology Progress Series*, 309, pp. 67–73.

Zintzen, V., Massin, C., Norro, A. & Mallefet, J., 2006, Epifaunal inventory of two shipwrecks from the Belgian Continental Shelf, *Hydrobiologia*, 555, pp. 207–19.

INDEX

Page numbers in *italics* refer to figures, those in **bold** refer to tables